July, 1976

HUMAN BIOLOGY IN HEALTH AND IN DISEASE

HUMAN BIOLOGY IN HEALTH AND IN DISEASE

SHIRLEY R. BURKE, R.N.
Community College of Allegheny County
Pittsburgh, Pennsylvania

John Wiley & Sons, Inc.
New York London Sydney Toronto

Library of Congress Cataloging in Publication Data:

Burke, Shirley R

Human biology in health and disease.

Includes bibliographies and index.

1. Pathology. 2. Anatomy, Human. 3. Human physiology. I. Title. [DNLM: 1. Biology. 2. Pathology. QZ4 B959h]

RB.B78 612 75-1472

ISBN 0-471-12371-4

Printed in the United States of America

10 9 8 7 6 5 4 3 2 1

To Allied Health students
past, present, and future

Preface

A basic understanding of the structure and function of the human body, and of some of the common disease processes affecting the various organ systems, is essential for students in the Allied Health professions. Students in these programs whose formal academic preparation is limited to one or two years need a core course in normal human biology that also introduces them to some of the common pathological conditions they will encounter in the practice of their specialties. This textbook is intended for these students.

Learning is not only more meaningful but also easier and more efficient for the student who is actively involved in the learning process. Thus, each chapter includes summary questions that the student will find useful both as starting points for class discussions and for independent study.

Laboratory Manual in Human Biology in Health and Disease, the manual that accompanies this text, provides the student with many opportunities to explore the subject matter through dissections of fresh tissues, microscopic examinations of tissues, and the study of X-ray films. The manual also includes exercises that will help the student to gain an understanding of the physiology of the organ systems. Suggestions for appropriate audiovisual aids are also listed in the manual. The learning process is greatly facilitated if the student finds in his own experience examples of the concepts he is trying to learn.

This textbook and its accompanying laboratory manual are the outgrowth of my experiences in teaching a variety of Allied Health students at Community College of Allegheny County in Pittsburgh. The illustrations are the work of Anthony Gall while he was a student at Community College of Allegheny College and of Gretchen H. Schoch, one of his art professors.

I am grateful to my colleagues Yvonne Chipman and Gail Rowles for their helpful comments and suggestions. I particularly thank the many students whose questions and suggestions have been so helpful. The enthusiasm and concern of these students for the welfare of others are a continuing inspiration.

Shirley R. Burke, R.N.

To The Student

You will find your course in Human Biology one of the most interesting courses you have had. It is about you, your family, and your friends. Most important, it is about those whom you desire to help, your patients and the physicians with whom you will work.

The first chapter of this text discusses the general organization of the body. In the second chapter you are introduced to some of the causes of disease and symptoms that are common to most disease processes. In the following chapters, a first chapter discusses the anatomy and physiology of a body system and the next chapter describes common procedures used to diagnose disorders of the system, common diseases of the system, and their usual treatment. The concluding chapters of the text are devoted to disorders that disrupt the function of many body systems. Your instructor will assign these chapters in the sequence that best meets your particular needs. This sequence is not necessarily the one that is used in the text.

You will notice that each chapter begins with a list of terms. Some of these terms are probably already familiar to you. If so, there is no need for you to look them up in the Glossary at the end of the text. Also found here are Appendixes of Common Medical Prefixes and Suffixes; of Medical Abbreviations; and of Weights, Measures, and Equivalents used in medical practice. If a term listed in the vocabulary section of the chapter is new to you, you should refer to the Glossary, as the term is used in that chapter but is not defined within the text of the chapter. If you have had little previous experience with medical terminology, you will be astonished to find how quickly what at the outset of your course sounds like a foreign language will become a part of your everyday speech. Because many medical terms are pronounced differently in different parts of the country, the Glossary makes no attempt to indicate the proper pronunciation of the term. You should pronounce the term in the way your instructor does or in the way you hear it pronounced when you are having your clinical experience.

At the conclusion of each chapter you will find a list of questions. These questions are not necessarily sequenced in the order in which the material was discussed in the chapter. In your clinical experience you will find you need this information structured in ways other than those in your text. If you can answer these questions, you have mastered the essential content of that chapter.

In your previous educational experiences you may have been required to memorize many facts. In this course, you are encouraged for the most part to discard this habit. It is better for you to understand one generalization and thereby predict three out of five facts correctly than it is to remember all five facts but not to understand them. By understanding general concepts, you will be able to deal with unpredictable situations. In your profession, you are going to meet many situations that are completely unexpected. This is one of the reasons that the Allied Health professions are so exciting.

The process of learning is self-accelerating: As you increase your store of information, you increase your capabilities of meeting your own needs as well as the needs of those you intend to serve. As you progress, you will increase your capacity to gain more information and your ability to adjust to the demands of your specialized field.

A clear exposition of the purposes and objectives of this text will assist you in the learning process. For this reason, you should at the outset consider carefully the following objectives. As you finish each set of chapters and each of the last four chapters, you should apply these objectives to assess your progress.

S.R.B.

The study of this book will enable you to:

1. Know the names of the structures in each of the organ systems.
2. Be able to describe the location of the parts of the body.
3. Know some of the parts of the various organs.
4. Know the functions of the organs.
5. Understand how the particular organ functions are accomplished.
6. Be able to predict many of the symptoms that result from malfunction of a particular organ or system.
7. Become familiar with common diagnostic procedures and how patients are prepared for the examinations.
8. Gain a knowledge of how the human body attempts to compensate in the presence of disease.
9. Therapeutically support some of the normal body defense mechanisms.
10. Become a skillful observer and an accurate reporter of your observations.

S.R.B.

Contents

HUMAN BIOLOGY IN HEALTH AND IN DISEASE

1 Body Organization

VOCABULARY

Alveoli
Autolysis
Endocrine
Enzyme
Epidermis
Exocrine
Inorganic
Intervertebral disc
Metabolism
Mitosis
Organic
Ovary
Pubic symphysis
Thyroid
Viscera

OVERVIEW

I CELLS
- **A.** Nucleus
- **B.** Cytoplasm
- **C.** Organelles

II TISSUES
- **A.** Epithelial
  - **1.** Simple squamous
  - **2.** Stratified squamous
  - **3.** Cuboidal
  - **4.** Columnar
  - **5.** Glandular
- **B.** Connective
  - **1.** Areolar
  - **2.** Adipose
  - **3.** White fibrous
  - **4.** Elastic
  - **5.** Neuroglia
  - **6.** Cartilage
  - **7.** Osseous
  - **8.** Blood
- **C.** Muscle
- **D.** Nerve

III MEMBRANES
- **A.** Mucous membrane
- **B.** Serous membrane
- **C.** Synovial membrane
- **D.** Cutaneous membrane

IV SYSTEMS
- **A.** Respiratory
- **B.** Circulatory
- **C.** Musculoskeletal
- **D.** Gastrointestinal
- **E.** Urinary
- **F.** Nervous
- **G.** Endocrine
- **H.** Reproductive

V BODY CAVITIES
- **A.** Thoracic
- **B.** Abdominal
- **C.** Pelvic
- **D.** Dorsal

VI Regions of the Body
VII Other Descriptive Terminology

CELLS

The structural and functional unit of all living matter is the cell. Most cells can be seen only under a microscope. Cells are made up of protoplasm, which is mainly water containing various organic and inorganic substances as well as several important organelles (little organs). The cell is surrounded by a membrane that determines to some extent what substances will enter the cell from the liquid cellular environment.

The nucleus of the cell contains DNA (deoxyribonucleic acid), which constitutes the hereditary material of the cell. It is the cellular nucleus that determines the work of the particular cell as well as the reproduction of the cell itself. The nucleus is surrounded by cytoplasm.

Located in the cytoplasm are the organelles. The most important of the organelles are the endoplasmic reticulum, ribosomes, mitochondria, lysosomes, Golgi complex, centrioles, microtubules, and microfilaments. We shall consider briefly the functions of each of these organelles.

The ribosomes are attached to the endoplasmic reticulum, in which protein synthesis takes place. Some of the endoplasmic reticula have no ribosomes; scientists believe that lipids (fatty substances) are synthesized in these endoplasmic reticula.

The mitochondrion is the organelle that produces the energy for cellular work, such as the manufacture of enzymes and hormones, and secures from its environment the raw materials needed by the cell. This energy is a result of the breakdown of a high-energy phosphate compound, ATP (adenosine triphosphate). The ATP is resynthesized from the utilization of foodstuffs, chiefly carbohydrates.

The lysosomes contain powerful enzymes and are considered the digestive organs of the cell. These organelles also function to remove damaged cells or damaged portions of cells. Rupture of the lysosomes will result in autolysis, or death of the cell.

The centrioles play a major role in cellular division. These organelles are located close to the nucleus of the cell and become active during cellular division, or mitosis.

The Golgi complex is concerned with the transport of enzymes and hormones that have been manufactured by the cell to the outside of the cell. Once outside of the cell, they enter the bloodstream where they can be taken where they are needed in the body.

Microtubules and microfilaments are believed to be involved, directly or indirectly, in attainment and maintenance of cell shapes. These organelles also probably contribute to the motility of the cell.

In addition to the cytoplasmic organelles, found in the cytoplasm are ribonucleic acid, some DNA, glycogen, lipids, proteins, and inorganic substances. Some cells—such as muscle and nerve cells—also contain fibrils, microscopic threadlike structures.

Several different types of cells exist, and each has a specialized function. Cells of the same type combine to make tissues. Membranes are combinations of different types of tissues.

TISSUES

Simple squamous epithelial tissue is made of a single layer of thin flat cells. This type of tissue is found in the lungs, where it facilitates the exchange of gases between the air sacs (alveoli) and the blood capillaries. This tissue also provides a smooth lining for the blood vessels and lymphatics and facilitates the exchange of gases, nutrients,and products of cellular metabolism between the capillaries and other body cells.

Stratified squamous epithelium is made of several layers of cells and is found in areas of the body needing protection against friction. For example, it lines the nose and mouth and is part of the epidermis of the skin.

Simple cuboidal epithelium is found in certain glands, for example, in the thyroid and on the surface of the ovary. Cuboidal cells secrete glandular products.

Columnar epithelial tissue is made of cells that resemble columns. There are four types of columnar cells. The simple columnar cells that line the gastrointestinal tract facilitate the absorption of nutrients, or foods. Ciliated (*cilia* means lashlike processes) columnar cells have these processes on their free surface which help to move substances toward the outside of the body. These cells are found in the respiratory tract and in parts of the reproductive system. Goblet columnar cells, which are found in the top layer of mucous membrane, secrete mucus to keep the membrane moist. The end organs of certain nerve fibers, such as the taste buds and the rods and cones of the retina, are a special type of columnar cells.

Glandular epithelium is found in both the endocrine and the exocrine glands of the body. These cells secrete substances such as digestive juices, hormones, and perspiration.

Connective tissues are primarily a supporting and binding type of tissue. Connective tissue can readily repair itself if injured, and it also helps to repair damage to other types of tissue.

Areolar connective tissue, the most common of all connective tissues, is thin and glistening. It connects the skin and other membranes to their underlying structures and exists as packing around blood vessels and organs.

Adipose tissue cells have the ability to take up fat and store it. They help to insulate and to protect as well as to provide a reserve food supply.

White fibrous connective tissue is very strong and flexible. Tendons and ligaments are made of white fibrous tissue.

Elastic connective tissue is found in the walls of blood vessels and other organs that must stretch and regain their original shape.

Neuroglia is a delicate type of connective tissue that supports nerve cells.

There are three types of cartilagenous connective tissue. Hyaline cartilage, which is found at the ends of bones in freely movable joints, is glossy and bluish white. It forms the preliminary skeleton of the embryo. The intervertebral discs and the pubic symphysis contain fibrous cartilage. Elastic cartilage is found in the external ear.

Osseous connective tissue is bone. Deposits of calcium and phosphate salts cause the intercellular material of osseous tissue to be hard.

Blood is liquid connective tissue. All cells are dependent upon blood to receive their nutrients and to transport the products of cellular metabolism.

There are three types of muscle tissue. Skeletal muscle is voluntary (controlled by your mind) and is attached to the skeleton. Visceral muscle is not under voluntary control (it is involuntary) and is found in the walls of blood vessels and in many of the organs of digestion. Cardiac muscle is found only in the heart. It also is involuntary.

Nerve tissue is highly specialized with properties of irritability and conductivity. This tissue transmits nerve impulses that assist in the integration of all physiological functions.

MEMBRANES

Membranes are composed of combinations of epithelial and connective tissue.

Mucous membrane lines cavities that open to the outside of the body, such as the respiratory and gastrointestinal tracts. Serous membrane lines closed cavities and secretes a slippery serous fluid for protection from friction. The pleura, which covers the lungs, and the peritoneum, which covers the organs of the abdominal cavity, are examples of serous membranes. Synovial membrane is tough because of the presence of much fibrous tissue. Synovial membrane lines the cavities of the freely movable joints. Dense fibrous membrane is tough and opaque. It is primarily a protective membrane and covers the brain and bones. It also makes up the sclera, the outer coat of the eye.

The cutaneous membrane is the skin. Its principle functions are as follows: It protects the underlying structures from injury and bacterial invasion; it helps regulate body temperature; as a sense organ, it keeps us aware of our environment; and it helps to some extent in the excretion of some body water and soluble wastes.

Tissues and membranes will be discussed in more detail in Chapter 3.

SYSTEMS

Each of the body systems performs a specific function; however, all of the body systems are greatly dependent upon the proper functioning of each of the other systems. For this reason, disease processes of one system well may be evidenced by malfunctioning of some of the other systems. For example, patients with certain types of heart disease may have great difficulty breathing. We shall discuss briefly each of the body systems and the major functions of each system.

The respiratory system is primarily concerned with the exchange of gases between the bloodstream and the microscopic alveoli of the lungs. The most important gases are oxygen and carbon dioxide.

The circulatory system has two divisions, lymphatic and blood. The lymphatic division is composed of lymph vessels, lymph nodes, and other lymphatic tissue. Its chief function is to return tissue fluid to the bloodstream and to filter this fluid, removing the foreign material and bacteria before they can enter the bloodstream. The blood circulation depends upon the pumping action of the heart, which sends blood containing nutrients and oxygen into the arteries and capillaries, where these products are delivered to the tissue cells. The capillaries return the products of cellular metabolism to the veins and then to the heart. If the products of cellular metabolism are waste products, they will be carried to the organs of excretion, chiefly the kidneys and lungs. If the products of cellular metabolism are substances such as enzymes and hormones needed for proper functioning of other cells, the circulatory system will deliver these products to the cells where they are needed.

The skeletal system is composed of bones and joints. The chief functions of the skeletal system are to provide a supporting framework for the body and a place for the attachment of muscles and ligaments; to protect the viscera; and to aid in the production of blood cells.

There are three different types of muscles, skeletal, visceral and cardiac muscle. Skeletal muscles are voluntary; they provide for locomotion and support of the body. Smooth or visceral muscles, which are involuntary, are found in many of the organs of digestion, where they move the products of digestion through the gastrointestinal tract. Smooth muscles are also found in blood vessels, where they function to regulate the diameter of the blood vessels. Cardiac muscle is also involuntary. This muscle is located only in the heart.

The gastrointestinal system processes the foods we eat and renders these foodstuffs into an absorbable form, so that they can enter the bloodstream and be distributed to the cells that need them. The gastrointestinal tract also aids in eliminating waste products.

The urinary system is composed of the kidneys, ureters, urinary bladder, and urethra. This system—which brings about the elimination of soluble waste products—is very important in maintaining the acid-base balance of the body. This balance is critical to health and will be discussed in Chapter 17.

The brain, spinal cord, and peripheral nerves (nerves extending to and from the brain and cord) make up the nervous system. This system not only is concerned with our thought processes but also regulates both voluntary and involuntary actions of muscles and glands.

The endocrine system works together with the nervous system in controlling and integrating all of the normal biochemical processes of the body. The pituitary gland, thyroid gland, parathyroids, pancreas, adrenals, gonads, pineal, and thymus are some of the major organs of the endocrine system.

The reproductive system functions to continue the species. Organs of the male and female reproductive systems differ both in their structures and locations.

BODY CAVITIES

The thoracic cavity is protected by the rib cage. The floor of the thoracic cavity is the diaphragm. This cavity contains the heart, lungs, bronchi, great blood vessels, the thymus gland, esophagus, and lymphatic ducts.

The abdominal cavity is inferior (meaning below) to the diaphragm and contains the organs of digestion, spleen, kidneys, adrenal glands, liver, gall bladder, and pancreas. The abdominal aorta, the largest abdominal artery, sends branching blood vessels to supply these structures with blood. The inferior vena cava and large portal vein receive the venous blood from these organs.

The pelvic cavity is inferior to the abdominal cavity. This cavity contains the urinary bladder, some of the organs of reproduction, and the distal part of the gastrointestinal tract.

The dorsal cavity contains the brain and spinal cord as well as the spinal cord nerve roots. These vital structures are well protected by the bones of the skull and vertebral column.

REGIONS OF THE BODY

Figures 1–1 and 1–2 show the regions of the body. You should carefully learn these regions of the body, as this knowledge will be of great help to you in understanding subsequent lessons. For example, the pectoral muscles are located in the pectoral region. The brachial artery, vein, and nerve are located in the brachial region.

OTHER DESCRIPTIVE TERMINOLOGY

You must also learn the meaning of terms such as lateral, medial, proximal, distal, inferior, superior, anterior or ventral, and posterior or dorsal. Medial means close to the midline; lateral means toward the side. For example, the heart is medial to the lungs, and the lungs are lateral to the heart. Proximal means closest to the point of attachment; distal means farther away. The elbow is proximal to the hand, and the hand is distal to the elbow. Inferior means lower and superior means higher. Anterior or ventral means in front and posterior or dorsal is in back. In applying any of these terms the body is always in the anatomical position, as shown in Figure 1–1.

The body as a whole, as well as individual organs, can be divided in different planes. Figure 1–3 illustrates the sagittal, coronal, and transverse or horizontal planes of the body.

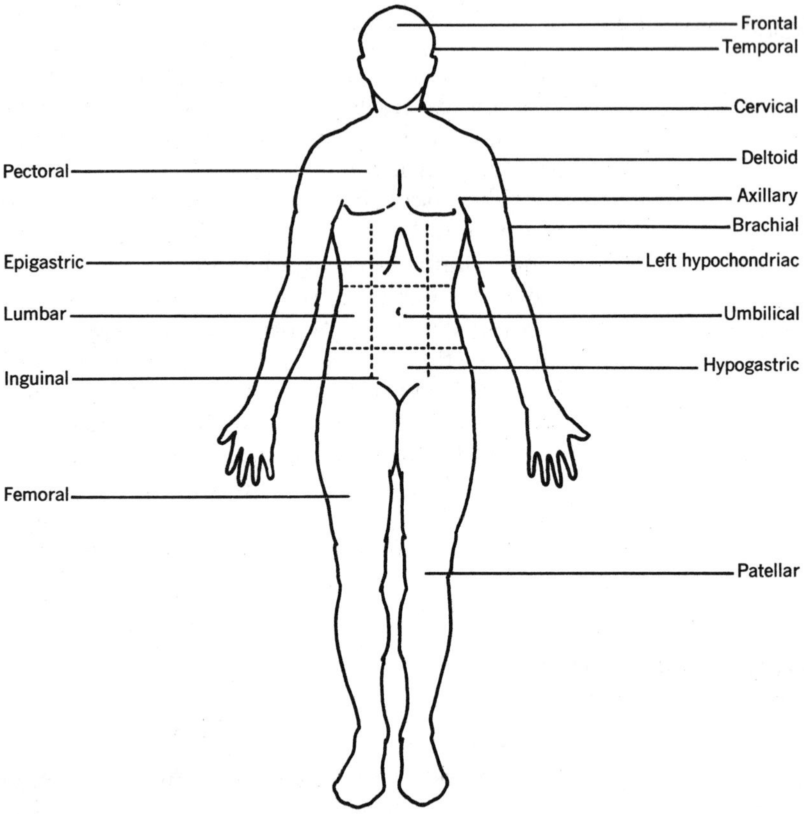

FIGURE 1-1
Anterior view of the human body in the anatomic position showing the regions of the body. Sometimes the regions of the abdomen are designated as the right and left upper quadrants and the right and left lower quadrants.

SUMMARY QUESTIONS

1. Name six cellular organelles and explain the function of each.
2. Where is simple squamous epithelial tissue located?
3. What is the function of stratified squamous epithelium?
4. Where are ciliated columnar cells located and what is the function of this type of cell?
5. What type of tissue is found in the tendons and ligaments?
6. Where is osseous tissue located and what are the functions of this type of tissue?
7. Name three types of muscle tissue and tell where each is located.
8. How do membranes differ in structure from tissues?
9. Where is mucous membrane located?

Seeing through the HUMAN BODY

IN COLORED ANATOMICAL TRANSPARENCIES

The accurate anatomical drawings, reproduced on transparent acetate film by "Trans-Vision®" process on the following pages, give you a unique opportunity to see through the human body. Each of the following anatomical plates (I-VIII) permits you to see a great many different parts of the human body from both their front and back views and in relationship to neighboring anatomical structures. By this unusual method of presentation you get a more accurate and intimate view of the interior of the body than you might otherwise ever obtain (at least outside a dissecting room).

The right-hand plates (II, IV, VI, and VIII) show you the front views, and the left-hand plates (I, III, V, and VII) show you the back views of the same organs. Each plate carries a list of a great many body parts visible on it. The body parts are numbered both on the drawings and in the list for quick identification.

An *alphabetical index* to the pictured and numbered body parts on all eight plates has been provided on the back of plate VIII, at the end of this special insert, for easy reference to the many body parts revealed in detail on the plates.

PLATE **I** shows the front skin of the body seen from its underside along with networks of veins which course between the skin and muscles. The veins are cut at points where they penetrate to deeper levels to join other veins returning blood to the heart.

PLATE **II** shows muscles of the face, neck, chest, and abdomen. On the figure's right side, muscles of the chest and abdomen are removed to show the front of the rib cage. Also shown are the thin membranous coverings, called fascia, of the thigh, arm, and forearm muscles.

PLATE **III** shows (from within) the cranial cavity, parts of the face, the rib cage, and muscles of the chest and abdomen. Note how the diaphragm separates the chest cavity from the abdominal cavity. As you follow the levels of the diaphragm on the succeeding plates, you will notice how much lower is its attachment on the back of the abdominal wall than on the front.

PLATE **IV** shows the brain and eyeball exposed and deeper muscles of the pharynx. Also shown are the three lobes of the right lung and the two lobes of the left lung. Below the diaphragm are the liver, gallbladder, stomach, and large and small intestines. Note the appendix hanging from the cecum.

PLATE **V** shows a mid-line (sagittal) section through the head revealing details of the brain, nasopharynx, and larynx. The lungs are cut through in a frontal section to show the space occupied by the heart between the two lungs. Below the diaphragm are back views of the abdominal organs whose front views appear on plate IV.

PLATE **VI** shows the septum that separates the two hemispheres of the brain, the nasal septum, the heart, and major blood vessels of the trunk and limbs. Now exposed below the diaphragm are the kidneys, pancreas, spleen, and duodenum.

PLATE **VII** shows the back views of the structures shown on plate VI. Note that the left kidney is located slightly higher than the right kidney and that the ureters empty into the bladder from behind.

PLATE **VIII** shows the left half of the head seen from a mid-line (sagittal) view, an inside view of the back of the rib cage, the front of the vertebral column and pelvis, and some of the major nerves emerging from the spinal cord. On the figure's left side are shown the deep muscles of the trunk, arm, and thigh. On the figure's right side these deep muscles are removed to show the underlying bones.

Drawings by ERNEST W. BECK, B.S., M.A., in consultation with Harry Monsen, Ph.D., Professor of Anatomy, College of Medicine, University of Illinois.

PLATE

I

This view shows the skin of the front of the body seen from its underside along with networks of veins which course between the skin and muscles. The veins are cut at points where they penetrate to deeper levels to join other veins returning blood to the heart.

Basilic vein (10)
Cephalic vein (27)
Cubital vein, median (35)
Epigastric vein, superficial (40)
Saphenous vein, great (120)
Thoracoepigastric vein (139)

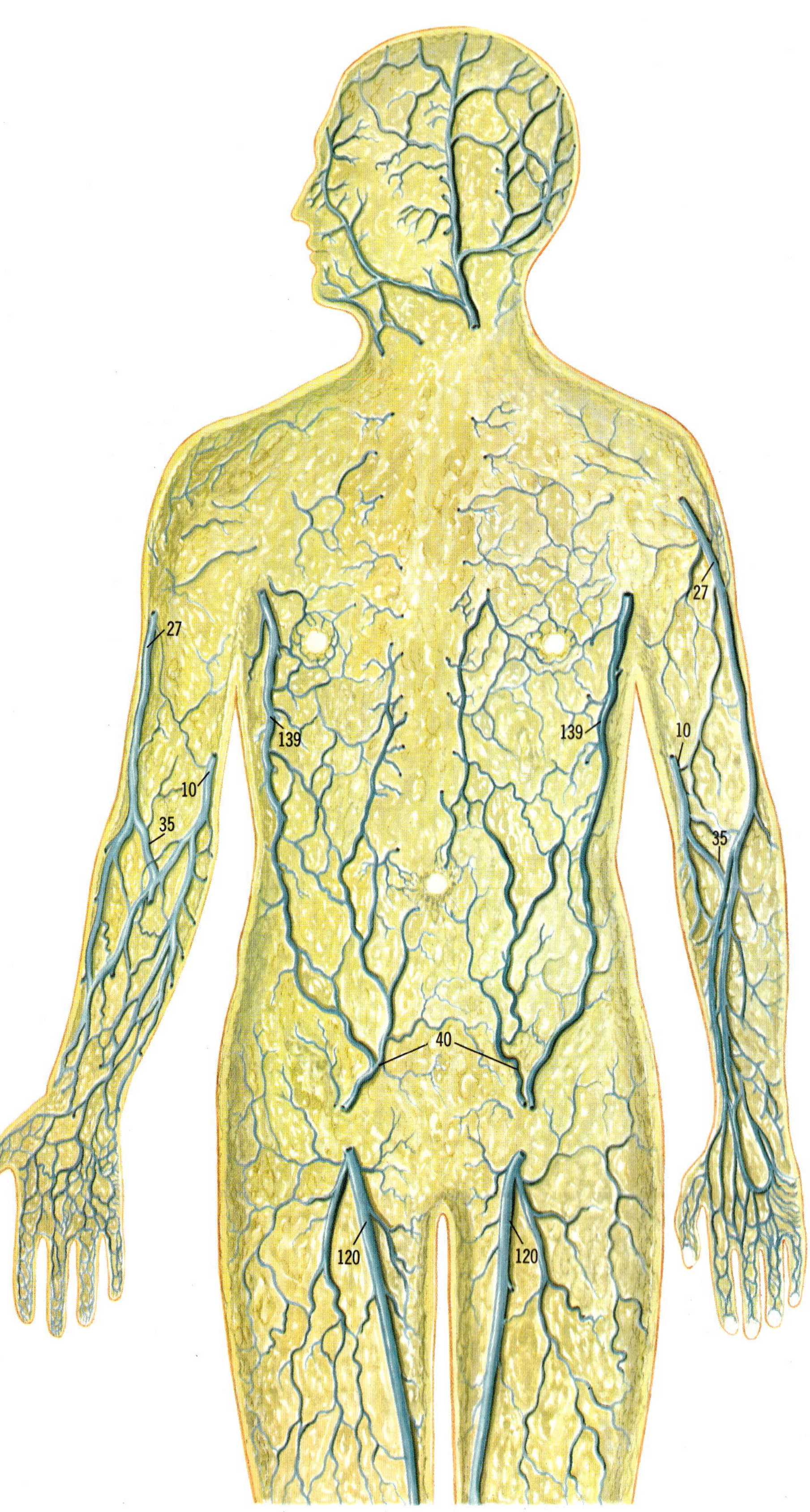

PLATE VIII

Adductor muscles (2)
Auditory (Eustachian) tube opening into pharynx (8)
Brachial nerve plexus (17)
Brachialis muscle (18)
Brain (22)
Carpal (wrist) bones (25)
Cerebellum (28)
Cerebrum (29)
Corpus callosum (33)
Deltoid muscle (36)
Diaphragm (37)
Femoral nerve (48)
Femur (49)
Forearm muscles (deep flexors of hand) (50)
Frontal bone (52)
Frontal sinus (53)
Gluteus medius muscle (56)
Humerus (60)
Iliacus muscle (61)
Ilium (62)
Intercostal (rib) artery, vein and nerve (64)
Intercostal (rib) muscles, external (65)
Intercostal (rib) muscles, internal (66)
Intervertebral disc (cartilage) (68)
Ischium (71)
Jugular vein, internal (72)
Larynx (75)
Mandible (78)
Maxilla (80)
Medulla oblongata (83)
Metacarpal (hand) bones (86)
Nasal turbinates (88)
Occipital bone (89)
Parietal bone (97)
Pituitary gland (101)
Phalanges (finger)bones (102)
Pons (103)
Psoas major and minor muscles (105)
Quadratus lumborum muscle (109)
Radius (111)
Rib (114)
Sacral nerve plexus (116)
Sacrum (117)
Scapula (122)
Sciatic nerve (123)
Sphenoid sinus (125)
Spinal cord (126)
Subscapularis muscle (135)
Thigh muscles (137)
b. Vastus intermedius
c. Vastus lateralis
d. Vastus medialis
Tongue (142)
Trapezius muscle (144)
Triceps brachii muscle (145)
Ulna (146)
Vertebral column (backbone) (153)

TRANS-VISION® MILPRINT, INC.
MILWAUKEE, WISCONSIN

INDEX

10. Name three functions of the cutaneous membrane.
11. Name the two divisions of the circulatory system and tell the function of each.
12. What are the functions of the skeletal system?
13. Name the structures of the urinary system.
14. What structures are located in the dorsal cavity?
15. Name the structure that separates the thoracic and abdominal cavities.
16. What organs are located in the pelvic cavity?
17. Where is the epigastric region?
18. Where is the hypogastric region?
19. Where is the hypochondriac region?
20. Name the region between the shoulder and elbow.
21. How do you use the terms proximal and distal?
22. How do you use the terms medial and lateral?

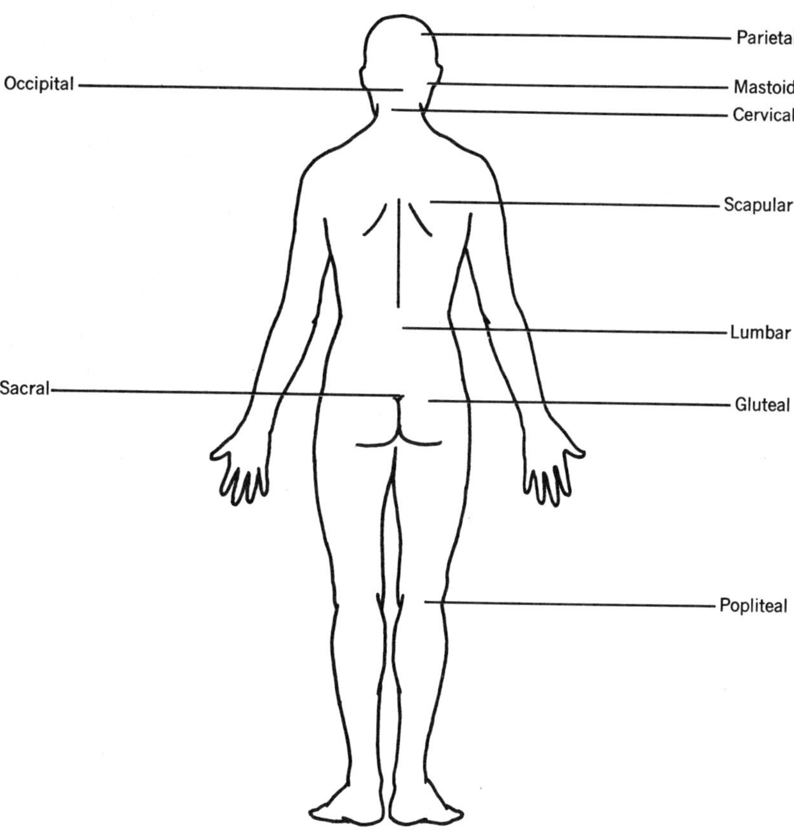

FIGURE 1-2
Posterior view of the human body and regions.

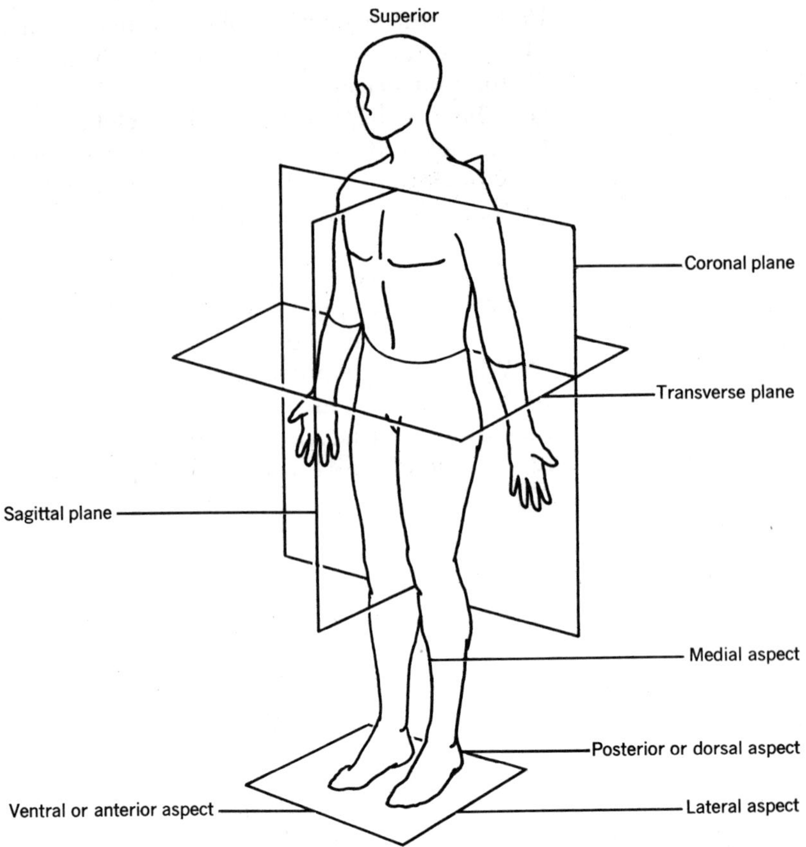

FIGURE 1-3
Planes of the body and terms of location and position.

23. Name the plane that divides the body into right and left sides.
24. Name the plane that divides the body into anterior and posterior parts.
25. Name the plane that divides the body into superior and inferior parts.

SUGGESTED READINGS

Brooks, S.M. *Integrated Basic Science.* 2nd ed. St. Louis, Mo.: The C.V. Mosby Co., 1970.

Chaffee, E., and E.M. Greisheimer, *Basic Physiology and Anatomy.* 2nd ed. Philadelphia, Pa.: J.B. Lippincott Co., 1971.

Nourse, A.E. *The Body.* Life Science Library. New York, N.Y.: Time-Life Books, 1964.

2 The Fundamental Processes of Disease

VOCABULARY

Achlorhydria
Adduction
Allergen
Anabolism
Anorexia
Antibody
Anticoagulant
Antihistamine
Aseptic
Aspirate
Bacteriostatic
Benign
Calculi
Catabolism
Catheter
Cilia
Contracture
Cutaneous
Dehydration
Dilitation
Enzyme
Epidermis
Epithelium
Eschar
Estrogen
Flexion
Genetic
Hematoma
Hypotension
Immunity
Inflammation
Interstitial
Intravenous
Malignant
Metastasize
Myocitis ossificans
Orifice
Osteoarthropathy
Palliative
Phagocytosis
Proximal
Radiation
Thrombus
Toxin
Vasodilatation
Vasospasm
Virulence
Viscera

OVERVIEW

I ETIOLOGY OF DISEASE

- **A.** Predisposing causes
  - **1.** Genetic
  - **2.** Dietary deficiencies
  - **3.** Inadequate Protection
  - **4.** Ignorance
  - **5.** Emotional problems
- **B.** Direct causes
  - **1.** Poisons
  - **2.** Biochemical alterations
  - **3.** Oxygen lack
  - **4.** Allergy and hypersensitivity
  - **5.** Neoplasms
  - **6.** Bacterial or viral invasion
  - **7.** Trauma
    - **a.** Cold trauma
    - **b.** Heatstroke and heat exhaustion
    - **c.** Burns

Pathology is the study of changes that take place in the body as a result of disease. They may be gross changes in structure such as atrophy or hypertrophy of an organ. In the case of atrophy, the part becomes visibly smaller than normal, and in hypertrophy its size is increased. These structural changes are also accompanied by microscopic changes. The study of the cell, which is the structural and functional unit of the body, is called cytology. The microscopic study of tissues is called histology. In many instances, structural changes result in alterations in the physiology or functions of the organs involved. Occasionally, there are functional changes without any apparent structural change, in which case the disease is said to be functional as opposed to organic.

ETIOLOGY OF DISEASE

PREDISPOSING CAUSES

There are many predisposing causes of illness. Essentially all are a result of a decreased body resistance. This decreased body resistance can be the consequence of many different factors or combinations of factors.

Our abilities and limitations are also a combination of many different factors, one of which is our heredity. That there are inherited predispositions to certain diseases has been known by scientists for many years. However, in the last decade great advances have been made in the field of genetics and genetic counseling. In subsequent chapters we shall discuss in greater detail some of the mechanisms of genetic transmission of both normal and pathological traits.

Improper diet, either in quantity or quality, can predispose to disease. Dietary deficiencies and the consequences of poor dietary practices are discussed in detail in Chapter 14. We shall point out here that the food we eat is the fuel used to accomplish the metabolic processes necessary to maintain health.

Inadequate protection obviously predisposes one to illness. Some of the obvious offenders in this area are improper or inadequate clothing, and improper sanitation. Every year many people also die or are injured because of dangerous cars, power tools, faulty heaters, wiring, or ventilation.

In recent years we have finally turned our attention to environmental factors, which affect our health and that of future generations. We are becoming increasingly aware of the devastating effects of overcrowding and noise. Such factors can predispose the population to many disabilities.

One of the most tragic predisposing causes of disease is the lack of information, or, worse still, inaccurate information. Health teaching is an important role for all members of the health professions. A well-informed person is better equipped to prevent disease and, if disease is present, to help the professionals in their roles of correcting the problem.

Many things that were once considered helpful or at least harmless are indeed injurious to our health. One of the most dramatic examples of how inaccurate information can prove disastrous occurred in early attempts to treat morphine addicts. During the Civil War, and for some time thereafter, morphine was considered safe for relieving pain. Not only was it given freely to those in acute pain, but it was also administered long into their convalescence. Many people became addicted to morphine. However, doctors finally recognized that, although often necessary in acute situations, morphine is not without serious harmful effects when used over prolonged periods. They then discovered that a new drug could be given to the addict—a drug that was quite effective in curing the morphine addiction. The name of the drug was heroin!

Before discussing some specific causes of disease, we shall consider one other predisposing cause, that of unhappiness. At best, this cause is vague, and perhaps not even scientifically well substantiated. It may have been this relationship between a person's outlook on life and his health that the Biblical psalmist recognized when he wrote "That which I feared has come upon me." Empirical evidence does exist that unhappy people are disease- and accident-prone. In a hospital setting, of course, patients are likely to be unhappy. Consider, however, your acquaintances and associates who are generally unhappy or who seem to derive little pleasure from their occupations and every-day activities. This group of people seems to be afflicted with all sorts of ailments with greater frequency and greater severity than are those of us who are more satisfied and optimistic about our lives. We do not insist that there is a cause and effect relationship, but merely that there appears to be a correlation. There are people with serious disabilities who are essentially optimistic. Happy people are much better able to cope with their unfortunate situations than are their unhappy counterparts.

DIRECT CAUSES OF DISEASE

Many different types of agents can directly cause structural, and or, functional changes in the body. When the cause of the disease (its etiology) can be identified, the management of the patient is generally directed toward removing this cause. If the etiology is unknown or if as yet we have no means of counteracting its cause, the treatment is more difficult and can usually only be of a palliative nature. Treatment then involves minimizing the symptoms, supporting the normal body defenses, and helping retard by various means or limiting the progression of the disease.

Poisons

Although poisons will be discussed more thoroughly in Chapter 23, these toxins are mentioned here because they are important agents in the etiology of disease. Many if ingested or inhaled can seriously damage the lining of the gastrointestinal or respiratory tracts. Others can cause lethal damage to the kidneys, liver, or other vital organs.

Biochemical Alterations

Biochemical alterations that can cause disease are mainly the result of pathology of the endocrine glands. These glands, together with the nervous system, control many important metabolic activities of the body. Clearly any pathology of these glands can seriously impair normal body functions. In Chapter 22 we shall explore this complex matter in greater depth.

Oxygen Lack

Whether a deficiency of oxygen is the result of an inability of the respiratory tract to oxygenate the blood properly or whether the circulatory system is unable to deliver adequate amounts of oxygen to the tissue cells, disease will result. Dyspnea, or labored breathing, may be subjective or it may be due to anoxia (the lack of oxygen). Ischemia is the lack of adequate amounts of oxygen in the tissue cells. If the ischemia is severe, necrosis (tissue death) will result.

Allergy and Hypersensitivity

Allergy is a result of altered reactions of the tissues in certain individuals on exposure to agents that are innocuous to most other people. Sensitivity is a term closely related to allergy; often the two may be indistinguishable. There are both an immediate and a delayed type of sensitivity. A delayed reaction appears from 24 to 48 hours after exposure. In this type, there are no circulating antibodies. Delayed sensitivity can be transferred only with whole cells. A positive tuberculin reaction is an example of delayed sensitivity.

Immediate sensitivity can be life threatening. In this type of sensitivity there are circulating antibodies in the serum, the fluid portion of the blood. The treatment of the allergic reaction is primarily symptomatic. The patient's symptoms may be relieved by an antihistamine drug, for example, pyribenzamine, or in the case of asthma, a drug such as aminophylline, which will cause bronchial dilitation and relieve the respiratory distress.

In process of developing hypersensitivity, a susceptible person is exposed to the allergen and develops an antibody to this allergen. Upon a later exposure to the allergen to which he has been sensitized, he will experience an antigen-antibody reaction, and the allergic signs and symptoms will develop.

The treatment of allergy is complex. Perhaps, the individual can be desensitized by taking small periodic doses of the allergen. In some forms of allergy, this can be accomplished only with great difficulty, and the allergic individual may have to avoid exposure to the allergen completely.

The types of substances that may cause allergic reactions in sensitive individuals include inhaled dust, molds, and animal dander. Certain foods or drugs may also be a source of an allergic reaction. Some people are sensitive to weeds, particularly poison ivy, when they come into contact with their skin. Insect bites or stings can also cause allergic reactions.

Neoplasms

At the present time there is not a great deal known about the causes of neoplasms or tumors. Some neoplasms are benign; others are malignant. Malignant tumors spread to other parts of the body by means of direct extension (they metastasize), via the bloodstream, or by way of the lymphatic system. Because malignant tumors undergo more rapid cellular division than do other cells of the body, they demand a large blood supply. The inability of the body to provide adequate blood supply can result in necrosis of the center of the tumor. Because both benign and malignant tumors may become quite large, some of the symptoms produced are a result of pressure on surrounding structures.

In the case of benign tumors, which do not metastasize (spread) to other parts of the body, surgery is usually the chosen treatment. Surgery is also used for treatment of malignant tumors, but in this case the surgical procedure is usually much more radical. Malignant tumors also are sometimes treated with radiation or with antineoplastic drugs. These drugs retard cellular division and decrease the growth of the tumor. Often, first the tumor is radiated in an attempt to reduce its size, and then surgery is performed. Radiation therapy following surgery may be continued for a prolonged period of time in the hope of destroying any malignant cells not removed at surgery. At the present time, immunotherapy for the treatment of cancer is a relatively new approach; the results of this type of therapy seem promising. Immunotherapy is based on the theory that the patient with cancer has a defect in his immunity system. Treatment is directed toward improving this resistance.

Bacterial or Viral Invasion

Certain types of microorganisms can cause disease when body resistance is decreased, when they invade parts of the body that normally they do not inhabit, or when their numbers or virulence is increased. Several factors influence the severity of the diseases caused by microorganisms.

Microorganisms normally inhabit many parts of the body. When they are in their usual habitat they usually do not cause disease. If, however, they invade other parts of the body, they will very likely cause disease. Infections caused by these resident bacteria are not communicable.

Communicable infections are caused by nonresident organisms that are transmitted either directly or indirectly from one host to another. These organisms enter the body through some portal: most organisms cannot invade the intact skin; however, they can enter the body through the respiratory tract, the gastrointestinal tract, or the genito-urinary tract.

As the number of microorganisms increases, the likelihood of an infection and the severity of that infection increase. The strength or virulence of the organism is also a factor in determining the extent of the infection. Virulence usually increases by rapid transfer of the organism through a series of susceptible individuals.

A final important factor in determining the extent of the infection is the defensive powers of the host. These powers include anatomical and chemical barriers, immunity, and the ability of the phagocytic cells to destroy the invaders. Normal body defenses will be discussed in detail later in this chapter.

Trauma

Injury or trauma is one of the more obvious causes of disease. There are many different types of traumatic agents. All result in varying degrees of damage to the cells and consequent disturbances in body structure and function. We shall discuss briefly a few general types of trauma. Later chapters describe more fully the result of trauma to the various organ systems.

Cold Trauma. Extreme changes in environmental temperature can result in serious cellular damage. Injury due to cold is much more severe if the cold is moist and if the exposure has occurred during prolonged periods of inactivity. If the patient also has some preexisting condition that interferes with normal blood flow to the part involved or if the blood circulation is diminished by tight clothing, the trauma is likely to be more serious than it would be otherwise. The very young and the aged are especially vulnerable to cellular damage from cold.

In response to cold injury, initially either blanching or erythema (unusual redness) of the affected part takes place. With pressure, the erythematous part blanches. Later, swelling and an abnormal collection of fluid in the part occur. This collection of fluid is called edema. Superficial blisters develop and may break down and form ulcers or even tissue death (gangrene) if the blood circulation is inadequate. In such cases, the trauma may be complicated by infection.

Three main aspects of treatment exist for patients who have suffered injury due to extreme cold. First, a relatively low environmental temperature should be maintained in order to reduce metabolism to the tissues involved, so that the available blood supply may be adequate to prevent gangrene. Second, maximum circulation must be maintained. Vessels in the traumatized area are likely to go into spasm (vasospasm), further impairing circulation and favoring blood clot formation, which may completely occlude (obstruct) circulation to the part. The vasospasms may be relieved by keeping the uninvolved parts of the body warm and thereby producing a reflex vasodilatation. Medications that favor vasodilatation, such as papaverine, may be helpful. To reduce the tendency toward clot formation it may be necessary to administer some anticoagulant, such as heparin. Finally, every effort must be made to prevent the wound from becoming infected, by the use of soft, nonirritating, sterile dressings and by the application of antiseptics. If infection does occur, specific antibiotic therapy is necessary.

Heatstroke and Heat Exhaustion. Heatstroke is due to prolonged exposure on a very hot day. The body's heat-regulatory center in the brain may lose control and the body temperature may be increased greatly. Patients do not perspire; they are weak and dizzy. Their skin is red, dry, and hot to the touch. If the condition is severe, convulsions may occur. The immediate treatment is to cool the patient with cold sponge baths and ice packs placed on the forehead and in the axilla, for sustained high fevers can result in brain damage.

Heat exhaustion is characterized by circulatory disturbances caused by excessive loss of salt and water from sweating. Such patients are pale and feel dizzy and faint. In addition to cooling a person with heat exhaustion, fluids and salts must be replaced.

Burns. In the case of trauma due to thermal burns, the severity of the damage depends upon several factors, mainly the depth of the burn and the extent of the surface area involved. Generally speaking, burns that involve only superficial tissues are not as serious. However, when large areas of the body are superficially burned, as in the case of sunburn, severe disruption of body physiology can take place.

In estimating the severity of a burn with respect to the depth of tissues involved, we describe the severity as being first, second, or third degree. In first degree burns, the skin is reddened, there are no blisters, but there may be some edema. A first degree burn usually peels, or desquamates, in three to six days and heals without difficulty. Second degree burns are characterized by blisters that contain fluid having a similar composition to that of blood plasma. Most of the epithelial layers of the skin, but not the deepest layers of the skin, are involved. The deepest layers contain the nerve endings (see Chapter 3 for a more complete discussion of the anatomy and physiology of the skin); for this reason, second degree burns are very painful. Second degree burns heal without scar formation in 10 to 14 days, if they do not become infected. In third degree burns the entire skin and subcutaneous tissues in the burned area are involved. These burns develop massive edema within a very short time following the burn. There will be necrosis of the involved tissues. This type of burn heals with scar formation.

The extent of surface area involved in a burn is also used to estimate the severity of the trauma. The "rule of nines" is used to determine the percentage of the surface area involved (see Figure 2–1). Severe first and second degree sunburns that involve as much

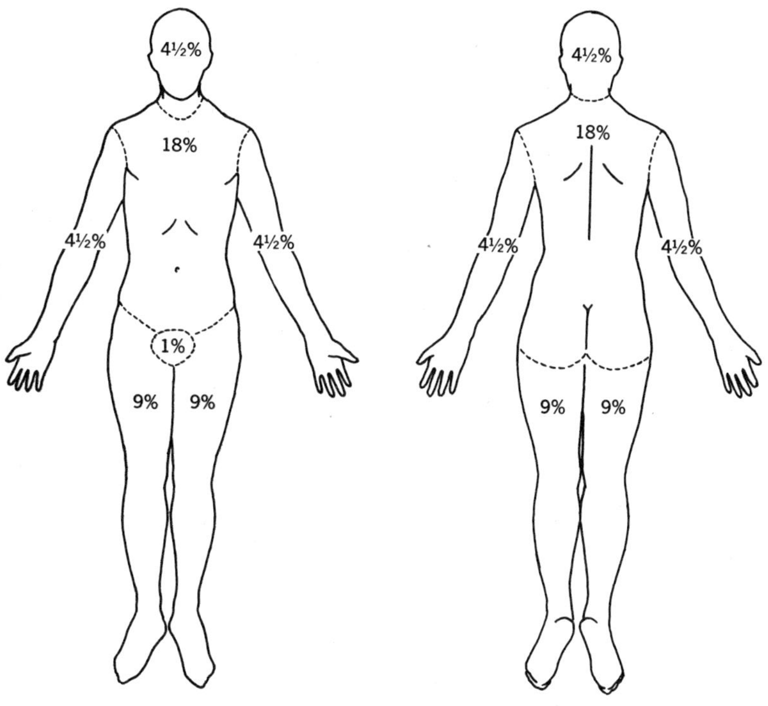

Region	Percentage of Body Surface
Head and neck	9
Right arm	9
Left arm	9
Anterior surface of trunk	18
Posterior surface of trunk and buttocks	18
Right leg and thigh	18
Left leg and thigh	18
Perineum and genitalia	1

FIGURE 2-1

Anterior and posterior views of the body divided into areas—9 percent or multiples thereof—for the purpose of quickly estimating amounts of burned surface. The Rule of Nines is usually attributed to Pulaski and Tennison.

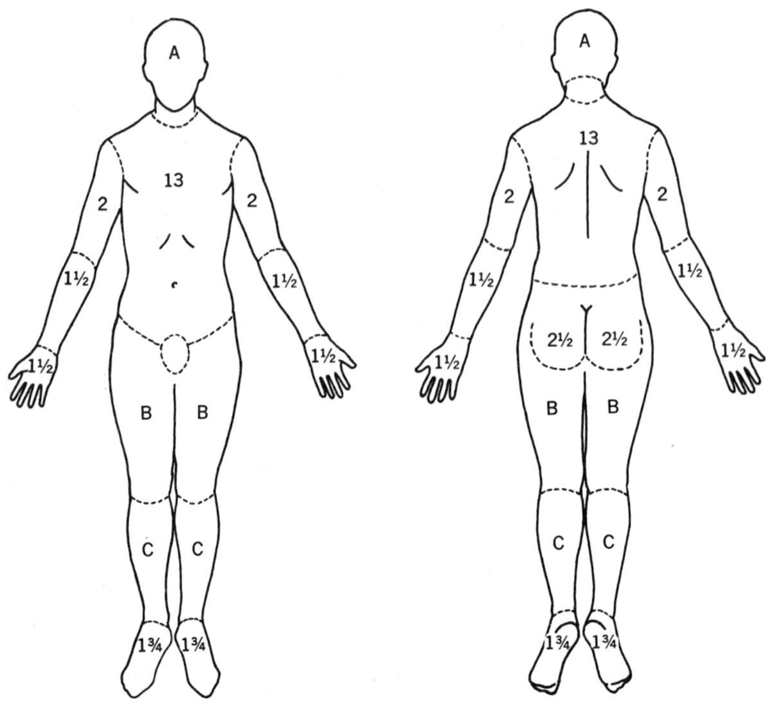

Areas	Age in Years					
	0	1	5	10	15	Adult
A – ½ of head	9½	8½	6½	5½	4½	3½
B – ½ of one thigh	2¾	3¼	4	4½	4½	4¾
C – ½ of one leg	2½	2½	2¾	3	3¼	3½

FIGURE 2-2
Body proportions change with growth. Shown here, in percentages, is the relationship of body area to whole body surface at various ages. This method of determining the extent of the burned area is attributed to Lund and Browder.

as 70 to 75 percent of the body surface cause more disturbances in body function than does a third degree burn of the finger tip. When 40 percent of the surface area is involved, regardless of the depth of the burn, there is cause for concern; if these happen to be mostly third degree, the trauma may be fatal. People rarely survive burns involving 80 percent of the body surface, particularly if the burns are mostly second and third degree burns.

In the case of electrical burns, the skin first turns white and then black. These burns are usually severe and may involve not only the skin but also blood vessels, muscles, tendons, and even bones. As a result of the electrical shock, the strong skeletal muscles may have such severe contractions that they cause fractures of bones. Patients may die from heart failure due to the electrical shock before any first aid is available. The first concern with electrical burns should be to maintain cardiopulmonary function and the second, to attend to the burns.

Regardless of the cause of burns, these injuries so severely disrupt the whole body physiology that patients with pre-existing health problems such as kidney disease, lung disease, or heart disease can have serious trouble with what otherwise might be a relatively mild burn trauma. Again, the very young and the elderly are especially vulnerable to burns.

Treatment. The current practice of first aid for a burn victim is to cover the area, ideally with a moist sterile covering. No two body surfaces should be in contact with each other while the patient is being transported to a treatment center. No ointments or salves should be applied. Although such lotions may temporarily help to relieve the pain by sealing off the burned area from the air, they may be a source of infection and will have to be scrubbed off when hospital treatment is instituted. This procedure may cause additional tissue damage.

If the patient is alert and is not in danger of vomiting, it may be helpful to give him sips of a mixture of fluid containing a teaspoon of salt and a half teaspoon of baking soda in a quart of water. This should be done only if more sophisticated treatment is likely to be delayed for some time. The mixture itself can be nauseating: If the patient aspirates (breaths in) the vomitus, the resulting respiratory tract damage may seriously complicate his recovery.

Under ordinary circumstances you would not become involved to any great extent in the direct care of badly burned patients. Since trauma due to burns may involve large numbers of victims and such massive numbers may overtax the resources of the community, we shall discuss in detail the management of burns.

The first 48 hours following a severe burn are critical. This initial shock phase is characterized by edema, which develops rapidly in the first 8 hours. There is severe dehydration in the unburned areas. Because so much blood plasma enters the burn edema, the circulating volume of the blood may be seriously jeopardized. Unless replacement is started quickly, irreversible shock may ensue. In the absence of respiratory distress, the most urgent need is that of intravenous fluids. Patients are very thirsty but must not be given tap water by mouth. Unlimited tap water may lead to water intoxication. They can, however be given electrolyte solutions (fluids containing salts) by mouth, but no more than 200 ml. (about three fourths of a cup) without a doctor's order. They should not be given any citrus fruit juices which contain large amounts of potassium. Because potassium from the damaged tissues passes rapidly into the bloodstream, patients probably have a dangerously high blood level of potassium during the first few days following their injury. Any oral fluids may cause gastric distention and vomiting, which will lead to even greater dehydration.

Patients should have a urinary catheter in place. The urinary output should be measured hourly; it should be about 30 to 50 ml. (approximately an ounce and a half) per hour. During the early phase of the burn treatment, intravenous fluids should be administered to maintain this amount of urinary output. During the first 48 hours any urinary output less than 10 ml. or over 80 ml. must be reported, as it may indicate that additional problems are developing.

Respiratory tract damage due to the inhalation of smoke and fumes particularly if the burn has occurred indoors, can be a serious complication of burns. Labored breathing is the most frequent symptom of respiratory tract damage. Coughing and the production of blood or smoke-tinged sputum are ominous signs. Damage to the respiratory tract almost always leads to infection, which will be evidenced by purulent sputum on about the third or fourth day following the burn.

If dressings are applied, they must be done with great care and, of course, under aseptic conditions. Dressings must be applied with uniform pressure, especially on the extremities. They must also be applied in such a way as to assure that the patient is in good body alignment and so that no two skin surfaces are in contact.

During the early period following the burn, movement of the burned parts should be discouraged, even if the burns are dressed. While eschar is forming, motion may crack this covering, delay healing, and provide a pathway for infection.

After about the second to the fourth day following the burn a reabsorption of the burn edema and a great increase in the urinary output will take place. The urinary output may total as much as 10,000 ml. over 24 hours. Fluid intake must be restricted, for one of the greatest dangers during this period is respiratory distress due to accumulation of edema fluid in the lungs.

In concluding a discussion of the critical aspects of the care and observations of acute burns, we shall mention the most common complications of severe burns. Shock, respiratory failure, and kidney failure are the most serious early complications. Later, infections, anemia, malnutrition, contractures, and gastrointestinal disturbances including peptic ulcers may occur. You may obtain a more thorough discussion of each of these complications in subsequent chapters.

Radiation Injury. Radioactive substances (some occurring naturally) give off alpha, beta, and gamma rays that can cause tissue damage. X-rays are man made gamma rays. Although the damage caused by gamma and X-rays is much more severe than that caused by the alpha and beta rays, all are potentially hazardous. The duration of the exposure to the rays and the distance between the exposed individual and the source of the radiation are important factors in determining the degree of radiation damage.

The body cells most vulnerable to radiation injury are those that most frequently undergo cell division. White blood cells and cells of the ovaries and testes are particularly sensitive to radiation. To protect from radiation injury it is necessary to limit the time of exposure, to maintain as much distance from the source of radiation as possible and to shield the body from the rays. In chapter 24 we shall discuss the effects of radiation more extensively.

Wounds of Bone and Soft Tissue

Injury to bones will be discussed in detail in Chapter 6. However, we shall point out here that fractures of bones can be caused not only by a direct blow but also by diseases of the bone that interfere with their ability to provide adequate structural supports. A compound fracture is one in which there are fragments of broken bone protruding through the skin. A simple fracture is one in which there may be some damage to the nearby soft tissue, but no bone protrudes through the skin. A comminuted fracture results in the shattering of the bone into many small fragments. Fragments of bone may be imbedded in the surrounding soft tissue. A greenstick fracture, which occurs almost exclusively in young children, is an incomplete break in the bone.

The following terms are used to described trauma to soft tissue. An abrasion is a superficial lesion that results from the scraping or rubbing off of skin or mucous membrane. Although this type of trauma is usually not serious, it is quite painful. Abrasions should be cleansed with soap and water. If bits of dirt or debris have entered the wound, they can best be removed by application of hydrogen peroxide. Stronger antiseptics such as iodine (particularly if the solution is not fresh and has become concentrated) may actually cause additional damage to the delicate abraded tissues.

A scab will form on the abrasion usually within a day. This scab provides protection to the wounded tissues while healing is occurring and therefore should not be removed unless infection occurs.

A laceration is a wound made by tearing. Again, the most important aspect of the care of the wound is cleanliness. Bleeding is not usually a problem of any major proportion. If there is considerable bleeding, it can usually be controlled by elevation of the part or by applying ice for a short period of time. Venous bleeding (bleeding from the veins) is easily controlled because venous pressure is low and veins are thin walled and collapse when they are opened. As artery walls are much thicker and do not collapse when cut, arterial bleeding is another matter. In this instance, a tourniquet should be applied close to the laceration for about ten minutes, at which time the bleeding should have stopped. If it has not, the wound will have to have surgical attention. Because the laceration is due to tearing, the edges of the wound are likely to be irregular. It may therefore have to be sutured if a healed scar is cosmetically undesirable or if the wound is extensive. If the wound is a minor laceration on a part of the body where appearance is not an important consideration, the edges of the wound can be brought closely together and taped securely in place with a bandaid. When adhesive tape is removed from the dressing of any wound, it should always be pulled toward the wound—first from one side and then from the other. Because children frequently fear the dressing change, first the tape can be moistened with a little mineral oil to allow it to be removed without undue pulling of the skin. If the dressing must be reapplied, the skin must be thoroughly cleansed before a new dressing can be applied. For people who are sensitive to adhesive tape, it may be helpful first to paint the skin where the tape is to be applied with tincture of benzoin, which will protect the skin.

A bruise, or ecchymosis, is due to the rupture of small superficial blood vessels in the wounded area. If there is a large amount of bleeding from these ruptured vessels, a hematoma may be formed.

Penetrating wounds that damage only a small area of skin but extend deep into the body tissues can be especially dangerous. Skin heals readily, but the underlying tissues heal far more slowly. If the wound becomes infected, it may have to be reopened to allow drainage. In the case of any wound that may be contaminated, the person should be given protection against tetanus: 250 units of Hypertet or 0.5 ml. of tetanus toxoid for an adult will provide temporary protection against this complication.

Wound Healing. Wound healing differs very little from one type of tissue to another. It is generally independent of the type of injury.

Let us consider some of the events involved in the healing of a wound. First, blood flows into the wound and fills the space with clots that help to unite the edges of the wound by retracting. Clot retraction usually begins about one half hour after the wound is incurred. In several hours, the clot loses fluid and forms a hard protective scab.

After the clot has formed, the injured and dying tissues produce necrosin, a substance that causes the nearby blood vessels to become more permeable; serum containing albumin, globulin, and antibodies leaks out of the vessels. These substances may be able to attack any microorganisms that are in the wound, and the fluid also provides a sustaining environment for the white blood cells that will appear about 6 hours later. The neutrophils, the first white blood cells to arrive, push a bit of their cytoplasm between the cells of the blood vessel wall and squeeze out. They may be able to ingest the organisms and digest most of the remains. If no bacteria are in the wound, the neutrophils rupture and release enzymes to attack the cellular debris so that it can be removed more easily. These white blood cells are attracted to the area because the injured tissues produce a substance called leukotaxin.

About 12 hours later monocytes arrive. These white blood cells have great phagocytic abilities. They can also produce enzymes that digest the fatty protective coverings of some bacteria. This inflammation fluid causes swelling and pain in the injured area. Heat is produced by the activity of the working cells and by extensive localized vasodilatation. The vasodilatation also causes redness in the inflammed tissues. Toward the end of the inflammatory period, fibroblasts appear and produce collagen, a protein that will become scar tissue. During the time that the collagen is being formed, an increasing number of capillaries migrate into the wound, so that the area is supplied with oxygen and the raw materials for protein synthesis. In the case of wounds of the skin, the final scar resembles the original tissue, except that it is denser; hair follicles, sweat glands, and sebaceous glands are usually absent. The amount of scar tissue formed depends mainly upon how much stress the wound receives and whether the wound was infected.

NORMAL BODY DEFENSES

INTACT SKIN

Most organisms cannot invade the intact skin. Fresh sweat, sebaceous secretions, and cerumen (a waxlike secretion in the outer ear canal) further discourage the growth of microorganisms. If these secretions are not fresh, not only are they unesthetic, but on decomposition they lose their antibacterial properties and actually provide nutrient media for a growing bacterial population.

Normal skin flora are toxic to the underlying tissues. Therefore, keeping their numbers at a minimum is obviously advantageous in the event that the skin is cut or abraded. The thickening of the epidermis at sites of friction, such as the palms of the hands and soles of the feet, is also a normal protective mechanism.

HAIRS IN THE NARES

The hairs in the nasal cavities also provide protection. They trap dust and other inhaled particles, so that they cannot go further down the respiratory tract to sites where the tissues are more sensitive.

MUCOUS MEMBRANES

Body orifices (openings) are protected by mucous membranes, which produce an antibacterial protective mucus. Organisms invading the respiratory, genitourinary, or gastrointestinal tract stimulate an increased production of this mucus, which is usually observable before other signs of infection, such as an increased body temperature.

Localized swelling of the mucous membrane in response to irritation helps to prevent the irritant from penetrating into deeper more vulnerable tissues. Heat is also produced in response to the irritant. Increased temperature may accelerate phagocytosis and the production of immune bodies.

In the respiratory tract the mucous membrane is ciliated. The microscopic cilia are very powerful; when stimulated by some irritant, they attempt to wave it toward the outside of the body.

The mucous membrane of the female vagina during the childbearing years of life is particularly effective in protecting against irritation and infection. During this period, the acid secretions of the vaginal mucosa are bacteriostatic and the mucosa is thick and resistant to irritation.

SALIVA AND TEARS

The saliva contains enzymes that are effective in providing a bacteriostatic environment. In the presence of dehydration or fever, the dry mouth predisposes patients to infections of the salivary glands. Tears contain lysozyme, a bacteriostatic enzyme. Tears constantly bathe the outer exposed surfaces of the eyes. This provides important protection, as these are the only visible living cells of the body.

ACID URINE

The normal acidity of the urine not only is antibacterial but also helps to keep the calcium salts in solution and to prevent the formation of calculi. Foods that favor increased acidity of the urine are meats, cereals, and cranberry juice. Most other fruits favor alkalinity of the urine, as do most vegetables.

GASTRIC JUICE

The hydrochloric acid of the gastric juice is very effective in destroying most bacteria. However, even with achlorhydria, ingested bacteria can usually be phagocytized by Kupffer's cells in the liver. Kupffer's cells are a part of the reticuloendothelial system, which combats invading microorganisms or toxins either by phagocytosis of the organisms or by forming antibodies to destroy them.

THE RETICULOENDOTHELIAL SYSTEM

In addition to Kupffer's cells in the liver, the reticuloendothelial system also includes cells in the spleen, bone marrow, and lymph nodes and other lymphatic tissue. All of these structures are important in the normal body defenses. Bone marrow and lymph nodes produce white blood cells to phagocytize bacteria. The lymph nodes also function to filter any bacteria or foreign material in the tissue fluid as it returns to the bloodstream.

Histocytes are located in almost all body tissues. The histocytes in infected tissues become monocytes that have great phagocytic properties. Circulating lymphocytes are probably spent cells that have already done their part in combating infection when they were in the lymph nodes. However, recent evidence suggests that the lymphocytes in the bloodstream can become histocytes, which in turn become monocytes. The lymphocytes in the bloodstream may also become plasma cells to help in the production of immunity. Circulating lymphocytes may become fibroblasts, which can make collagen (a protein) and build scar tissue.

TEMPERATURE CONTROL AND FEVER

The remarkable stability of body temperature in the healthy adult, regardless of activity and environment, gives us a convenient and reasonably reliable measure of the relative health of the individual. The control center for the regulation of body temperature is located in the hypothalamus in the brain. The temperature of the blood circulating through the hypothalamus causes this center to set into operation mechanisms either to conserve heat or to dissipate excess heat.

If the temperature of the blood circulating through the hypothalamus is lower than it should be, the following changes will take place to conserve heat and to increase heat production: Vasoconstriction in the skin decreases heat loss by radiation and conduction. The adrenal glands increase their production of epinephrine and norepinephrine to cause further vasoconstriction and to increase metabolism and therefore heat production. Shivering also increases the production of heat, as does an increased production of thyroxin by the thyroid gland.

If a patient who has a fever is pale (vasoconstriction is present) and feels cold (heat production is not sufficient to satisfy the temperature control center in the hypothalamus), his fever is no doubt increasing. On the other hand, once the compensatory mechanisms have conserved and produced sufficient heat to increase the temperature of the blood above the requirements of the temperature control center, the patient becomes flushed (vasodilatation is occurring), his skin feels warm, and he is perspiring (all help to eliminate the excess heat). You can be certain that his temperature is now decreasing.

Some authorities believe that an elevated temperature may enhance the inflammatory process, in which case it may be helpful in eliminating the cause of the fever. There are, however, several detrimental effects of fever, particularly if it continues over a prolonged period of time—for example, weight loss from the increased metabolism, increased heart work, and the loss of valuable body water and salts. If the fever is very high for a long period of time, brain damage may occur.

Although the reason is obscure, aspirin is usually effective in reducing fever. Aspirin, however, does not lower the body temperature in an individual who does not have a fever. For very high fevers sponge baths and ice packs may be necessary procedures to reduce the fever.

The normal oral temperature of an adult is 37°C (98.6°F). Upon arising in the morning, the healthy adult usually has an oral temperature of around 36°C. Babies and young children who are ill usually will have much higher temperatures than does a feverish young adult. From middle age on, fevers of 101°F are usually indicative of much more serious illness than would be the case for a young child.

SHOCK

Shock may be caused either by loss of vascular volume or by an excessive vasodilatation, which may greatly increase the size of the vascular bed. The loss of vascular volume may be the result of hemorrhage, inadequate intake of fluids with excess losses of fluids from perspiration, gastrointestinal loss of fluids as in severe diarrhea, or plasma loss as in burns. Surgical shock can occur even though there has been no significant bleeding during the surgery. In this case, the stress of the surgery has caused the capillaries to become excessively permeable and vascular fluid has been lost to the interstitial tissue spaces.

Regardless of the cause of shock, the normal body compensatory mechanisms are the same. Vasoconstrictor substances are produced to decrease the size of the vascular bed so that vital centers, the heart and brain, are supplied, while the less important structures receive only minimal supplies of blood. As a result of this compensatory mechanism, because there is decreased metabolism in the peripheral tissues, the patient in shock appears pale and his skin is cool.

The low blood pressure characteristic of shock causes less fluid to be filtered out of the capillaries in order to conserve vascular volume. Actually, if the blood pressure is very low, the capillaries will increase the return of tissue fluid to the vascular compartment and thereby increase vascular volume but cause some dehydration to the peripheral tissues. In this case, the patient will be thirsty.

The kidneys also help to compensate for the decreased vascular volume. Less urine will be produced, so that water can be saved for the vascular compartment.

With such effective compensatory mechanisms, you might wonder why shock sometimes leads to deeper and deeper shock that becomes irreversible even with vigorous treatment. The smooth muscles of the blood vessel walls may themselves be too ischemic to contract and to cause vasoconstriction. The blood flow becomes sluggish in these vessels and tends to clot. Clotting will decrease the cardiac output of blood, which will cause the shock to become more severe.

The aged individual, whose blood vessel walls have lost much of their elasticity, is much more dependent upon the pumping action of the heart to perfuse blood through his circulatory system.

These patients are much more likely to suffer irreversible shock with blood pressure levels that would not be particularly dangerous to a younger patient.

Measures must be taken to treat the cause of the shock; however, in addition to specific therapy, simple first aid measures can be life saving. The patient should be kept flat with his legs elevated. Although light blankets may be used to decrease chilling, no excessive heat should be applied. Heat will cause vasodilatation and increase the metabolism of the peripheral tissues, both of which will oppose the normal body compensatory mechanisms. If the patient is conscious, oral fluids should be given liberally. If the shock is severe, intravenous fluids must be given. Drugs that cause vasoconstriction and elevate the blood pressure are sometimes prescribed. Some doctors believe that such drugs should not be used to treat shock.

PAIN

Pain is a major protective mechanism. It gives us warning that something is wrong so that we can take measures to help correct the difficulty. Diseases such as cancer which are not evidenced by pain until the process is well advanced, are much more difficult to manage than is a disease such as acute appendicitis, which gives pain warning early in the disease process.

Cutaneous pain is mediated mostly by superficial pain nerve endings. If these superficial nerves are destroyed, the pain is not as severe. Abrasions are often more painful than is a deep cut.

Visceral sensation is not well localized because there are so few visceral nerve fibers. For pain to be felt, a large area of the organ must be stimulated. Visceral sensation may take either of two pathways to the brain, which interprets the sensation. It may go over the parietal pathway (parietal means body wall), in which case the sensation will be felt in the skin covering the organ involved. If the nerve impulse travels over a visceral pathway, it enters nerve fibers a few segments above the organ involved; therefore, the pain will be experienced higher than the actual location of the stimulus.

Several types of stimuli may cause visceral pain. Although cutting the skin will cause pain, an incision into viscera does not result in pain. Ischemia is probably the most important stimulus that causes visceral pain. This type of pain is provoked by movement and relieved by rest. In the case of a perforated peptic ulcer, the pain is caused by chemical irritants. In addition to this severe pain, the abdominal muscles contract and the abdomen becomes quite rigid. Stretching of the viscera also produces pain. This type of pain is similar to the pain of ischemia, because the blood vessels in the distended viscera are compressed. Spasm also causes compression of blood vessels and decreases blood supply. The increased activity of the organ in spasm increases the need for blood supply and also causes an increased production of acid metabolites, which cannot be readily removed because of the diminished blood supply.

For pain to be perceived, the brain must be functioning, the pathway to the brain must be intact, and the pain receptors must respond to the stimulus. Each of these three factors is important in different types of anesthesia. In general anesthesia, the brain cannot perceive the pain. When a nerve is blocked, the pathway to the brain is interrupted. A local injection of procaine renders the pain receptors insensitive to the pain stimulus.

IMMOBILITY

Immobility may be caused by paralysis, pain, trauma, limited joint action, or restriction of activity for either medical or psychiatric reasons. Regardless of the cause of the inactivity, it has devastating effects on all of the organ systems if it is prolonged.

With prolonged immobility, a desalting of the bones and a weakening of the structural supports can result, particularly if there is a deficiency of estrogen, as in women past their menopause. Calcium is withdrawn from the bone and excreted by the kidney. An increase in the dietary calcium is contraindicated because the bones cannot use it. If excess dietary calcium is taken, kidney stones may form, particularly if the urine is more alkaline than is normal. The excess calcium may be deposited in muscle and cause myositis ossificans. It may also be deposited in joints and cause osteoarthropathy.

About 45 percent of the body weight is muscle. Thus, it is not surprising that weakness due to immobility comprises a major factor in any illness. In addition to inactivity causing muscle weakness, the muscles can atrophy. As little as one or two months of immobility can reduce muscle mass by as much as 50 percent. With prolonged immobility of about a year, the muscle fibers may become infiltrated with fat and fibrous tissue. Atrophy of skeletal muscles decreases muscle size, functional movement, and strength of contraction, and it causes impaired coordination.

Contractures can develop as a result of prolonged immobility. Atrophy and shortening of the muscles may involve the joints. Flexors and adductors are the strongest muscles. Therefore, the deformed extremity will be in a flexed and adducted position unless constant attention is given to positioning that favors the extensors and abductors and unless the part is frequently exercised in as full a range of motions as is possible.

Lessened muscle activity decreases the circulation to the skin and other soft tissues. Prolonged pressure also decreases blood supply and nerve impulses. Under these circumstances, decubital ulcers (bed sores) may develop, particularly over bony prominences like the base of the spine. Prevention of this complication depends upon positioning and massage to improve circulation. Once the ulcer has occurred, treatment is difficult, because all wound healing depends upon a good blood supply to the affected part.

The effects of immobility on the cardiopulmonary system can be serious. Even after a few days of bed rest some loss of muscle tone occurs. Upon arising for the first time following a period of bed rest, the patient is likely to have orthostatic hypotension; this decreased blood pressure is due to an increase in the diameter of the blood vessels and also to the fact that when the patient gets out of bed the nervous system may not be able to adjust rapidly enough to regulate vascular diameters properly. Thus, the patient may become light-headed or may faint.

With prolonged bed rest there is a decreased venous return of blood to the heart, which favors clot formation (thrombus). The most likely place for the thrombus to form is in the large veins of the legs. If the thrombus begins to move (becomes an embolus), it may be fatal when it reaches the vessels of the lungs.

Decreased respirations while a patient is immobile not only decrease venous return to the heart but also decrease chest expansion. These conditions can lead to complications such as hypostatic pneumonia and retained mucus, which may also favor respiratory infections.

With immobility, tissue catabolism increases and anabolism decreases. Both of these conditions can cause anorexia and malnutrition. Gastric and intestinal distention with constipation is also a common complication of immobility. If the constipation is severe enough, fecal impaction can lead to bowel obstruction.

In the urinary system prolonged immobility can lead to stasis of urine and urine retention. As mentioned earlier, the kidneys may not be able to eliminate bone minerals and stone formation may occur, particularly if the urine is alkaline and dehydration with a decreased urine volume exists.

Many psychological problems are also associated with prolonged periods of immobility. Motor function constitutes one of the most important aspects of human behavior. Even much of our thought processes are concerned with planning some action. Immobility and isolation decrease motivation, problem-solving ability, and learning.

Three major types of behavioral changes may occur in the immobilized patient. The patient's inability to manipulate himself or his environment can result in frustration, anger, and fear. Perceptual changes also occur when the person has a reduction in both the quantity and the quality of information available to him. This type of sensory deprivation can cause stressful interpersonal relationships. The third type of psychological problem occurring with prolonged immobility is that of role change. Our society values highly both youth and activity. Whether or not there is actually a decrease in the person's social status is not relevant. Relevant is the fact that he feels that his status had deteriorated and that those contributions he can make to his family and society are of little value. These three problems are very difficult for both the patient and his aides. If a patient can be helped through the stage of frustration and anger into a stage of accepting his disability and focusing his attention on what he can do rather than on what he cannot do, he has made great progress toward a full and rich life in spite of a serious disability.

SUMMARY QUESTIONS

1. What is pathology?
2. What do you call the study of the cells?
3. What is histology?
4. What is atrophy?
5. What term would you use to describe labored breathing?
6. What is ischemia?
7. What is a neoplasm?
8. What symptoms may result from cold trauma?
9. Differentiate between heatstroke and heat exhaustion.
10. How is the severity of a burn determined?
11. List the appropriate first aid measures for treating a severely burned person.
12. How does an abrasion differ from a laceration?
13. What is a hematoma?
14. Why are penetrating wounds particularly dangerous?
15. List the structures of the reticuloendothelial system and explain the function of this system.
16. How can you tell when a patient's fever is decreasing?
17. What is normal body temperature?
18. What two conditions can result in shock?

19. Explain the normal body compensatory mechanisms of shock.
20. What type of patient is most likely to suffer from irreversible shock and why is this patient most vulnerable?
21. What first aid measures are appropriate in the treatment of shock?
22. List several complications that may occur with prolonged immobility.

SUGGESTED READINGS

Bowden, M. L., and I. Feller. "Family Reaction to a Severe Burn," *American Journal of Nursing,* February 1973.

Cole, W. H., and Pueston, C. B. *First Aid Diagnosis and Management.* 6th ed. New York: Appleton-Century-Crofts, 1965.

Gardner, A. F. *Paramedical Pathology.* Springfield, Ill.: Charles C Thomas, 1972.

Jose, D. G. "The Cancer Connection with Immunity and Nutrition," *Nutrition Today,* March/April 1973.

Lilijedahl, S. O., and G. Birke. "The Nutrition of Patients with Extensive Burns," *Nutrition and Metabolism,* Vol. 14, Supplement, 1972.

Romano, T., and D. R. Boyd. "Illinois Trauma Program," *American Journal of Nursing,* June 1973.

Schiffers, L. J. *Healthier Living.* 3rd ed. New York: John Wiley, 1970.

Stern, L. D. "Changing Concepts in the Management of Burns," *The Professional Medical Assistant,* May/June, 1973.

Wollmer, R. *The Conquest of Pain.* New York: Alfred A. Knopf, 1961.

3 Tissues and Membranes

VOCABULARY

Distal
Embryo
Intercellular
Ligaments
Periphery
Subcutaneous
Tendons
Vasoconstriction
Vasodilatation

OVERVIEW

I TISSUES
- **A.** Epithelial
  - **1.** Simple squamous epithelium
  - **2.** Cuboidal epithelium
  - **3.** Columnar epithelium
  - **4.** Stratified epithelium
  - **5.** Glandular epithelium
- **B.** Connective
  - **1.** Areolar
  - **2.** Adipose
  - **3.** Dense fibrous
  - **4.** Cartilage
  - **5.** Osseous
  - **6.** Blood
- **C.** Muscle
  - **1.** Skeletal
  - **2.** Visceral
  - **3.** Cardiac
- **D.** Nerve

II MEMBRANES
- **A.** Mucous
- **B.** Serous
- **C.** Synovial
- **D.** Fibrous
- **E.** Cutaneous

Cells are the structural and functional units of all living organisms. There are different types of cells. Although all cells are composed of protoplasm that is surrounded by the cell membrane, the characteristics and structure of cells vary depending upon the specialized function of the particular cell. Figure 3–1shows a typical cell and the various parts of the cell. Most of the structures within the cell can be seen only with an electron microscope.

TISSUES

Tissues are made of cells that are similar to one another in structure and in intercellular substance. Each type of tissue is specialized for the performance of specific functions. One can observe the structural differences of the various tissues with an ordinary light microscope.

EPITHELIAL TISSUE

In general, epithelial tissues cover the body surface and line the body cavities. These tissues primarily protect the underlying structures, secrete fluids, and absorb substances needed by the body.

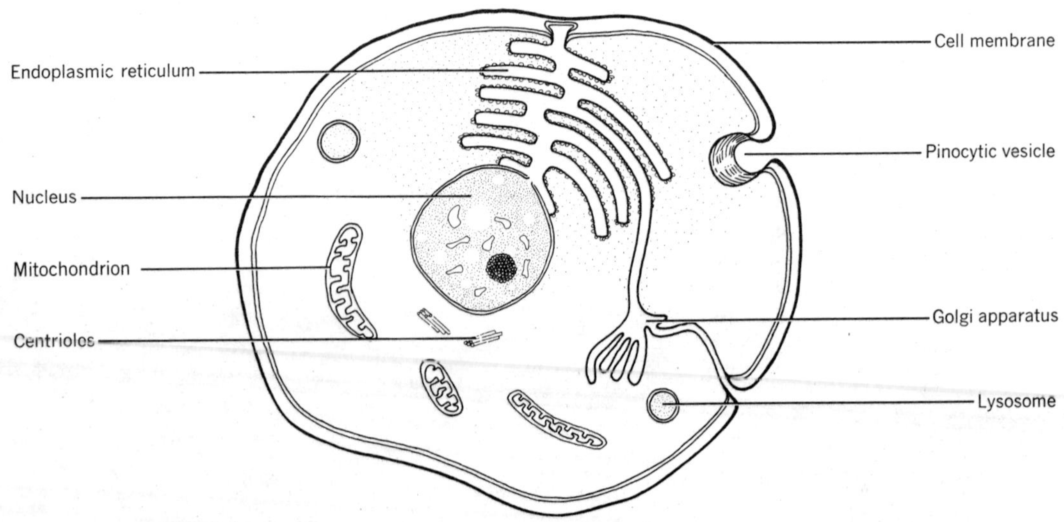

FIGURE 3-1
Diagram of a typical cell.

Simple Squamous Epithelium

Simple squamous epithelial tissue consists of a single layer of flat, platelike cells that are fitted closely together. Simple squamous epithelium lines the blood and lymph vessels as well as the air spaces in the lungs. It provides for exchange of nutrients, gases, and the products of cellular metabolism between the bloodstream and other cells of the body (see Figure 3–2).

Cuboidal Epithelium

The cells of cuboidal epithelial tissue are shaped like cubes (see Figure 3–2). This type of tissue is found in glands and in parts of the kidney.

Columnar Epithelium

Cells of columnar epithelium are tall and narrow. One area in which this type of tissue is found is the lining of the small intestine, where the cells are specialized for the absorption of the products of digestion. Some columnar cells, called goblet cells, produce a protective mucus. Other columnar cells have cilia (hairlike processes) on their free surface; the cilia move particles and secretions across the surface of the tissue. Some of the columnar cells of the epithelial lining of the respiratory tract are ciliated. Figure 3–2 shows the different types of columnar cells.

Stratified Epithelium

Stratified epithelium, consisting of more than one layer of cells, serves to protect the underlying structures. As the surface cells are brushed off by friction, new cells are pushed to the surface by the deeper layer of the tissue. This type of tissue makes up the outer layer of our skin and is also found lining the mouth.

Glandular Epithelium

Glandular epithelium is specialized to secrete a variety of different substances. Exocrine glands secrete substances through ducts onto the surface of the body, for example, sweat, or into hollow organs such as the stomach. Endocrine glands pour their secretions directly into the bloodstream.

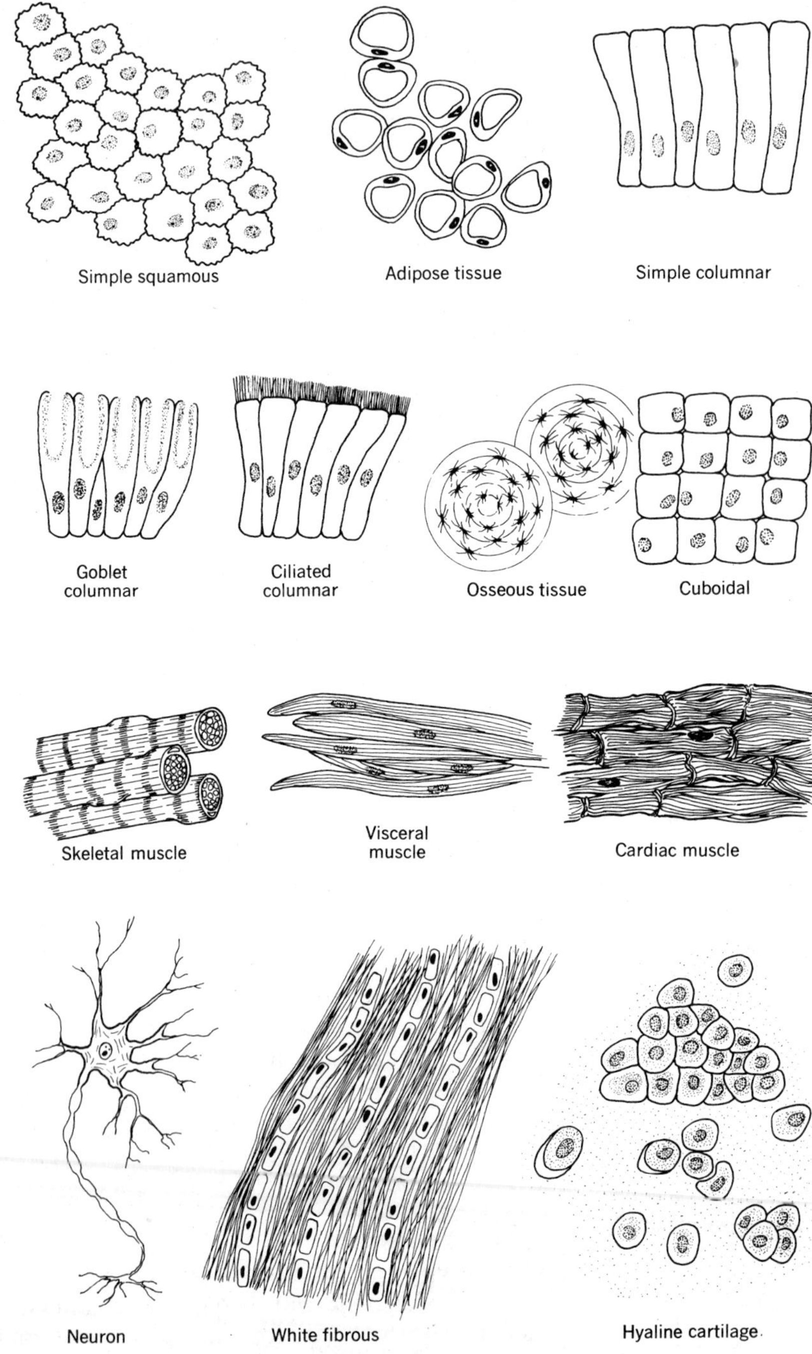

FIGURE 3-2
The microscopic appearance of various body tissues.

CONNECTIVE TISSUES

Connective tissues support, protect, and bind together other tissues. The particular function of the connective tissue may depend upon the cells of the tissue or upon the characteristics of the substance found between the connective tissue cells.

Areolar Connective Tissue

Areolar connective tissue is widely distributed throughout the body. It lies beneath most of the epithelial tissues and serves as packing around blood vessels and nerves. It is thin and glistening.

Adipose Tissue

Adipose tissue has the ability to store fat droplets (see Figure 3–2). It is found beneath the skin; around the kidneys, where it provides support and protection; in joints, where it serves as padding; and in the yellow marrow of long bones. Adipose tissue is a reserve food supply, supports and protects various organs, and provides insulation against heat loss.

Dense Fibrous Connective Tissue

Tendons and ligaments are dense fibrous tissue. This very strong tissue is particularly suited to connect bones together or to connect muscles to bones. Dense fibrous tissue is also found covering muscles.

Cartilage

The intercellular material of cartilage is a firm jelly-like substance. There are three types of cartilage: elastic, hyaline, and fibrous. Elastic cartilage, which is flexible, is found in the external ear. Hyaline cartilage is found at the ends of bones in freely movable joints and in part of the nasal septum, and connects the ribs to the breast bone. The skeleton of the embryo is formed from hyaline cartilage. Fibrous cartilage which is not as firm as hyaline cartilage but is very strong, is found in the discs between the vertebrae.

Osseous Connective Tissue

Osseous tissue is bone (see Figure 3–2). The intercellular substance of osseous tissue in an adult is mostly calcium phosphate salts. In a young child there is relatively less of the mineral salts and more protein in the bone; for this reason the bones of children are somewhat more flexible than those of an adult. Osseous tissue provides a supporting framework for the body and also provides bony protection for some organs such as the brain.

Liquid Connective Tissue

Liquid connective tissue is blood. The intercellular material of this tissue is fluid. Blood serves as a vital transport system for distributing nutrients, gases, and the products of cellular metabolism throughout the body. For a detailed discussion of the composition of blood see Chapter 7.

MUSCLE TISSUE

As we discussed in Chapter 1, there are three types of muscle tissue: skeletal, visceral, and cardiac. In subsequent chapters we shall present a more detailed discussion of muscle tissue.

NERVE TISSUE

Nerves are composed of nerve cells or neurons and supporting cells called neuroglia. Figure 3–2 shows the general structure of a neuron. These cells are highly specialized and serve to transmit impulses to and from various parts of the body. Sensory impulses originate in the periphery and go to the brain for interpretation to keep us advised of conditions in our environment. Motor impulses originate in the central nervous system and result in muscle or glandular activity.

MEMBRANES

Membranes are combinations of tissues. The membranes that are of particular importance to your understanding of the human body are mucous, serous, synovial, fibrous, and cutaneous.

MUCOUS MEMBRANE

Mucous membrane is a combination of epithelial and connective tissues. Mucous membranes line the body cavities that open to the outside of the body, the respiratory tract, the gastrointestinal tract, and the genitourinary tract. All mucous membranes contain some goblet cells that secrete mucus for lubrication and protection.

SEROUS MEMBRANE

Connective and epithelial tissues make up serous membranes, found lining closed cavities of the body. Examples are the pleura, which covers the lungs and lines the thorax; and the peritoneum, which is found in the abdominal cavity. Serous membrane secretes a slippery fluid that protects against friction.

SYNOVIAL MEMBRANE

Freely movable joint cavities are lined with synovial membrane. This type of membrane is composed entirely of connective tissues. The cells in the surface layer secrete synovial fluid to provide lubrication.

FIBROUS MEMBRANE

Fibrous membranes are strong, protective membranes composed entirely of connective tissues. Examples of this type of membrane are the dura mater (covering the brain), the periosteum (covering the bones), and the outer covering of the eye, which is called the sclera.

CUTANEOUS MEMBRANE OR SKIN

The outer layer of the skin is stratified squamous epithelium called the epidermis. The inner layer, or dermis, is connective tissue. Although these two layers are firmly attached to each other, they are quite different in their characteristics (see Figure 3–3).

The epidermis has no blood vessels. The outermost layer of cells of the epidermis are dead cells that are constantly being worn away and replaced by the living cells in the deeper layer of the epidermis. The deep layer of the epidermis produces melanin, which is responsible for the skin color. Exposure to sunlight increases the production of melanin and therefore "sun tan." Freckles are irregular patches of melanin.

The dermis is the inner layer of skin. Conelike elevations on the surface of the dermis, or papillae, are the structures that are responsible for fingerprints.

Elastic fibers in the dermis allow for extensibility and elasticity of the skin. There are considerably more elastic fibers in the dermis of a youth than of an elderly person. This fact and the disappearance of fat from the subcutaneous tissue result in the characteristic wrinkled appearance of the skin in elderly people.

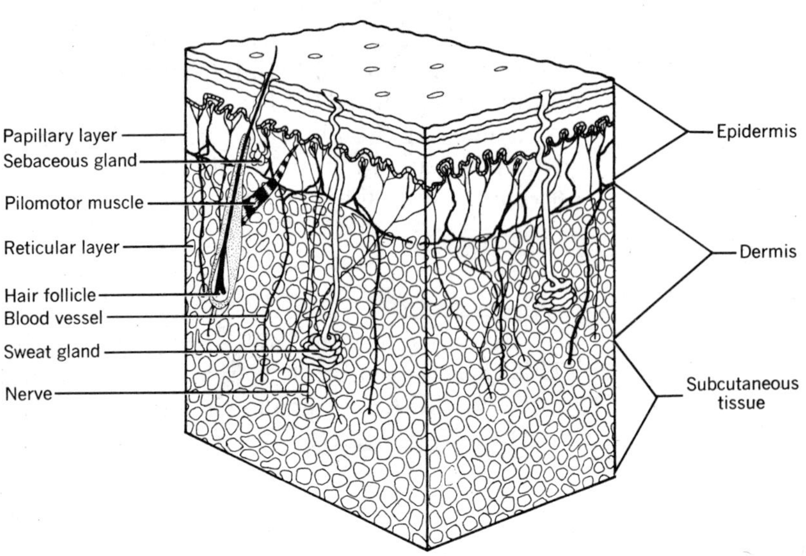

FIGURE 3-3
A three-dimensional diagram of the skin.

Certain cutaneous glands deserve special mention, because they produce secretions important to the functions of the skin. Sebaceous glands are found almost everywhere on the surface of the body except on the palms of the hands and the soles of the feet. Sebaceous glands produce sebum, an oily secretion that helps to keep the skin soft and prevents the hair from becoming dry and brittle.

Sweat glands in the dermis pour their secretions onto the surface of the skin through ducts that pass through the epidermis. The evaporation of perspiration is one mechanism by which the body is able to maintain a stable internal temperature in the presence of high external temperatures.

Ceruminous glands are found in the skin of the passages leading into the ear. These glands secrete cerumen, a waxy substance that helps to protect the ear drum. If the amount of cerumen is excessive, it may interfere with hearing.

Hair is present on most of the skin surfaces. However, it is most abundant on the scalp. The hair grows from a follicle deep in the dermis. Because the hair that we see is not living tissue, cutting or shaving has no effect on its growth.

Fingernails and toenails protect the distal parts of our fingers and toes. The nails are most firmly attached at the base of the nail. This crescent-shaped area is called the lunula.

In summary, the function of the skin is to protect the underlying structure from bacterial invasion and from drying out. It serves as a sense organ in that it contains nerve endings that respond to touch, pain, and changes in temperature. The skin helps to regulate body temperature by the evaporation of perspiration and by the vasoconstriction or dilatation of blood vessels in the skin. If the internal body temperature increases, the vessels of the skin dilate and the warm blood flows more freely near the cooler surface of the body, where the heat can be dissipated by radiation and conduction. Although not its major function, the skin does help in the elimination of some of the body wastes.

SUMMARY QUESTIONS

1. Name the types of cartilagenous tissues and tell where each is located.
2. Of what type of tissue is bone composed?
3. Where are mucous membranes located?
4. Name the glands of the skin and discuss the function of each.
5. What type of membrane lines the closed cavities of the body?
6. What is the function of adipose tissue?
7. Where is dense fibrous connective tissue found in the body?
8. What is the function of stratified epithelium and where is it found in the body?
9. Name the various types of muscle tissue and discuss the function of each type.
10. What are the functions of the cutaneous membrane?
11. Give some examples of fibrous membranes.

SUGGESTED READINGS

Brady, R.J. *A Programmed Approach to Anatomy and Physiology.* 2nd ed. Washington D.C.: Robert J. Brady Co., 1968.

Schiffers, L.J. *Healthier Living.* 3rd ed. New York, N.Y.: John Wiley, 1970.

Snyder, J. "Be a Soft Touch All Winter," *Today's Health,* January 1973.

4 Diseases of the Skin

VOCABULARY

Allergen
Analgesic
Antihistamine
Dysfunction
Edema
Erythema
Exacerbation
Exudate
Macule
Malaise
Remission
Salicylate
Serous
Steroid
Tranquilizer
Vesicle

OVERVIEW

I Acne
II Seborrheic Dermatitis
III Eczema
IV Urticaria
V Contact Dermatitis
VI Psoriasis
VII Impetigo
VIII Warts
IX Herpes Simplex
X Herpes Zoster
XI Furuncles and Carbuncles
XII Paronychia
XIII Sebaceous Cysts
XIV Prickly Heat
XV Dermatophytosis
XVI Corns and Calluses
XVII Infestations and Bites
XVIII Lupus Erythematosis
XIX Scleroderma

There are a variety of diseases of the skin with which you should be familiar. Many of these diseases are of unknown etiology and as a result treatment is usually directed toward relief of the symptoms. Fortunately, although skin diseases cause discomfort, most are self-limiting.

One of the most common symptoms associated with diseases of the skin is itching, or pruritus. Excessive warmth, rough prickly fabrics, and emotional stress aggravate the itching. Frequently it is more noticable at night, probably because the patient's attention is not occupied. Scratching not only increases the severity of the itch but may denude the skin and make the patient more susceptible to infection. Antipruritic lotions such as calamine are helpful. Starch baths (one pound of cornstarch in a tub of warm water) may also help to relieve the itching. The patient should be instructed not to use soap when taking the starch bath, and to pat the skin with a soft towel rather than to rub it when drying.

ACNE

Acne characteristically occurs during adolescence. It is usually self-limiting and will disappear in the late teens or early twenties. However, 8 or 10 years is a long time for a young person who longs to be attractive and popular to endure the pimples and blackheads of acne.

Frequent, thorough washing of the affected areas with a mild soap is helpful because it helps to reduce the oiliness and at least temporarily reduces the number of bacteria on the skin causing the infections. The blackheads and pimples or pustules should not be picked at or squeezed because unskilled manipulation of these lesions can result in secondary infections and perhaps permanent scarring.

The doctor may order steroid creams to be applied to the affected areas. Before applying any medication, the patient should wash his hands and the affected area thoroughly. Short treatments with ultraviolet light, administered by the doctor, may be beneficial because this treatment will lessen the oiliness and reduce the number of infection-causing bacteria. Excessive use of ultraviolet light can be harmful. It may be helpful to avoid certain foods such as chocolates, nuts, and fatty foods. Some doctors completely restrict dietary caffine.

If the acne has caused scarring, it can be made less conspicuous by dermabrasion. Under anesthesia the outermost layers of the skin are removed by sandpaper or a rotating wire brush. Following this dermabrasion the skin feels raw and sore. Some serous exudate and crusting will take place, but the patient should not wash the area for five or six days and avoid picking or touching it until sufficient healing has occurred.

SEBORRHEIC DERMATITIS (DANDRUFF)

Dandruff frequently occurs with acne; however, it is not limited to the adolescent period of life. The symptoms are an oily scalp, itching, irritation, and the formation of greasy scales. Severe or prolonged seborrheic dermatitis may cause the premature loss of hair.

Dandruff usually responds well to frequent shampooing (two or three times a week) with tincture of green soap, brushing the hair, and massaging the scalp. Although the condition improves with treatment, the symptoms frequently return if the regimen of treatment is interrupted.

ECZEMA

Eczema is characterized by vesicles (blisters) on reddened itchy skin. The blisters burst and weep; crusts form later from the dried fluid. Because the condition is usually aggravated by emotional stress, tranquilizers may be ordered. Antihistamines and steroids are also helpful in the treatment of eczema. Wet dressings and starch baths may relieve some of the symptoms.

URTICARIA (Hives)

Urticaria is usually the result of an allergy. However, the condition is sometimes related to emotional stress. The most familiar example of urticaria is a mosquito bite. In severe cases, there will be wheals (rounded, white elevations) and itching of the entire body. Steroids and antihistamines are used to treat urticaria.

CONTACT DERMATITIS

The most common causes of contact dermatitis result from contact with poison ivy, poison sumac, or poison oak. Prompt cleaning of the exposed skin with soap and water followed by application of alcohol may prevent or lessen the reaction to these allergens. The symptoms include redness, itching, blisters, and edema. Scratching spreads the lesions. Antipruritic lotions and wet dressing help to relieve the symptoms.

Many people are allergic to cleaning agents, cosmetics, other chemicals, and certain metals. In such cases, it is well to avoid the particular allergen. The doctor may be able to desensitize the patient to the allergen by giving repeated small doses of the substance.

PSORIASIS

Psoriasis is characterized by patchy erythema and scales. It most frequently occurs in young adults and middle-aged people. Its cause is unknown. Ointments may be used to soften the scales; sometimes, ultraviolet light treatments are also helpful. Low fat diets are usually recommended. Steroids, antihistamines, and tranquilizers may also be used, as in the case of many other skin diseases.

IMPETIGO

Impetigo usually is the result of infection by streptococci or staphlococci. It is more common in infants and children than in adults. Because this condition, unlike most other skin diseases, is contagious, care must be taken in treating and handling the lesions.

The symptoms of impetigo are erythema and vesicles that rupture and cover the skin with a sticky yellow crust. The crust must be removed with soap and water or mineral oil before antibiotic ointments are applied. The condition can usually be cured in a few days. However, if it is severe in a newborn infant, it may be fatal.

WARTS

Warts are caused by a virus. Susceptibility to warts varies considerably from one individual to another. Sometimes warts disappear spontaneously. If they do not, nitric or sulfuric acid may be applied deep into the root of the wart, after the thick horny outer tissue has been pared away. Although most warts are not painful, plantar warts that occur on the soles of the feet may cause considerable discomfort.

HERPES SIMPLEX (Cold Sores)

Herpes simplex is also caused by a virus. Many healthy people harbor this virus and under circumstances such as emotional stress or respiratory infections the herpes simplex may appear.

The condition is characterized by blisters on inflammed skin, usually around the mouth. Although there is no specific treatment and the lesions subside in about a week, topical applications of tincture of benzoin can help to relieve the pain and burning. The patient should be told that when the benzoin is first applied the burning will be more severe, but that subsequent applications will not be painful.

HERPES ZOSTER (SHINGLES)

The cause of herpes zoster is a viral infection. Patients have fever and malaise. After a few days, erythema and vesicles appear along the course of a nerve. Although there is no specific treatment for this painful condition and it can last for months, analgesics and calamine lotion may be used for symptomatic treatment.

FURUNCLES AND CARBUNCLES

A furuncle is a boil. A carbuncle is a large, swollen erythematous lesion, often on the back of the neck. The carbuncle is very painful and has several opening through which pus drains. Both conditions are usually the result of staphylococcal or streptococcal infections.

Hot, moist compresses help to localize the infection and to promote drainage. It may be necessary to incise and to drain the lesion surgically. Antibiotics are usually ordered.

Because furuncles and carbuncles frequently occur in patients who have diabetes mellitus, the patient should have blood and urine examinations to rule out this endocrine disease. If diabetes is present, it must also be treated.

PARONYCHIA

A paronychia is an infected hangnail. Often, the patient needs only to soak his finger frequently in warm water. If this treatment is not successful, the base of the nail will have to be surgically removed.

Painful as such a procedure might seem, it usually is not. The nail will grow back in about 3 to 4 months.

SEBACEOUS CYSTS

Sebaceous cysts are the result of the blockage of a duct of a sebaceous gland. Usually, the swelling is small and a doctor can simply incise and drain the cyst in his office. Occasionally, the cyst becomes very large and unsightly. In such cases, it will have to be removed in a hospital operating room. The doctor usually will discharge the patient immediately after the surgery and instruct him to return to his office in 4 or 5 days to have the dressing changed. If a drain was placed in the wound, it is removed at this time.

PRICKLY HEAT

Prickly heat is a rash of bright red pimples, usually contracted during hot weather. The discomfort of prickly heat may be relieved by cool baths using mild soap. The patient should apply a light coating of cornstarch powder. He should be cautioned that too much powder will cause caking, particularly in the folds of skin, which will further aggravate the condition. He should also wear absorbant underclothing.

DERMATOPHYTOSIS (Athlete's Foot)

Athlete's foot is a fungus infection. Susceptability to the infection varies considerably from one person to another. Usually it first affects the skin between the toes; the skin is red, cracked and sore. It may spread to other parts of the feet, the hands, axillae, and the groin. Because it is an infectious condition, care must be taken to avoid transmission from one person to another by means of contaminated towels and toilet articles. Persons with the infection should avoid going barefoot in dormitory bathrooms and gym locker rooms.

Antifungal agents such as Desenex powder or ointment are usually effective in treating dermatophytosis. The patient must take particular care to dry between his toes. He should avoid footwear that causes his feet to perspire and change his shoes and socks frequently.

CORNS AND CALLUSES

Corns and calluses are usually the result of friction caused by poorly fitted shoes. Corns are hard, raised areas that are often painful. Calluses are flat, thickened patches. The only effective treatment is to relieve the pressure or friction. Keratolytic agents such as salicylic acid help to remove the thickened hard skin. Rubbing cream may also soften the calluses.

INFESTATIONS AND BITES

Pediculosis is caused by lice. The lice may infest the hair of the scalp, body hair, or pubic hair. Symptoms are itching and irritation of the skin. Scratching denudes the skin and makes it more susceptible to infection. A variety of ointments, powders, and lotions containing benzyl benzoate or benzine hexachloride are effective in the treatment of pediculosis. If the patient continues to have close contact with others who harbor the parasites, the condition will recur.

Scabies is caused by a mite. The itching of this condition is intense. The mite burrows deep into the skin. This condition can be readily transmitted from one person to another by close personal contact. The treatment of scabies is thorough bathing followed by the application of ointments or lotions that contain benzine hexachloride or benzyl benzoate. After the treatment, the patient should completely change his clothing.

LUPUS ERYTHEMATOSIS

Lupus erythematosis is a rare condition characterized by erythematous, macular (raised) lesions. The itchy lesions frequently appear on the face in a butterfly pattern. As well as skin lesions, many patients with lupus have dysfunction of kidneys, joints, lungs, and heart.

Although there is no specific treatment for lupus, steroids help to relieve the symptoms and salicylates such as aspirin relieve joint pains and reduce fever. There are remissions and exacerbations of this disease. Prognosis is guarded and ultimately the disease may be fatal.

SCLERODERMA

Scleroderma is a systemic disease involving not only the skin but also muscles, bones, heart, and lungs. The skin becomes smooth, hard, and tight. Although the disease is progressive, there are periods of remissions and exacerbations. In severe cases, the disease is usually fatal.

There is no specific treatment for scleroderma. However, symptomatic treatment with ointments, massage, and heat may help relieve the stiffness and inelasticity of the skin. Steroids may provide temporary relief.

SUMMARY QUESTIONS

1. What is pruritus?
2. What measures are helpful in relieving pruritus?
3. What is pediculosis?
4. How is a paronychia treated?
5. What measures can you recommend to someone who suffers from acne?
6. How do you treat dermatophytosis?
7. How do you prepare a starch bath?
8. What should you do if you come into contact with poison ivy?
9. What causes a sebaceous cyst?
10. Differentiate between herpes simplex and herpes zoster.

SUGGESTED READINGS

Edelstein, M. L., *et al.* "Malignant Melanoma, A Selective Review of the Biology, Histochemistry and Diagnostic Cytology," *The American Journal of Medical Technology,* October 1973.

Johnson, D. F. *Total Patient Care, Foundations and Practice.* 3rd ed. St. Louis, Mo.: C. V. Mosby Co., 1972.

Stewart, W. D., *et al. Synopsis of Dermatology.* St. Louis, Mo.: C. V. Mosby Co., 1970.

5 The Musculoskeletal System

VOCABULARY

Afferent
Brachial
Condyle
Cortisone
Efferent
Embryo
Glucocorticoid
Leukocyte
Platelet
Proximal
Steroid
Temporal region

OVERVIEW

I COMPOSITION OF BONE
- **A.** Protein
- **B.** Inorganic salts

II GROSS STRUCTURE OF BONES
- **A.** Classification of bones according to their shape
- **B.** Bone marrow
- **C.** Blood supply of bone

III BONE FORMATION
- **A.** Membranous bone formation
- **B.** Cartilagenous bone formation

IV THE APPENDICULAR SKELETON

V THE AXIAL SKELETON

VI ARTICULATIONS OR JOINTS
- **A.** Classification of joints according to amount of motion
- **B.** Angular movements, circumduction, and rotation

VII ORGANS OF LOCOMOTION
- **A.** Types of muscle tissue
- **B.** Physiology of contraction
- **C.** Skeletal muscles
- **D.** Visceral muscles
- **E.** Cardiac muscles

VIII INNERVATION OF MUSCLES

IX PHYSIOLOGY OF CONTRACTION
- **A.** All or None Law
- **B.** Muscle tone

X PRIME MOVERS AND ANTAGONISTS

Bones, which are composed of mainly osseous connective tissue, have many important functions. They provide a supporting framework for the body, protect the viscera, and provide a place for attachment of muscles. Some bones contain red bone marrow, which forms red blood cells and some white blood cells. This marrow also aids in the destruction of old, wornout red blood cells.

COMPOSITION OF BONE

The chief organic constituent of bone is collagen, which is a protein. About two-thirds of bone is inorganic calcium phosphate. The source of the protein and the inorganic salts is the food we eat. Because calcium phosphate is the primary ingredient for proper bone density, we shall consider some of the factors involved in the metabolism of this inorganic salt.

Vitamin D is essential for the absorption of these minerals into the blood vessels of the intestine. Although it can be synthesized by the skin on exposure to sunlight, the best sources of vitamin D are fish-liver oils, eggs, milk, and butter. Because it is a fat-soluble vitamin, the action of vitamin D is dependent upon proper fat metabolism. Its action is opposed by cortisone and other glucocorticoids. For this reason, patients on long-term steroid therapy may have decreased bone density and are predisposed to pathological fractures. Sex hormones (estrogen and testosterone) favor the deposition of calcium into the bones. Postmenopausal women who have low levels of estrogen may also have diminshed bone density.

The enzyme alkaline phosphatase is the catalyst needed in the regulation of bone and blood levels of calcium. Hormones from the thyroid and parathyroid glands are also important in the regulation of bone and blood levels of calcium. These hormones will be discussed in detail in Chapter 21.

Mechanical stress also favors bone formation, whereas inactivity favors the desalting of bone. Patients with fractures of the weight-bearing bones are frequently put in traction until they can be mobilized. Traction supplies the mechanical stress that will favor bone healing while the patient is inactive. Traction also helps to keep the bone fragments in correct alignment. Without traction the strong leg muscles may go into spasm and cause overriding (overlapping) of the fractured ends of the bone.

Bone is covered with periosteum, a dense, fibrous membrane. Osteoblasts are located in the deep layer of the periosteum. These osteoblasts are the bone-building and bone-repairing cells. Beneath the periosteum is compact bone. This type of osseous tissue, as its name implies, is very dense. Cancellous bone is located in areas that are not subjected to great mechanical stress and where the greater weight of compact bone would be a problem. Cancellous bone is light and spongy.

GROSS STRUCTURE OF BONES

Bones can be classified according to their shape. Long bones are located in the extremities. The two ends of the long bones are called the epiphyses. Each epiphysis is covered with articulating hyaline cartilage. There is cancellous bone in the epiphysis. The shaft of the bone (or diaphysis) is compact bone and covered with periosteum. In the center of the bone is the medullary cavity that contains yellow bone marrow (see Figure 5–1).

Short bones are located in the wrist and ankle. These are mostly cancellous bone, with a thin sheet of compact bone around the cancellous bone. Flat bones are formed of two plates of compact bone, with cancellous bone between these plates. The skull, ribs, scapula, and sternum are flat bones. Irregular bones are located in the vertebrae and the face. They are mostly cancellous bone with a compact cover.

Sesamoid bones are extra bones formed within certain tendons. The patella (kneecap) is a sesamoid bone. Excluding the sesamoid bones, because their numbers vary, there are 206 bones in the human body.

BONE MARROW

Yellow bone marrow is located in the medullary cavity of long bones. Primarily composed of adipose tissue, it serves as an area of fat storage.

Red bone marrow is found in all cancellous bone of children. In the adult, it is located only in the vertebrae, hips, sternum, ribs, cranial bones, and proximal ends of the femur and humerus. Red bone marrow is composed of many cells supported by a highly vascular, delicate connective tissue. The red bone marrow forms red blood cells, granular leukocytes, and platelets. It also destroys old red blood cells and some foreign materials.

BLOOD SUPPLY

Bone is a quiet organ, not requiring wide variations in its blood supply; its normal needs for blood are relatively minimal. All bones have many microscopic periosteal arteries. The long bones also have a nutrient, or medullary artery. This artery enters the medullary cavity through a tunnel in the shaft of the bone.

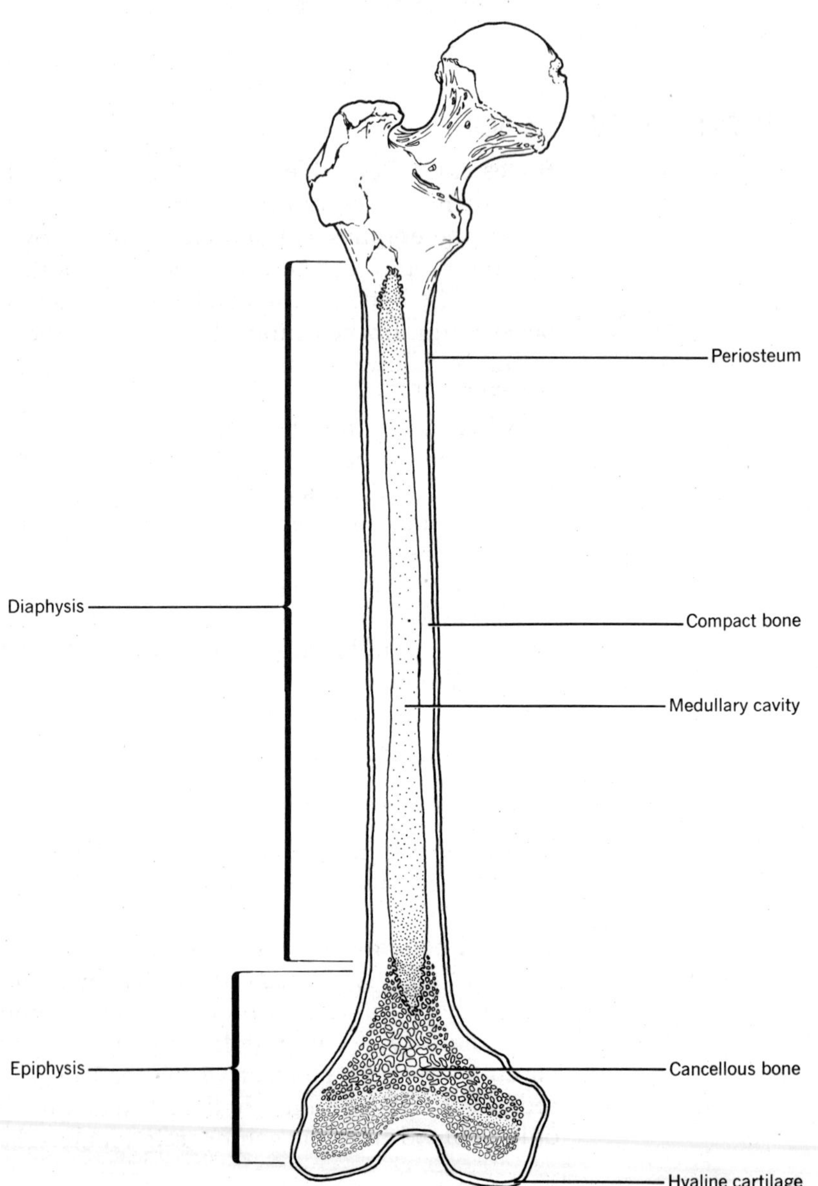

FIGURE 5-1
A transverse section of a long bone.

BONE FORMATION

Bone formation usually begins in the third month of fetal life. The bones of the skull begin as fibrous membrane, and the rest of the bones of the body are formed from hyaline cartilage.

In membranous bone formation, the fibrous membrane already has the shape of the bone to be formed. There is an ossification center in the middle of the membrane, and ossification (laying down of the inorganic salts) begins in the center and radiates toward the periphery. The ossification process is not complete at the time of birth. The fibrous membrane that will become cranial bones forms the fontanels or soft spots of the infant's head. These fontanels allow for molding of the skull and an easier passage through the birth canal. The fontanels and the open joints between the cranial bones also allow for growth of the skull.

The cartilagenous bone of the embryo is formed from hyaline cartilage. Slightly different patterns of ossification exist for different types of bones. Long bones have three centers of ossification, one at each end of the forming bone and one in the center of the shaft. Short bones have only one ossification center in the middle of the forming bone; ossification proceeds toward the periphery. As the bone develops, the cartilage cells degenerate and the osteoblasts (the bone-building cells) move in, causing the deposition of the calcium salts. This process continues until only a small strip of cartilage remains. Finally, usually about the age of 18 to 20 years, the bones are completely ossified. Young bones are less brittle than older bones, because the former have relatively more collagen than they do inorganic salts.

Long bones grow in circumference by the activity of osteoblasts in the deep layer of the periosteum. Osteoblasts are bone-building cells. Osteoclasts dissolve bone in the center of long bones and keep the medullary cavity open; they assure that the bones are reasonably strong but not too heavy. After the osteoblasts have completed their initial function of bone formation, they become maintenance cells, or osteocytes. Osteocytes help in the exchange of calcium salts betwen bone and blood.

THE APPENDICULAR SKELETON

SHOULDER GIRDLE

The shoulder girdle includes the scapula and the clavicle. The glenoid cavity, an indentation in the scapula, receives the head of the humerus. The clavicle articulates with the sternum or breast bone and extends laterally to the shoulder.

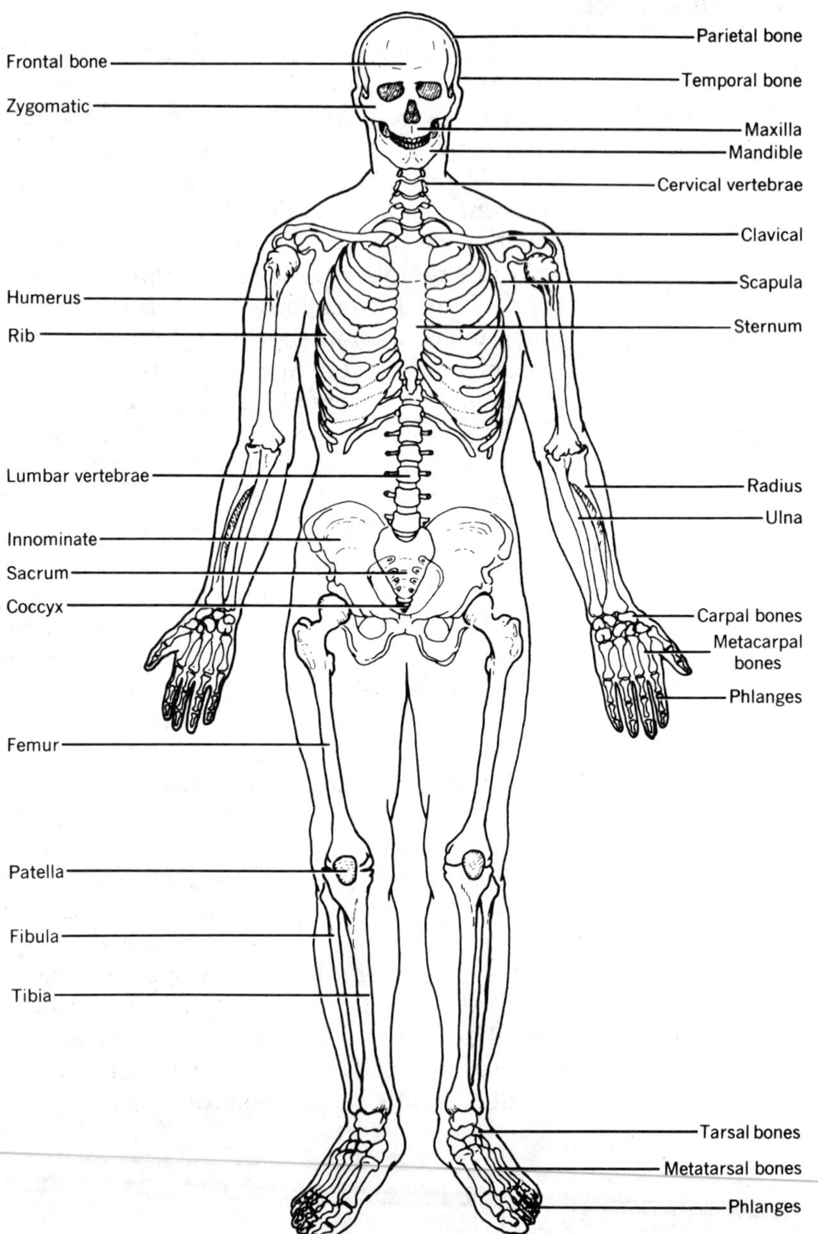

FIGURE 5-2
Articulated skeleton. The appendicular skeleton is made up of the bones that form the shoulder girdle, pelvic girdle, and those of the extremities. The other bones comprise the axial skeleton.

ARM AND HAND

The humerus is the bone in the brachial region. The forearm has two bones, the ulna on the medial aspect and the radius on the lateral aspect.

There are eight carpal bones in the wrist. The palm of the hand has five metacarpal bones; there are 14 phalanges in the fingers, two in the thumb and three in the other fingers.

PELVIC GIRDLE

The pelvic girdle is made up of two innominate bones (hip bones), the sacrum and coccyx. It protects the urinary bladder, some reproductive organs, the lower colon, and the rectum. In the female it is broad and roomy with a large outlet, whereas in the male it is narrow and the pubic arch is much smaller.

The innominate bone begins as three separate bones that fuse into one and form the socket (acetabulum) holding the head of the femur. The ilium is the upper, flared portion of this hip bone, the ischium is the lower and strongest part, and the pubis is the place where the two hip bones unite anteriorly to form the symphysis pubis.

LOWER EXTREMITY

The femur (thigh bone) is the largest and heaviest bone of the lower extremity. There are two bones in the lower leg. The tibia is the larger of the two and is on the medial aspect. The fibula is a slender bone on the lateral aspect of the leg. There are seven tarsal bones in the ankle. The calcaneus, which forms the heel, is the largest of these tarsal bones. Five metatarsals form the arch and the ball of the foot. There are 14 phalanges in the toes.

THE AXIAL SKELETON

BONES OF THE SKULL

The cranium (skull) is formed by eight bones. The frontal bone forms the forehead and the roof of the orbits of the eyes. The frontal bone contains the frontal sinuses, which are air spaces lined with mucous membrane. Secretions from these sinuses drain into the nose. The two parietal bones form the roof and sides of the head. The occipital bone is the base of the skull. The foramen magnum is an opening in the occipital bone which serves as a passageway for the spinal cord. The two temporal bones are located in the temporal region. These bones house the very important structures concerned with equilibrium and hearing. The mastoid sinuses, which are also located in the temporal bone, drain into the middle ear.

The ethmoid bone is a small bone, part of which forms the upper area of the nasal septum. It also contains sinuses that drain into the nose. Just posterior to the ethmoid is the sphenoid bone. The body of the sphenoid contains the sphenoid sinus, which drains into the posterior part of the nasal cavity. All of these sinuses help to give resonance to the voice and lightness to the skull.

BONES OF THE FACE

Two nasal bones form the bridge of the nose: the lacrimal bones, which are on the medial wall of the socket of the eye; and the turbinates, or conchae, which are scroll-like bony projections from the lateral walls of the nasal cavity. The vomer forms the posterior and lower part of the nasal septum. Two palatine bones form the posterior part of the roof of the mouth. The two zygomatic bones are the prominent bones of the cheek. The maxillary bone is a fusion of two bones that forms the anterior part of the roof of the mouth; it has alveolar processes, or sockets, that support the upper teeth. The maxillary bone also contains a pair of sinuses that drain into the nasal cavity. The mandible, which is the only movable bone of the skull, is the lower jaw bone. The hyoid bone is a horseshoe-shaped bone fastened by muscles in the neck below the mandible.

SUTURES

The joints of the cranial bones are called sutures. These joints are synarthritic joints and are immovable. The sagittal suture is located between the two parietal bones. The coronal suture is found between the frontal bone and the parietal. The lambdoidal lies between the parietal and the occipital bone. The squamosal suture is the junction between the temporal, sphenoid, and parietal bones.

FONTANELS

As mentioned earlier, the fontanels are composed of fibrous membrane, in which ossification is incomplete at the time of birth. The anterior fontanel, which is the largest, is located at the junction of the sagittal and coronal sutures. The posterior fontanel lies between the sagittal and lambdoidal sutures. The anterolateral fontanels are located at the junction between the frontal, parietal, sphenoid, and temporal bones. The posterolateral fontanels are located at the junction of the parietal, occipital, and temporal bones.

Within a few months after birth only the anterior fontanel remains palpable. This, the last to close, usually does so after about 16 months.

VERTEBRAL COLUMN

The vertebral column supports the head and trunk and provides protection to the spinal cord and nerve roots. In the adult, there are 26 bones in the vertebral column; however, in the child, there are 33 vertebrae, as those of the sacrum and coccyx have not fused.

There are seven cervical vertebrae. The first cervical vertebra is the atlas, which articulates (joins) with the condyles of the occipital bone of the skull. The second is the axis, from which arises the odontoid process (a toothlike projection) to articulate with the atlas.

There are 12 thoracic vertebrae, all of which have facets for articulation with the ribs. They are larger than the cervical vertebrae, and you can easily palpate (touch) their spinous processes in the midthoracic region of your back.

The five lumbar vertebrae are the heaviest of the vertebrae and have short thick processes.

The sacrum consists of five fused sacral vertebrae; it forms the posterior part of the pelvic girdle.

A fusion of four vertebrae forms the coccyx. Between the sacrum and the coccyx is the sacrococcygeal joint. This joint is somewhat more movable in the female than in the male.

Between each pair of vertebrae are intervertebral discs that are fibrous cartilage with soft centers. These discs serve as shock absorbers and cushions.

THORAX

The thorax is made up of a cage of bones and cartilage, covered with muscles and skin. The floor of the thorax is the diaphragm. The chief function of this bony cage is to protect the thoracic viscera.

The sternum, which is approximately six inches long, is located in the anterior midline of the thorax. It consists of a fusion of three parts: the manubrium, the body or middle portion, and the xiphoid, which is the inferior portion.

There are twelve ribs on each side of the thorax; they are all attached to the thoracic vertebrae. The first seven pairs, called the true ribs, are attached directly to the sternum by separate costal cartilages. Each of the next three pairs, called the false ribs, is attached to the cartilage of the rib above it and thence to the sternum. Because they have no anterior attachment, the last two pairs of ribs are called the floating ribs.

ARTICULATIONS OR JOINTS

Joints can be classified according to the amount of motion they permit. Synarthrotic joints are immovable joints such as the sutures in the skull.

Amphiarthrotic joints have limited motion. The bones that form these joints are united by fibrous tissue and cartilage. Examples of the amphiarthrotic joints are the symphysis pubis, the sacroiliac joint, and the vertebral joints.

Diarthrotic joints, which are freely movable, contain a cavity surrounded by binding tissue that is lined with synovial membrane. This membrane secretes fluid for lubrication.

Several types of movements can be accomplished at diarthrotic joints. Angular movement includes flexion and extension—decreasing and increasing the angle that the bones make to each other. Hyperextension is extension beyond the anatomical position. Abduction is an angular movement in which the body part is moved away from the midline of the body. Adduction is moving the part toward the midline.

Circumduction is a movement in which the distal end of the extremity inscribes a circle while the shaft inscribes a cone. Only at ball and socket joints, that is, the shoulder and the hip, can circumduction be accomplished.

Rotation is a revolving or twisting without angular movement. In internal rotation, the extremity is turned toward the midline. In external rotation, the part is turned away from the midline.

Some special types of movement take place only in the wrist and ankle. Supination occurs when the palms are turned forward or up. In pronation the palms are turned down. When the plantar surface of the foot is turned toward the midline, the feet are said to be inverted. Eversion concerns turning the plantar surface away from the midline. Plantar flexion is accomplished by standing on the toes; dorsiflexion takes place when the toes are pointed upward.

ORGANS OF LOCOMOTION

TYPES OF MUSCLE TISSUE

There are three different types of muscle tissue. Skeletal muscles, which are attached to the skeleton, are under a person's voluntary control. For this reason, they are sometimes called voluntary muscles. Visceral muscles, which are involuntary, are located in the organs. Cardiac muscle, also involuntary, is located only in the heart.

Skeletal muscle is capable of contracting very rapidly and powerfully. There must be nerve impulses to bring about contraction of skeletal muscles. Efferent nerve fibers from the brain and spinal cord send impulses for contraction, and afferent fibers from the muscles to the central nervous system inform the brain of the degree of contraction taking place.

Visceral and cardiac muscle also have efferent and afferent nerve supplies. In the case of the visceral muscle, these nerves are less important than in the skeletal muscles because both visceral and cardiac muscle have automaticity, that is, the ability to contract without nerve supply. In the case of the cardiac muscle, although it has automaticity, the efferent nerve impulses control the rate of contraction according to the needs of the body. In visceral and cardiac muscle, the afferent impulses are concerned with sensations of pain, spasm, and stretch.

PHYSIOLOGY OF CONTRACTION

When a muscle is contracting, the muscle cells become shorter and thicker. The contraction causes the part to move. The chemistry of contraction results in the production of heat, which is one of the factors helping to maintain normal body temperature.

In addition to chemical changes, electrical changes also take place during muscle contraction. Electrical changes have considerable clinical significance because they can be measured and recorded to give valuable diagnostic information. In the case of cardiac contractions, the electrical changes are measured by an electrocardiograph. Electromyography measures the electrical changes taking place during the contraction of skeletal muscles.

All muscle cells abide by the All or None Law. Each muscle cell when stimulated gives its best response or it does not contract at all. The strength of the contraction of the entire muscle depends upon the number of cells stimulated and the condition of the muscle.

Muscle tone is a steady, partial contraction that is probably present at all times in healthy muscles. Muscle tone is lowest when

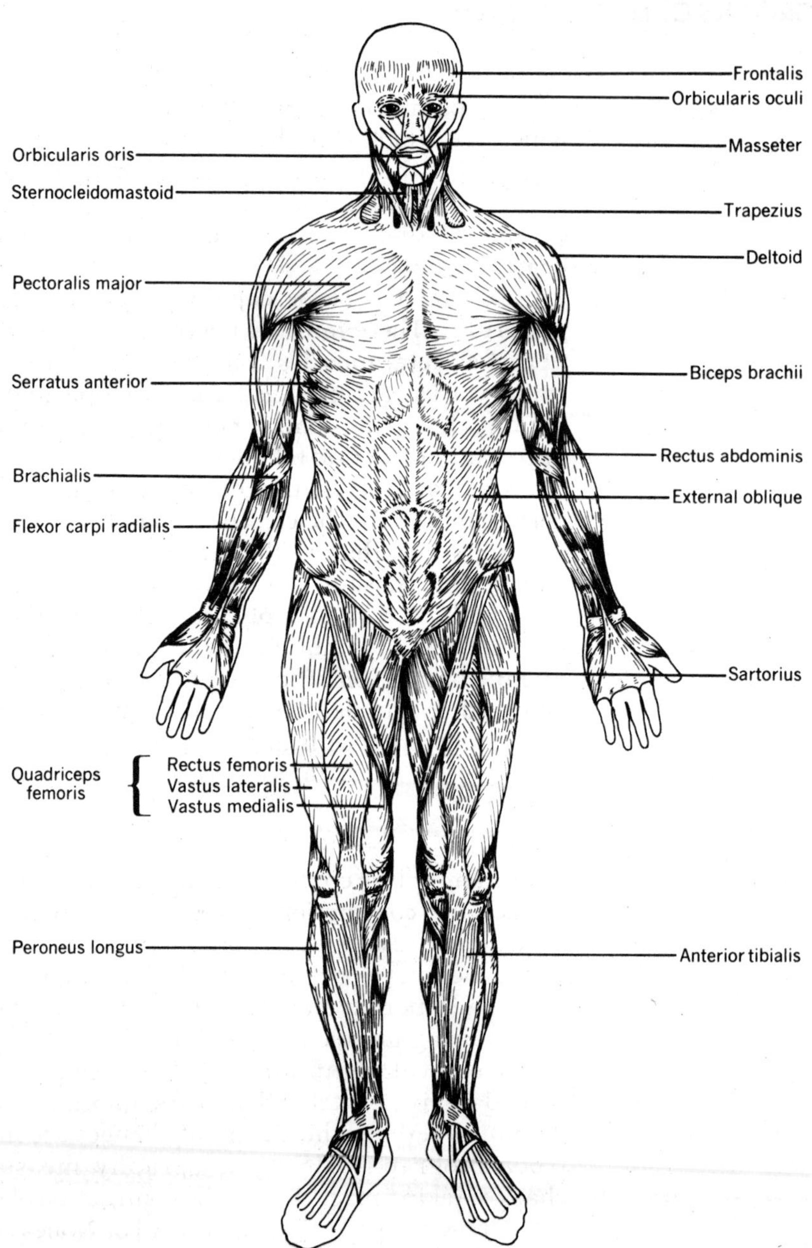

FIGURE 5-3

An anterior view of the superficial muscles of the body. Refer to the anatomical transparencies to see the structure and location of the deeper muscles.

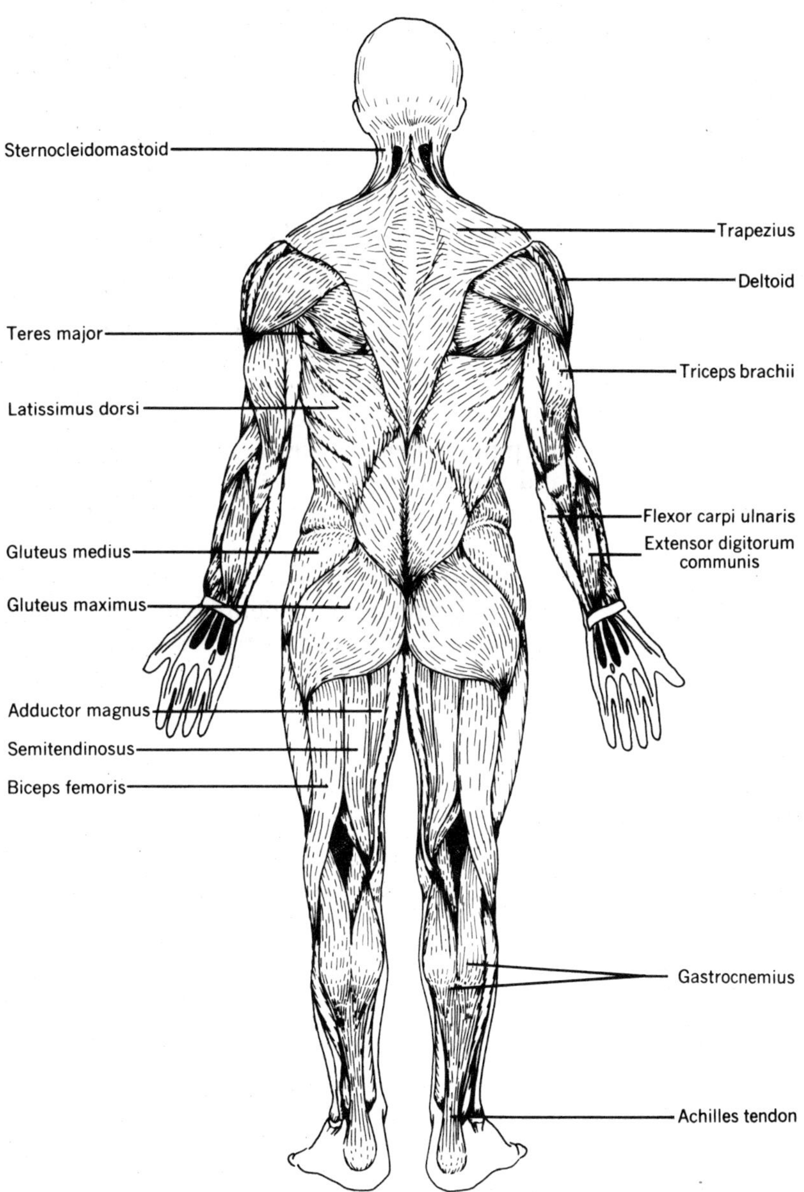

FIGURE 5-4
A posterior view of the superficial muscles of the body.

we are asleep or ill. When a person has been ill and has had even a relatively short period of bed rest, the muscle tone can be markedly diminished. Because the patient is generally unaware of this, he should be assisted when he is allowed out of bed for the first time.

SKELETAL MUSCLES

Skeletal muscles are sometimes named according to their action, such as flexors or extensors. They are also named according to their location, the direction of their fibers, the gross shape of the muscle, or the number of divisions that the specific muscle has.

The muscles with which we will be concerned are called prime movers. When a prime mover contracts, its antagonist must relax. The antagonist therefore does the opposite of the prime mover. For example, when the elbow is flexed, the biceps is the prime mover and the antagonist—the triceps—must relax. In order to extend the elbow, the triceps becomes the prime mover and the biceps must relax.

Table 5–1 includes some of the important muscles of the body. If you are interested in detail concerning muscles, you should consult some of the anatomy texts listed in the recommended readings at the end of the chapter.

SUMMARY QUESTIONS

1. What factors favor the deposition of calcium into bone?
2. Where is the periosteum located?
3. Name the end portions of long bones.
4. Where is red bone marrow located?
5. Where is yellow bone marrow located?
6. How are the bones of the skull formed?
7. What are the bone-building cells called?
8. What cells of the bone help in the exchange of minerals between blood and bone?
9. Where is the humerus located?
10. Name the three parts of the innominate bone.
11. Where is the foramen magnum located?
12. Name the thigh bone.
13. Where is the acetabulum?
14. Name all of the sinuses that drain into the nose.
15. Name the bone at the base of the skull.
16. Name the bone in the temporal region.

TABLE 5-1

Muscle	Location	Function
Muscles of Facial Expression		
Orbicularis oculi	Circles eyelids	Closes eyelid
Levator palpebrae superior	Posterior part of eye orbits and eyelids	Opens eyelid
Oculi recti (4)	Superior part of orbits	Rolls eye upward
	Inferior part of orbits	Rolls eye downward
	Lateral and medial parts of orbits	Turn eye side to side
Oculi obliques (2)	Superior lateral side of eyeball	Turns eyeball downward and laterally
	Inferior lateral side of eyeball	Turns eyeball upward and laterally
Orbicularis oris	Circles the mouth	Purses lips
Masseter	From zygomatic bone to mandible	Closes jaw
External pterygoid	From sphenoid to mandible	Opens mouth
Movement of the Head		
Sternocleidomastoids	From sternum and clavicle to the temporal bone	Flex and rotate head
Semispinalis capitus	From thoracic vertebrae to occipital bone	Extends head
Movement of the Shoulder		
Trapezius	From occipital bone to scapulae	Raises and pulls shoulders back
Pectoralis minor	From shoulder girdle to ribs	Depresses shoulder forward
Movement of the Upper Extremity		
Pectoralis major	Pectoral region	Flexes and adducts anteriorly
Latissimus dorsi	Lateral aspect of the back	Extend and adduct posteriorly
Deltoid	Deltoid region	Abducts arm
Biceps brachii	Anterior aspect of brachial region	Flex forearm and supinate hand
Triceps brachii	Posterior aspect of brachial region	Extend forearm
Anterior forearm muscles	Anterior aspect of the forearm	Flex wrist and pronate hand
Posterior forearm muscles	Posterior aspect of the forearm	Extend wrist and supinate hand
Muscles of Inspiration		
Diaphragm	Floor of thoracic cavity	Increases vertical diameter of thorax

Muscles of the Abdominal Wall		
External obliques	Superficial layer of abdominal muscles	Compress viscera and aid in forced expiration. Flex and rotate vertebral column
Internal obliques	Beneath external obliques	Same action as the external obliques
Transversus	Beneath internal obliques	Compresses viscera and aids in forced expiration
Rectus abdominus	From symphysis pubis to sternum	Depresses thorax, flexes and rotates vertebral column
Muscles of the Back		
Sacrospinalis	From sacrum to occipit	Extends vertebrae
Muscles of the Pelvic Floor		
Levator ani	Pelvic floor	Support pelvic viscera
Coccygeus	Pelvic floor	Supports pelvic viscera
Rectal sphincter	Surrounds anus	Keeps anus closed
Movement of the Lower Extremities		
Gluteal group	Gluteal region	Some muscles of this group extend the thigh and rotate thigh outward. Others abduct and internally rotate the thigh
Adductor group	Anterior and medial aspect of the thigh	Adduct the thigh
Hamstrings	Posterior aspect of the thigh	Flex lower leg
Quadriceps femoris	Anterior aspect of femoral region	Extends lower leg
Anterior tibial	Anterior aspect of lower leg	Dorsiflexes and inverts the foot
Posterior tibial	Posterior aspect of the tibia	Inverts and plantar flexes
Gastrocnemius	Prominent muscle of the calf	Adducts and inverts the foot, plantar flexes

17. Name the lower jaw bone.
18. Name the first and second cervical vertebrae.
19. Classify joints according to the amount of motion they permit.
20. What is circumduction and at what joints can this type of movement be accomplished?
21. What is the difference between abduction and adduction?
22. Differentiate between the three types of muscle tissue.
23. What is dorsiflexion?
24. What is the All or None Law?
25. What muscle flexes the forearm?
26. What muscle extends the thigh?
27. Contraction of the sternocleidomastoid causes what movement?
28. What muscle contracts to extend the head?
29. What muscle abducts the arm?
30. Name the muscles of the abdominal wall.
31. What types of muscle tissue have automaticity?
32. Contraction of what muscle will extend and adduct the arm posteriorly?

SUGGESTED READINGS

Brooks, S. M. *Integrated Basic Science.* 3rd ed. St. Louis, Mo.: C. V. Mosby Co., 1970.

Chaffee, E., and Greisheimer, E. M. *Basic Physiology and Anatomy.* 2d ed. Philadelphia, Pa.: J. B. Lippincott Co., 1969.

Nourse, A. E. *The Body,* Life Science Library, New York, N.Y.: Time-Life Books, 1964.

Tokay, E. *Fundamentals of Physiology: The Human Body and How It Works.* New York, N.Y. Barnes & Noble, 1967.

6 Disorders of the Musculoskeletal System

VOCABULARY

Anaerobic
Analgesic
Atrophy
Chronic
Crepitus
Cyanosis
Ecchymosis
Edema
Embolus
Epigastric
Exacerbation
Exudate
Gynecology
Hydrocortisone
Macerated
Orthopedics
Prednisone
Remission
Resuscitation
Steroid
Trauma
Urology

OVERVIEW

I FRACTURES
- **A.** Types of fractures
- **B.** First aid
- **C.** Types of reductions of fractures
- **D.** Casting and cast care
- **E.** Traction
- **F.** Immobilization of fractured ribs
- **G.** Complications of fractures
- **H.** Crutch walking

II SPRAINS AND DISLOCATIONS
III RHEUMATOID ARTHRITIS
IV OSTEOARTHRITIS
V LOW BACK PAIN
VI OSTEOMYELITIS
VII HERNIAS

FRACTURES

Probably the most common of the disorders of the musculoskeletal system is that of broken bones. Although the most frequent cause of fractures is that of trauma, some are pathological fractures, that is, the result of a bone disease that cause a weakening of the structural supports so that bones break with little or no actual trauma. Pathological fractures may occur because of such conditions as cancer of the bone, osteoporosis (porous bones), or complication from prolonged steroid therapy.

Fractures are classified as follows (see Figure 6–1):

Simple fracture: A break in the continuity of the bone does not produce an open wound in the skin. This type of fracture is also called a closed fracture.

Compound or open fracture: Fragments of the broken bone protrude through a wound in the skin. This type of fracture is likely to be complicated by infection.

Greenstick fracture: The bone bends and splits, but it does not break completely. This type of fracture occurs primarily in children.

Comminuted fracture: The bone is broken in several places and splinters of bone may be embedded in the surrounding soft tissue.

Spiral fracture: The bone has been twisted apart. This kind of fracture is relatively common in skiing accidents.

Impacted fracture: The broken ends of bone are jammed into each other.

Silver-fork fracture: The bone is fractured at the lower ends of the radius, so-called from the shape of the deformity that it causes.

Depressed skull fracture: A fracture of the skull causes a fragment of bone to be depressed below the surface.

Complete fracture: The fracture line goes all the way through the bone.

FIRST AID

The most important first aid care of fractures is prevention of movement of the affected parts, thereby protecting the surrounding soft tissues from additional trauma. This immobility can be accomplished by splinting the injured part in the position in which the fracture occurred. The splint, which should be applied before the patient is moved, must include the joints above and below the fracture site. The splint should be padded so that the soft tissues are protected uniformly.

If the fracture is a simple fracture, bleeding is not usually a

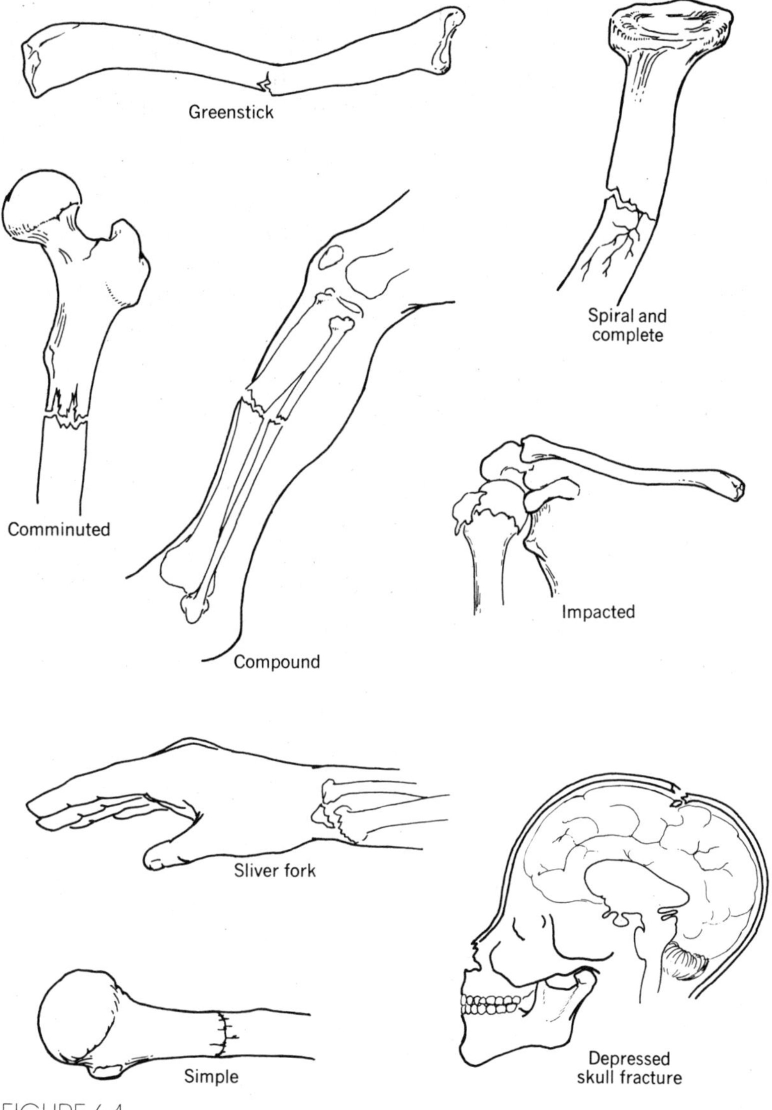

FIGURE 6-1
Types of fractures.

major problem. However, a great deal of bleeding may accompany a compound fracture; measures must be taken to control the bleeding and to combat shock. Aside from controlling bleeding and treating shock, haste in first aid in a case of fracture is unnecessary and can be perilous. Those who have incurred back injuries may be spared serious spinal cord damage by being moved carefully on a firm board. If this procedure is not practical, the patient should be left exactly as he is. Only bleeding and shock should be treated until skilled attendants with proper equipment arrive.

Following fractures, particularly those of the femur, the muscles surrounding the fracture go into spasm, which causes the broken ends of the bone to override (overlap). In this case, the injured extremity is noticeably shorter than the other. Muscle spasm is also harmful because it causes additional pain. Except in the case of a compound fracture, traction should be applied manually by steady pulling of the intact part distal to the fracture. The amount of traction should be just enough to support the extremity in such a way that motion is minimized. The traction should be used before splinting, and it must be continued until a splint is applied which can maintain the pull. If the patient must be transported without a splint, manual traction should be maintained continuously until the patient reaches the hospital.

In the case of a compound fracture, the open wound should be covered with a sterile dressing. If none is available, dirt should be kept away from the wound. Although this type of fracture should be splinted, traction should not be applied.

TYPES OF REDUCTION

After the doctor has X-rayed the patient and has a clear view of the type of fracture and the amount of displacement of the bone segments, he will reduce the fracture (replace the parts in their normal position). He performs a closed reduction by gently manipulating the part, redirecting it to its normal position. After he has reduced the fracture, he will immobilize the patient in this position by a bandage or cast.

An orthopedic surgeon performs an open reduction by making a surgical incision and then exposing and realigning the bone. This type of reduction is used if soft tissue such as blood vessels and nerves lie between the ends of the broken pieces of bone, and or if the patient has suffered a comminuted fracture. Following an open reduction, the part is immobilized in this position either by external fixation such as casting, or by internal fixation, in which one of a variety of types of metal plates, screws, or rods is affixed to the bone.

CASTING AND CAST CARE

Casts are made from bandages that are impregnated with plaster of Paris. Two rolls of the bandage are placed in a bucket of tepid water. The rolls are stood on end so that air can escape and the water can penetrate all of the folds with a minimum loss of plaster. When no more bubbles come from the roll, the technician removes the first roll and gently presses the extra water out of it and gives it to the doctor. A third roll is then placed in the water so that it will be ready to apply by the time the second roll has been used. Because the longer the roll stays in the water, the more plaster it loses, timing is important. Vigorous squeezing of the roll will also cause loss of plaster. After the cast has been applied, the doctor will take a second set of X-rays to ensure correct setting.

Sometimes stockinette or padding is placed on the part before the case is applied, especially over bony prominences. If this is done there can be no wrinkles in the padding.

While the cast is drying, it must be left uncovered so that water can evaporate from it. However, no drafts should fall directly on a patient in a wet cast. When a wet cast is moved, it should be supported with the palm of the hand rather than the fingertips. A thumb print on a cast can leave an indentation that may cause pressure on the patient's skin and develop a sore. The cast should dry on plastic-covered pillows to prevent its flattening and to elevate the part, decreasing the amount of swelling that may develop.

After the cast is dry, the roughened edges of the cast should be covered with adhesive tape. If the cast is lined with stockinette, the end of the stockinette should be pulled over the edges of the cast and taped to the cast. If this is not done, the edges of the cast will crumble and cast crumbs may get under the cast and irritate the skin.

Casts should be kept clean. They can be washed with scouring powder and a little water. If the cast is likely to be soiled with urine or feces, it should be protected with plastic. Although drawing and writing on casts is not harmful, the entire cast should not be painted over. The paint will interfere with the porosity of the plaster, and the skin under the cast may become macerated.

A new type of cast made of meshlike material is lighter in weight than the conventional plaster of Paris cast, yet it affords as much support. After it is applied, it is baked under infrared light. In addition to its light weight, this type of cast is not adversely affected by moisture and allows a patient who wears it to shower.

When an extremity is casted, the fingers or toes should be observed frequently during the first few days. Any swelling, discoloration, or abnormal temperature of the digits should receive immediate attention. Such symptoms may indicate that the cast is too tight and that the blood supply is impaired. Localized areas of pain under the cast may indicate a pressure area that can result in serious tissue destruction.

Casts are removed by a mechanical cast cutter. The patient should be assured that although the machine is noisy it will not hurt him.

After the cast has been removed, the patient will have both pain and stiffness. The skin under the cast may be covered with crust composed of exudate, oil, and dead skin. Oil and warm baths will soften and remove this crust after a few days.

An extremity that has been in a cast for a prolonged period of time will be weak and will need continued support. An elastic bandage can be put on a leg; an arm may be supported with a sling. These weak muscles must be exercised not only to increase their strength but also to improve circulation to the part. A doctor or physiotherapist can help teach the patient active and progressively graded exercises so that his muscles and joints will regain full range and strength of motion.

TRACTION

Traction is sometimes used, particularly for fractures of leg bones. The traction may be used as a temporary measure to help in disengaging the bone fragments until other measures can be taken to immobilize the part, or it may be maintained until healing takes place.

Suspension traction involves using weights and pulleys. Usually about 8 to 10 pounds of weight will be necessary for a fractured femur of an adult. This weight must be applied continuously. A box at the foot of the patient's bed can support his foot on the uninvolved side and keep him from sliding down in bed (see Figure 6–2). Clearly, if the patient does slide down in bed and the weights rest on the floor, no traction is being applied to his leg. The patient must use a firm mattress, or perhaps a board placed under the mattress.

As in the case of a patient whose extremity was casted, the patient who has been in traction for a prolonged period of time must use exercises to help regain muscle strength and range of motion once the traction is removed.

IMMOBILIZATION OF FRACTURED RIBS

Fractured ribs are usually immobilized by securely applying an elastic bandage to the chest or by strapping the chest with adhesive tape. Many people are sensitive to adhesive tape, particularly when it is left on for long periods of time. A generous application of tincture of benzoin to the skin before the tape is applied may help

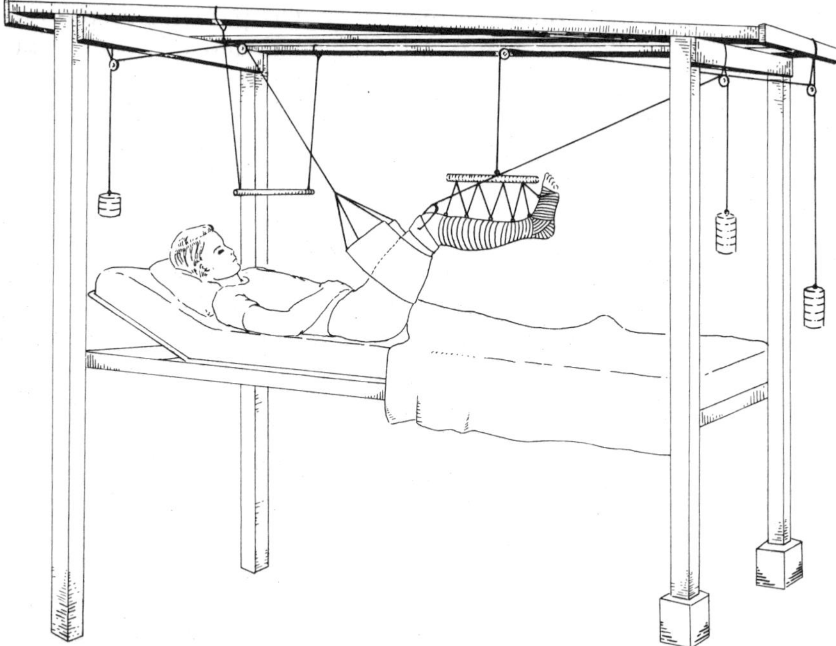

FIGURE 6-2
A type of skeletal traction.

prevent adhesive tape reactions, which can be painful. Better still, the tape can be applied sticky side out, one layer on top of another. Once the chest is taped in this manner, talcum powder should be applied to the sticky outer surface so it will not adhere to the patient's clothing. All of this tape can be removed simply by cutting it off with bandage scissors. Tape that has adhered to the skin, can be removed by applying mineral oil to the outer surface of the tape, waiting a few minutes, and then removing the tape. The oil penetrates the tape so that it can be removed without discomfort.

COMPLICATIONS OF FRACTURES

Delayed union or nonunion of a fracture can be caused if the part has been inadequately immoblized or if soft tissue is lodged between the bone fragments. Poor circulation and infection can also result in delayed union.

Although fat embolus is not a common complication of fractures, it does occur particularly in severe fractures and can be life threatening. In such cases, it usually happens in the 12 to 24 hours following the injury. The patient's respirations will be rapid, his pulse rate will increase, and he will be pale or cyanotic. This emergency situation may necessitate resuscitation (artificial support of respirations and heart action).

Infection, particularly in the case of a compound fracture, can be a very serious complication. Wounds are deep and may be infected by the gas-gangrene bacillus, an anaerobic organism. Symptoms of this infection are fever, pain, a thin watery exudate that is foul-smelling, and local puffiness and discoloration. Crepitation due to gas bubbles in the subcutaneous tissues may be felt under the skin. Tetanus is another infection that may complicate compound fractures. Both tetanus and gas-gangrene are treated by opening the wound so that air can enter and the wound can be irrigated. Appropriate antibiotics are also used.

For other complications, you should refer to Chapter 2, which discusses the complications of immobility.

CRUTCH WALKING

The patient must have correctly fitting crutches and should be instructed in their proper use. Crutches that are too long may exert pressure on the brachial nerve plexus in the axilla and cause nerve damage more disabling than the fracture. Even if the crutches are properly fitted, crutch palsy can develop if the patient frequently rests his weight on the axillae rather than on his hands.

The patient should usually do exercises to strengthen his arm and shoulder muscles in preparation for crutch walking. He can do push-ups or lift weights while lying in bed.

Several types of gaits are used by patients who must walk on crutches. In a swing-through gait, the patient advances the two crutches simultaneously and swings his weight through the crutches landing beyond them. Although this method is fast, it should be discouraged unless the patient is in a hurry, because it can lead to atrophy of the muscles of the legs and hips. Teenagers probably should not be taught this method unless they recognize its dangers and will use it carefully and infrequently.

In the two-point gait, the patient advances the left foot and the right crutch simultaneously, and then advances the right foot and the left crutch. This gait is the most normal type of crutch walking.

A patient who is allowed little or no weight bearing should use a three-point gait. He should advance both crutches at the same time while he uses the strong leg to stand on and swings the strong leg through the crutches. The injured leg can go with the crutches or it can stay in line with the strong leg.

To use a four-point gait, the patient advances one at a time, the right crutch, left leg, left crutch, and the right leg. This method allows equal but partial weight bearing on each limb and therefore is used for patients who have arthritis, cerebral palsy, or poliomyelitis.

If the patient uses a single crutch, he should place it on the unaffected side. If you are assisting a disabled person, you should walk on his unaffected side. When assisting such a patient on a stairway, if you can not walk along side the patient, you should walk behind him as he goes up the stairs and in front of him as he goes downstairs. It is more dangerous and difficult for a disabled person to go downstairs than it is for him to walk upstairs.

SPRAINS AND DISLOCATIONS

Sprains are injuries to the ligaments surrounding a joint. There is swelling, pain, and loss of motion in the affected joint; there may also be ecchymosis in the area. The part should be X-rayed to rule out the possibility of a fracture. Treatment of sprains consists of elevation of the part and application of an elastic bandage. An injection of procaine (a local anesthetic) and hydrocortisone into the joint will help relieve the pain. After application of an elastic bandage, the patient can continue to use the affected part.

Dislocations occur when the articular surfaces of a joint are no longer in contact. Dislocations usually are a result of trauma; however, they may occur as a result of joint disease. The patient experiences pain, loss of function of the joint, and malposition (the joint is not in the normal position). A dislocation is treated by manipulation of the joint until the parts are in their normal position. Immobilization with elastic bandages, a cast, or a splint for several weeks will allow the joint capsule and surrounding ligaments to heal.

RHEUMATOID ARTHRITIS

Rheumatoid arthritis is an inflammatory disease of the joints characterized by chronicity, remissions, and exacerbations. This disease has an insidious onset. The patient notices that a joint or two is stiff when he wakes up in the morning. Slowly some joints—usually those of the fingers—become sore, red, and swollen. Later, other joints become involved. As the disease progresses, the patient suffers fever and malaise. He also has a low tolerance for temperature changes and any kind of stress. Rheumatoid arthritis can be seriously disabling.

The treatment of rheumatoid arthritis is mostly palliative. Salicylates, usually in the form of aspirin, are given. Because large doses of aspirin may cause erosions of the mucosa of the stomach, patients should have milk and crackers before taking the aspirin. They should be instructed to watch for evidence of gastrointestinal bleeding such as black, tarry stools. They should drink at least 1500 ml. of fluids daily to lessen the possibility of renal calculi (stones). Steroids such as prednisone or cortisone are also helpful, but these medications are usually resorted to only after other measures have failed to give relief. Some of the untoward side effects of prolonged steroid therapy include fluid retention, peptic ulcers, poor wound healing, adrenal atrophy, and even emotional problems.

Exercises and physiotherapy can be helpful for the patient with rheumatoid arthritis; however, during acute exacerbations of the disease, it may be necessary to immobilize the painful joints to provide local rest. Splints or bivalved casts may be used for this immobilization. A bivalved cast is made by casting the part; after the cast is thoroughly dry, it is cut so that the cast can be removed for bathing or other treatment. When the cast is reapplied, the two parts are held together with elastic bandages.

OSTEOARTHRITIS

Osteoarthritis is a slowly, but steadily progressive disease that has no remissions and no systemic symptoms. It is usually a disease of late middle age and old age. Although the affected joints usually suffer little or no swelling, they are painful and stiff, particularly in the morning and during damp weather. Local rest of the affected joints, warmth, and aspirin are helpful. Short periods of moderate exercise, five or six times a day, which do not strain the patient may be prescribed. The affected joints may be injected with a steroid preparation. This form of steroid therapy does not have as severe side effects as the steroid preparations that are taken orally.

LOW BACK PAIN

Low back pain is a common complaint and an important cause of loss of work time as well as a source of much aggravation to both the sufferer and those who are trying to help him. Because the real cause of the problem is frequently hard to identify, the ailment is

difficult to treat. The pain may be the result of urological diseases, gynecological problems, psychosomatic illness, or an orthopedic condition. The most frequent orthopedic causes of low back pain are osteoarthritis, lumbosacral strain, unstable lumbosacral mechanism, or herniated intervertebral discs.

Acute lumbosacral strain is treated with bed rest, muscle relaxants, local heat, and analgesics to relieve the muscle spasm. The patient with an unstable lumbosacral mechanism may also be treated with measures to relieve muscle spasm and exercises to increase the strength of the paravertebral muscles. Sometimes a brace or a lumbosacral belt is prescribed.

A herniated intervertebral disc (slipped disc) causes severe back pain that radiates down one leg. The pain recurs and is particularly severe when the patient strains, coughs, or lifts a heavy object, or even a light object, if he uses poor body mechanics. The pain is caused by pressure on nerves, usually in the lumbar region. The spongy center of the intervertebral disc herniates and causes this pressure.

The herniated intervertebral disc may be treated conservatively with bed rest on a firm mattress or possibly with traction. Patients should practice good body mechanics and avoid twisting when they roll from side to side in the bed. Quick motions should be avoided. Occasionally, if conservative measures are unsuccessful, a doctor must resort to surgery. He may do a laminectomy, that is, removal of the posterior arch of the vertebra to expose the cord. The herniated disc is then surgically removed.

Sometimes a spinal fusion is necessary in order to treat a herniated intervertebral disc. In this type of surgery, bone is taken from some other area, such as the iliac crest, and is grafted onto the vertebrae. Two or more vertebrae are united by means of this graft. Although mobility of the joint is permanently lost, the patient becomes accustomed to a permanent area of stiffness.

OSTEOMYELITIS

Osteomyelitis is an infection of bone. Although it can result from an infected compound fracture, it more frequently is a blood-borne infection, starting at a focus of infection elsewhere in the body. The patient has severe pain, fever, and local swelling. There is a serious danger that osteomyelitis can become chronic. Treatment includes antibiotic therapy and surgical drainage of the pus. Antibiotics are put directly into the surgical wound, which is kept open for drainage. The affected part may be immobilized in a position utilizing

the help of gravity in draining the pus. Rest and good nutrition will facilitate healing by increasing the patient's resistance.

HERNIAS

Probably the most common type of muscle disorder, hernias result when weakened muscles allow the viscera underlying the weakened muscle to protrude. The most common hernias are those that occur in the normally weak places of the abdominal wall—inguinal ring, femoral ring, and umbilicus—and in the diaphragm, at the place where the esophagus enters the stomach.

With the exception of the diaphragmatic hernia, surgery is usually the chosen treatment. Trusses are usually not recommended because an uncorrected hernia is vulnerable to serious complications. An incarcerated hernia cannot be reduced (the protruding viscera cannot be pushed back into the abdominal cavity). Edema of the protruding structures and constriction of the opening through which the bowel has emerged make it impossible for them to return to the abdominal cavity.

Strangulation of the bowel can occur if the incarcerated hernia has not received prompt surgical attention and the opening through which the viscera is protruding obstructs the blood supply to the loop of bowel. This condition can lead to necrosis of the trapped loop of bowel and will necessitate a bowel resection (the surgical removal of the loop of bowel).

A diaphragmatic or hiatus hernia is the protrusion of a part of the stomach through a defect in the wall of the diaphragm at the point where the esophagus passes through the diaphragm. Patients with this type of hernia have heartburn, belching, and epigastric pressure after eating. They may help their distress by eating frequent, small meals and by avoiding any food for about three hours before retiring; they might also sleep in semisitting position, and avoiding bending over. If a patient is obese, he should be put on a weight reduction diet. A patient with a chronic cough must be treated, because coughing puts additional strain on the weakened area.

Surgical treatment of a hiatus hernia involves thoracic surgery to replace the stomach or other organs protruding into the abdominal cavity and to repair the defect in the diaphragm. Such surgery is more major than that required for other uncomplicated hernias.

Occasionally, incisional hernias occur following abdominal surgery. Obese or elderly patients and those who suffer from malnutrition are more prone to developing incisional hernias than are

other surgical patients. Surgical incisions that have been infected and in which wound healing has been impaired may also predispose the patient to an incisional hernia.

A variety of other diseases are characterized by muscle weakness and poor coordination. These diseases, such as multiple sclerosis, myasthenia gravis, and myotonia, will be discussed in Chapter 11.

SUMMARY QUESTIONS

1. Differentiate between a simple fracture and a compound fracture.
2. What type of patient is most likely to have a greenstick fracture?
3. What type of fracture is most likely to become infected?
4. Discuss first aid measures that are appropriate to use for a patient who has fractured his leg.
5. Discuss open and closed reductions of fractures.
6. What should you observe when a person has had a cast applied to an extremity?
7. What is the difference between internal and external fixation of a fracture?
8. Discuss complications of fractures.
9. How might nerve damage result from the improper use of crutches?
10. Discuss the different types of gaits used with crutches.
11. Discuss the treatment of sprains and dislocations.
12. How does rheumatoid arthritis differ from osteoarthritis?
13. Discuss several aspects of the treatment of a patient with a herniated intervertebral disc.
14. What is osteomyelitis and how is it treated?
15. What are the most common types of hernias?
16. Why are trusses usually not recommended for a patient who has a hernia?
17. How is a hiatus hernia treated?
18. What type of patient is most likely to develop an incisional hernia?

RECOMMENDED READINGS

Compere, E. L., *et al. Pictorial Handbook of Fracture Treatment.* 5th ed. Chicago, Ill.: Year Book Medical Publishers, 1963.

Gardner, A. F. *Paramedical Pathology.* Springfield, Ill., Charles C Thomas, 1972.

Jordana, A. "Protection from X-Ray Exposure," *The Professional Medical Assistant,* May/June 1973.

Krause, M. V. *Food, Nutrition and Diet Therapy.* Philadelphia, Pa. W. B. Saunders Co., 1966.

Robbins, S. L., and Angell, M. *Basic Pathology.* Philadelphia, W. B. Saunders Co., 1971.

Soika, C. "Combating Osteoporosis," *American Journal of Nursing,* July 1973.

7 The Circulatory System

VOCABULARY

Agglutinate
Amino acid
Anastamosis
Antibody
Anticoagulant
Collateral circulation
Endothelium
Filtration
Glucose
Glycogen
Hypertension
Mediastinum
Oriface
Osmosis
Parasympathetic
Parietal
Sympathetic nerves
Varicosed veins
Visceral
Ulnar

OVERVIEW

I HEART
- **A.** Location and description
- **B.** Pericardium
- **C.** Structure of the heart
  - **1.** Myocardium
  - **2.** Endocardium
  - **3.** Heart chambers
  - **4.** Heart valves
- **D.** Blood supply
- **E.** Nerve supply
  - **1.** Intrinsic nerve supply
  - **2.** Exlrinsic nerve supply
- **F.** Electrocardiogram

II PHYSIOLOGY OF CIRCULATION
- **A.** Cardiac output
- **B.** Blood flow
- **C.** Resistance to blood flow
- **D.** Velocity
- **E.** Blood pressure
  - **1.** Arterial
  - **2.** Venous

III ARTERIAL CIRCULATION
- **A.** Pulmonary
- **B.** Systemic

IV CAPILLARIES

V VENOUS RETURN

VI COMPOSITION OF BLOOD
- **A.** Plasma
- **B.** Formed elements
  - **1.** Red blood cells
  - **2.** White blood cells
  - **3.** Thrombocytes

VII THE LYMPHATIC SYSTEM

The general function of the circulatory system is to transport fluid with dissolved substances and particles to and from all parts of the body. There are two divisions of this system: the lymphatic, which helps to return tissue fluid to blood; and the blood. The blood division is a closed circuit; thus, pathology in any part, be it a blockage in the flow or a leak, will lead to easily predictable symptoms.

THE HEART

We shall now consider some of the major structures of the blood circulatory system. The heart is the pump for the blood. It is located in the thoracic cavity, medial to the lungs, in the mediastinum. The sternum is anterior to the heart, and four thoracic vertebrae are posterior. The base of the heart is directed upward and to the right, and its apex is directed downward and to the left of the midline.

The heart is surrounded by a sac called the pericardium. This sac has two layers, the parietal and the visceral. Between the layers is a small amount of fluid that serves as a lubricant. The visceral pericardium, also known as the epicardium, is closely attached to the cardiac muscle or myocardium. The lining of the heart is called the endocardium, which is endothelial tissue continuous with the lining of the blood vessels.

The upper chambers of the heart, or the atria, have thin walls and a smooth inner surface. The septum between the right and left atria has a scar on it which is called the fossa ovale. This fossa, the foramen ovale in fetal life, functioned to allow some blood to flow directly from the right atrium to the left—bypassing the nonfunctioning lungs of the unborn child. The right atrium receives venous blood from the superior and inferior vena cavae and also from the coronary sinus. The coronary sinus receives its blood from the coronary veins, which are returning blood from the capillaries of the myocardium. The right atrium sends its blood to the right ventricle. The left atrium receives blood from four pulmonary veins and sends blood to the left ventricles.

The lower chambers of the heart are the ventricles, which have much thicker walls than the atria. Their inner surfaces are irregular, because they contain several papillary muscles as well as stringlike structures, or chordae tendineae, that are attached to the papillary muscles and to the valve flaps that are located between the upper and lower chambers. The chordae tendineae prevent the valves from turning inside out when the ventricles contract and force the blood upward. The right ventricle sends its blood to the pulmonary artery and thence to the capillaries of the lungs to be oxygenated and to eliminate carbon dioxide. The left ventricle, whose walls are normally about three times as thick as the walls of the right ventricle, forms the apex of the heart. Pulsations of the apex can be heard between the fifth and sixth ribs, about 2 inches below the left nipple in the male. The function of the left ventricle is to pump blood into the aorta and to all parts of the body (see Figure 7–1).

HEART VALVES

The valves of the heart, which are made of tough fibrous tissue, all function to prevent the backflow of blood. Between the right atrium and ventricle is the tricuspid valve. It has three triangular flaps that are attached to the chordae tendineae. Between the left atrium and ventricle is the bicuspid, or mitral valve. When the ventricles contract, blood is forced upward and closes these valves. Semilunar valves prevent the blood from flowing back into the ventricles once it has been pumped out into either the pulmonary

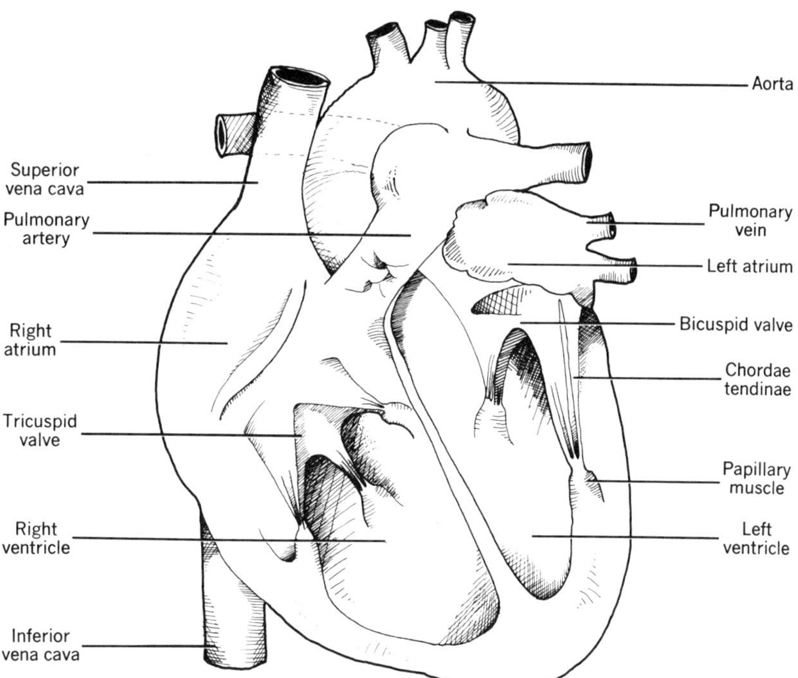

FIGURE 7-1

A schematic drawing of the internal structure of the heart and the major vessels entering and leaving the chambers of the heart.

artery or the aorta. Both the aortic and pulmonary semilunar valves are composed of three halfmoon-shaped pockets that catch the blood and balloon out to close the orifices. It is the closure of the heart valves that makes the heart sounds. When the atrioventricular valves close, the first sound (lupp) is heard; when the semilunars close, the second sound (dupp) is heard. Abnormal heart sounds therefore may be indicative of some valvular pathology.

BLOOD SUPPLY TO THE HEART

The blood supply to the myocardium travels via the coronary arteries. Right and left coronary arteries are the first branches off the ascending aorta. They go to a rich capillary network throughout the myocardium, from there to coronary veins, and then to the coronary sinus and right atrium. Occlusion, or blockage of coronary blood supply, may result in a heart attack if collateral circulation is not adequate.

INTRINSIC NERVE SUPPLY

The heart has an intrinsic nerve supply that is very important because the heart is able to contract without any efferent or motor nerve fibers. Structures involved in this intrinsic supply are the sinoatrial (SA) node, the atrioventricular (AV) node, and the bundle of His (see Figure 7–2). The SA node is the pacemaker and initiates the beat. Located about the place where the superior and inferior vena cavae enter the right atrium, it sends electrical impulses via the atrial myocardium to the AV node located just below the coronary sinus in the septum. The AV node sends the impulses to the bundle that branches throughout the walls of the ventricles. Digitalis, a drug that slows and strengthens the heart beat, is commonly used in the treatment of some heart diseases. An overdose of digitalis may produce heart block because the bundle will not transmit impulses to the ventricles. In such a case, the ventricles will beat at their own intrinsic rate, which is about 35 beats per minute. This rate will not alter with rest or exercise as does the normal heart rate.

EXTRINSIC NERVE SUPPLY

The extrinsic nerve supply of the heart is responsible for altering the rate and force of the cardiac contraction to meet the needs of the body. Efferent nerves going to the SA node to initiate contraction originate in the cardiac center in the medulla of the brain. The vagus, a parasympathetic nerve, will slow the heart rate; branches

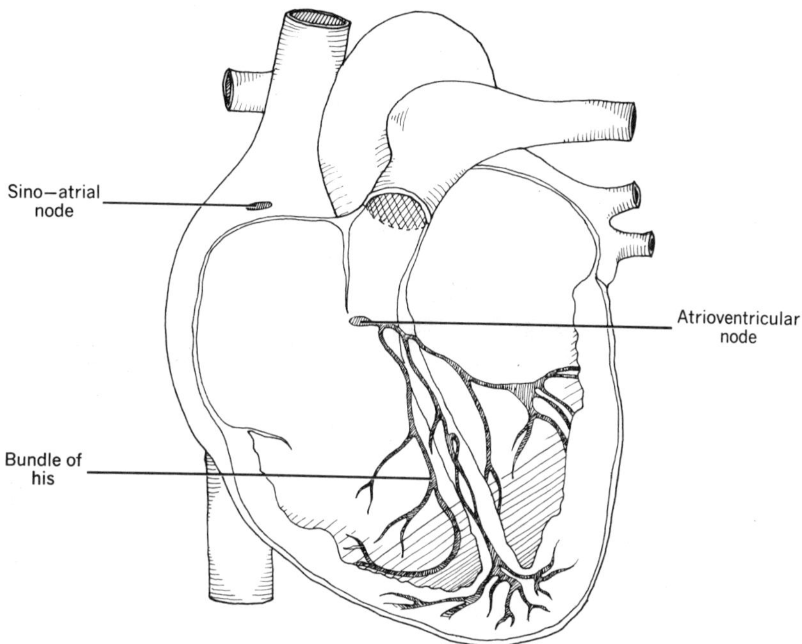

FIGURE 7-2
The SA node initiates the cardiac cycle. Impulses from the SA node are transmitted through the atrial myocardium to the AV node and then to the bundle of His, which has fibers extending throughout the walls of the ventricles. When these fibers are stimulated, the ventricles contract.

from the sympathetic nerves will increase its rate depending upon the needs of the body. You should see Chapter 11 for a more detailed discussion of the nervous system.

Afferent, or sensory, nerve fibers are stimulated by oxygen lack. If coronary blood flow is inadequate, the cells of the myocardium are not receiving sufficient oxygen. Substernal pain usually radiates down one arm, usually the left, or elsewhere in the body. When the walls of the right atrium are stretched because of an increase of blood, other afferent nerves are stimulated. These fibers send impulses to the medullary center to cause it to increase the rate and strength of contraction.

ELECTRICAL CHANGES

Electrical changes take place during the cardiac cycle; these changes can be measured with an electrocardiograph, or ECG (see Figure 7–3). The p-wave occurs when the impulse has been received by the SA node. The QRS complex occurs when the impulse

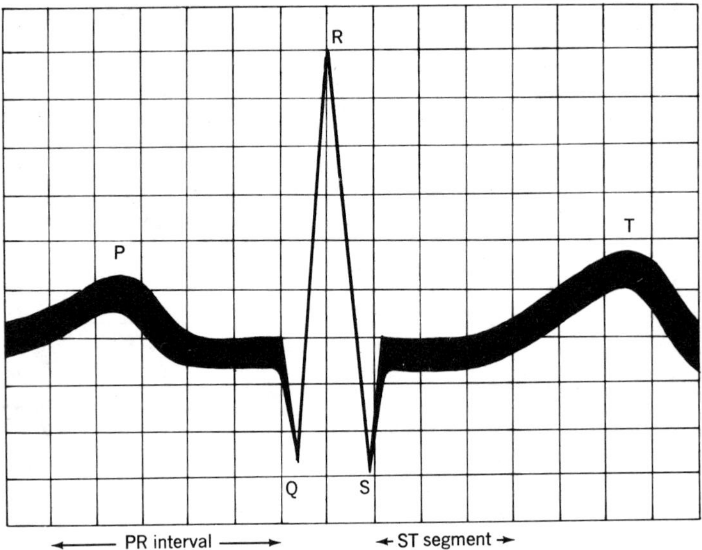

FIGURE 7-3
A normal ECG tracing.

is passing through the ventricles and the T-wave indicates ventricular rest, or diastole. The PR interval is the time it takes for the impulse to cross the atria and AV node to reach the ventricles. This time should be about 0.12 to 0.2 second. If the interval is too short, the impulse has reached the ventricles through a shorter than normal pathway. If the interval is too long, a conduction delay in the AV node has probably taken place. The QRS complex takes about 0.12 second. The entire cardiac cycle is about 0.8 second if the heart is beating at 75 beats per minute. The systole, or contraction, of the ventricles takes 0.3 second, and the diastole takes 0.5 second. Because the heart abides by the All or None Law, the only way the rate of contraction can be increased is to shorten the diastole; when stimulated, it gives its best contraction or none at all. The significance of this increase in rate with the shortening of the diastole is that greater work is being required of the heart and less rest is being provided. Figure 7–4 will help you understand the relationship between the contraction of the ventricles, the heart sounds, and the electrical changes taking place during the cardiac cycle.

CIRCULATION

Circulation depends primarily on the action of the heart, the condition of the blood vessels, and the viscosity of the blood. It can,

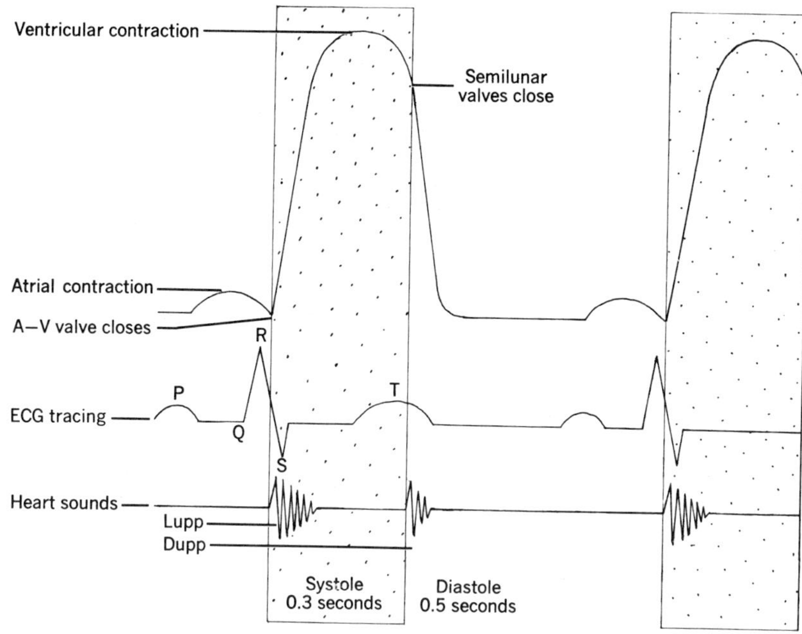

FIGURE 7-4

A schematic drawing showing the relationship between the closure of the heart valves, heart sounds, systole, diastole, and an ECG tracing.

however, be influenced by temperature, by body size and activity, and by some drugs. An increased body temperature will increase the rate of the heart beat, as will increased activity. Generally speaking, the smaller the individual the more rapid the heartbeat. Drugs that enhance the activity of the sympathetic nervous system will increase the heart rate, whereas parasympathetic agents and sedatives will decrease the heart rate.

CARDIAC OUTPUT AND STROKE VOLUME

Cardiac output is the rate at which blood is pumped by the heart. Stroke volume is the amount of blood pumped by each beat of the heart. At a rate of 75 beats per minute, 70 milliliters of blood are ejected by each ventricle per beat. This amount, which reaches about 5 quarts a minute, can easily be doubled in exercise. A weak heart must pump faster to make up for a low stroke volume. A well-trained athlete has a relatively slow pulse because he has a good stroke volume. Young elastic vessels can tolerate a great increase in stroke volume. As you get older and your vessels lose their elasticity, an increased stroke volume may cause your weakened vessels to rupture.

BLOOD FLOW

Blood flow is the quantity of blood moving through a vessel or vessels at a given period of time. Normally, at rest, the vessels of the skeletal muscles contain only about 15 percent of our blood; however, during exercise, the muscles can command as much as 75 percent of the blood. The blood always flows from a higher pressure to a lower pressure. This pressure is highest in the aorta and falls greatly by the time it reaches the capillaries. From the capillary bed, the pressure continues to drop as the blood flows into the veins. The drop on the venous side is not so great as that on the arterial side.

RESISTANCE TO FLOW

Friction causes resistance to blood flow. When blood vessels are constricted, resistance to flow increases. The thicker or more viscous the blood, the greater the resistance to the flow. Following hemorrhage—when there has been an increased return of tissue fluid to the vascular compartment—the blood will decrease in viscosity, the flow rate will increase and accordingly the pulse will be rapid and weak.

VELOCITY

The velocity of the flow of blood is the distance that it travels in a given period of time. Measurement of this velocity is called circulation time. One of two substances—either decholin, which has a bitter taste, or ether—is injected into an arm vein. The time that it takes the substance to get to the capillaries of the tongue (and be tasted) or to the lungs from which the exhaled air will have the odor of ether is measured. Normal values for the arm-to-tongue circulation time are about 10 to 16 seconds; and for the arm-to-lung circulation time, about 4 to 8 seconds. A prolonged circulation time suggests some increased resistance to the flow of blood.

BLOOD PRESSURE

Arterial blood pressure is determined by cardiac output and resistance to the flow of blood. It falls progressively from the time the blood leaves the left ventricle until it returns to the right atrium. In the aorta the pressure is normally about 120 mm. Hg (millime-

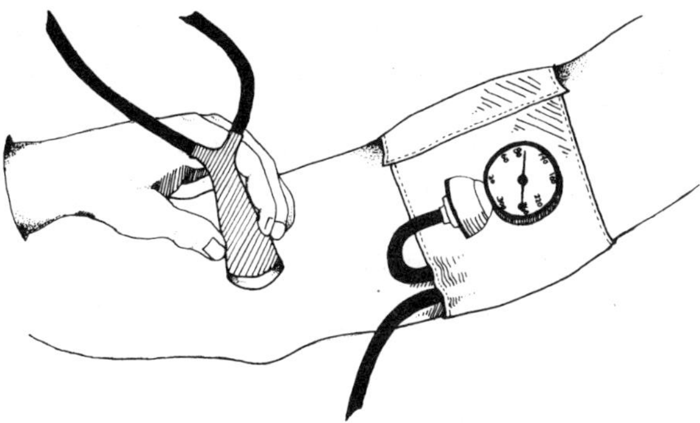

FIGURE 7-5
Measurement of the blood pressure in the brachial artery.

ters of mercury); by the time it gets back to the right atrium it is near 0 mm. Hg. Systolic pressure measures the force of the ventricular contraction. When you take a person's blood pressure, you wrap a blood pressure cuff around his upper arm and place a stethoscope over the brachial artery, which is on the medial aspect of the antecubital fossa. You then pump the cuff up either until the gage reads higher than you expect the pressure to be or until you can no longer feel a radial pulse. You should deflate the cuff slowly and listen for the first sound. This will be the systolic pressure, normally about 120 mm. Hg. You then continue deflating the cuff slowly until the sounds disappear. This moment will indicate diastolic pressure, which in a healthy young adult is usually about 80 mm. Hg. (see Figure 7–5). You record this blood pressure as 120/80.

Diastolic pressure, the measure of the peripheral resistance to blood flow, represents the force that must be overcome by the left ventricle before any blood can enter the aorta. For this reason, it is usually considered of greater clinical significance than is the systolic pressure. Mean pressure (approximately the arithmetic average of the systolic and diastolic pressures) is more important than either the systolic or diastolic alone, because it represents the average rate at which the blood is circulated.

Venous pressure is also measured. Pressure in an arm vein is normally about 10 to 0 mm. Hg. However, because it is measured with a water manometer, not a mercury manometer, water values are about 60 to 100 mm. The pressure is measured by inserting a needle into an arm vein. A manometer is attached to the needle, which has a three-way stopcock. The arm of the stopcock is arranged so that the venous blood will flow into the manometer. The height of the column of blood indicates the venous pressure.

If frequent venous pressure readings are required, central

venous pressure (CVP) measurements are used. A polyethylene catheter is passed through a vein and into the entrance of the right atrium. This catheter is attached by means of a three-way stopcock to a manometer and a continuous intravenous infusion. When a reading is to be made, the infusion is stopped and fluid is allowed to enter the manometer. The fluid rises in the manometer and slowly falls. The level to which it falls and starts fluctuating is the CVP reading.

ARTERIAL CIRCULATION

Arteries carry blood from the heart to the capillaries. They are more elastic than veins and do not collapse when cut. The arteries are generally deeper and have smaller diameters than their corresponding veins. Capillaries are the microscopic connections between the arterioles and venules. Veins differ from arteries in that they are thin-walled and collapse when cut. Most veins have valves to prevent the backflow of blood.

The pulmonary circulation arises from the pulmonary artery and branches down to capillaries that surround the alveoli, in which the blood gases (oxygen and carbon dioxide) are exchanged. Once the blood is oxygenated and most of the carbon dioxide has been removed, the blood is returned to the left side of the heart via the pulmonary veins. Blood pressure is low in the pulmonary circuit, and no fluid is filtered out into the alveoli. However, fluid may be filtered out into the alveoli of people with pulmonary hypertension; their respirations will sound moist. These people suffer from pulmonary edema.

The systemic circulation begins in the aorta. Off the arch of the aorta lead three branches: the left common carotid; the left subclavian; and the innominate, which gives rise to the right common carotid and right subclavian. A short cord, the ligamentum arteroisum, extends from the undersurface of the aortic arch to the pulmonary artery. In fetal life, this structure was the ductus arteriosus, which permitted blood to flow from the pulmonary artery into the aorta and thus to bypass the pulmonary circuit.

Circulation to the head and neck begins with the common carotid arteries, which subdivide into the external carotid supplying the superficial structures of the head, and the internal carotid supplying the deeper structures, which feeds into the circle of Willis at the base of the brain. The circle is an important anastamosis (joining) of vessels that provides for collateral circulation to this vital area in the event of some blocking of the blood supply to the

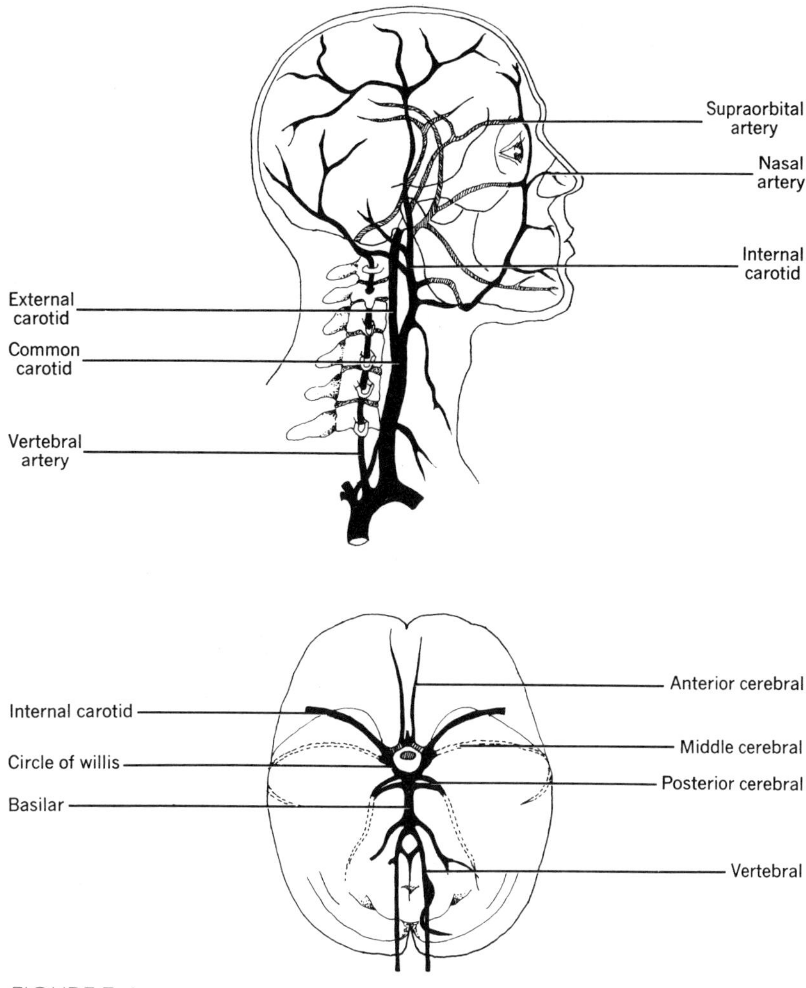

FIGURE 7-6
Blood supply to the head.

brain. The vertebral arteries also feed into the circle of Willis (see Figure 7–6).

The vertebrals are branches off the subclavian arteries. You can feel pulsations in these subclavian arteries just above the clavicles. The subclavian changes its name as it proceeds toward the upper extremities: It first becomes the axillary and then the brachial, which divides to form the radial and the ulnar arteries. You can feel pulsations in the radial artery on the lateral, distal anterior aspect of your forearm. The radial and ulnar arteries anastamose to form the volar arch in the hand. This arch gives off small digital arteries to the fingers (see Figure 7–7).

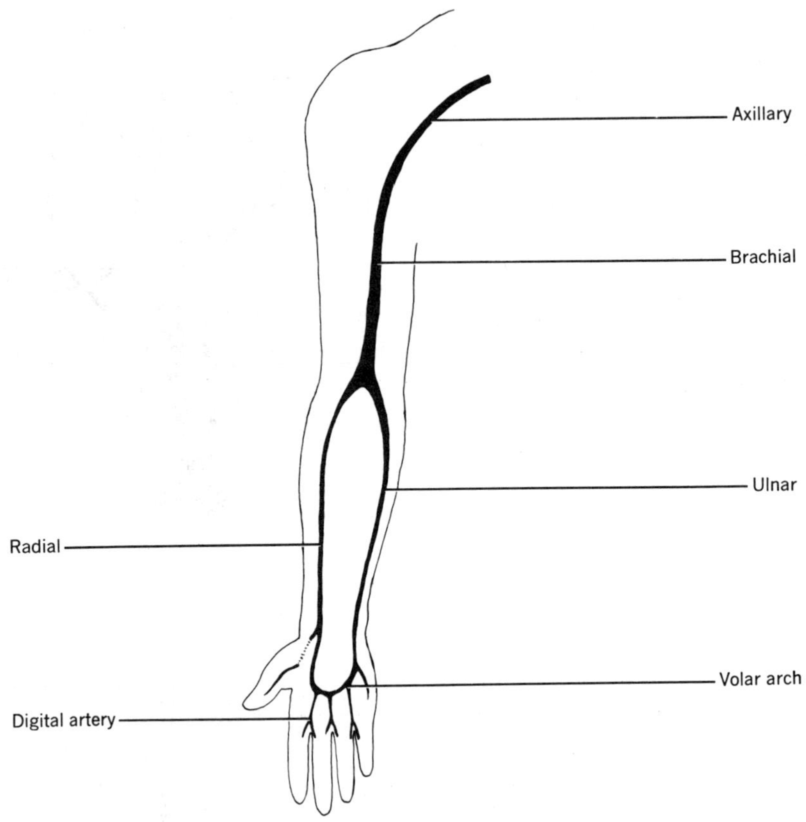

FIGURE 7-7
Blood supply to the upper extremity. These arteries branch from the subclavian artery.

Two major visceral arteries come off the thoracic aorta: the bronchial, which is a nutrient artery to the lungs; and the esophageal. The parietal branches are the intercostals, supplying the intercostal muscles, and the superior phrenic, supplying the upper surface of the diaphragm.

The first visceral branch off the abdominal aorta is the celiac, lying just below the diaphragm (see Figure 7–8). This artery feeds into a complex anastamosis that has several branches to supply the stomach, spleen, pancreas, liver, gall bladder, and the first part of the small intestine. Other visceral branches of the abdominal aorta are as follows: the superior mesenteric, which supplies the small intestines and the first half of the large bowel; the suprarenals; the renals; the inferior mesenteric, which supply the last half of the large bowel; and the spermatics or ovarians. The parietal

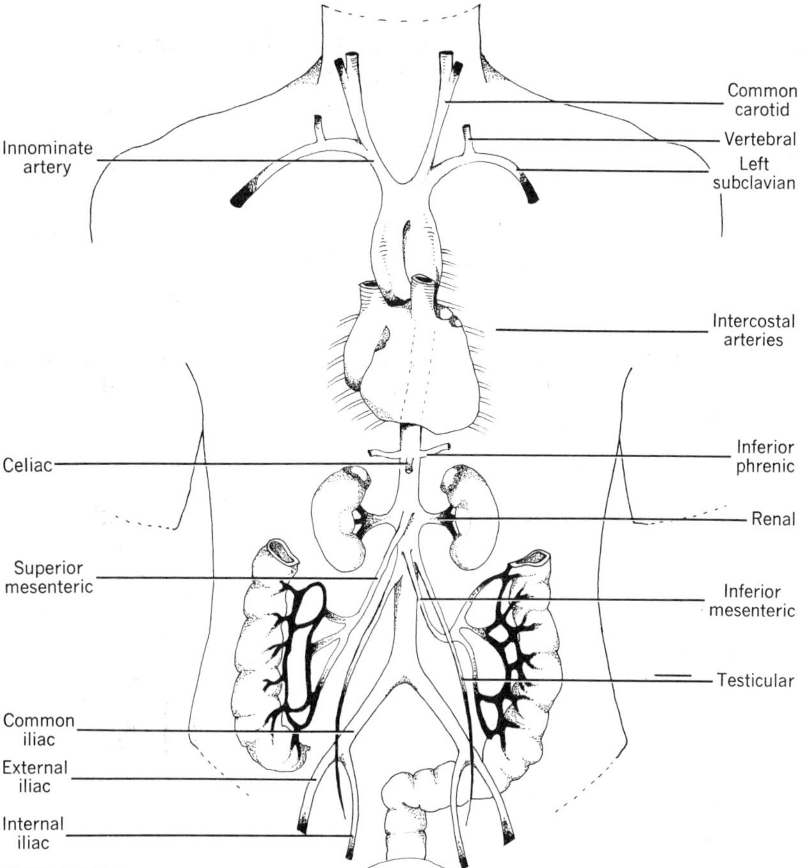

FIGURE 7-8

Major branches off the aorta.

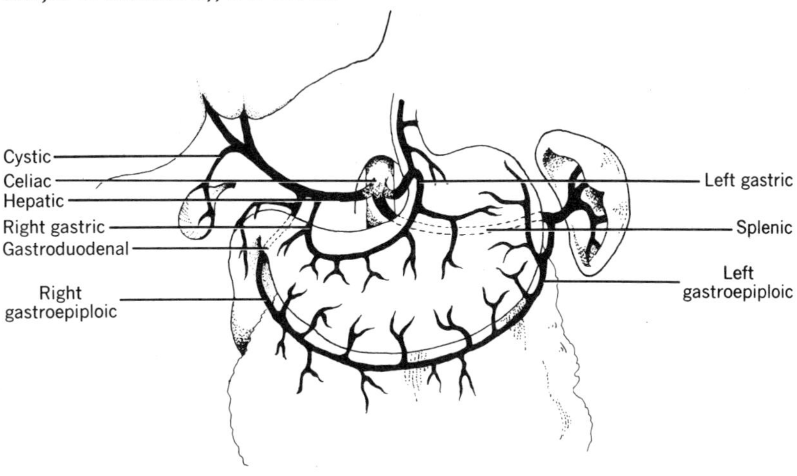

FIGURE 7-9

Branches of the celiac artery.

branches of the abdominal aorta include four lumbars and a middle sacral (see Figure 7–8).

About the level of the fourth lumbar vertebra, the aorta divides to form the common iliac arteries. These in turn divide to become the internal iliac, supplying the pelvic viscera, the external genitalia, and buttocks; and the external iliac, which goes down the lower extremity and becomes the femoral. The femoral also divides to become the anterior and posterior tibial. The anterior tibial extends down to the foot where it becomes the dorsalis pedis artery. The deep plantar artery, one of the branches of the dorsalis pedis, descends into the sole of the foot where it unites with a branch of the posterior tibial artery and forms the plantar arch.

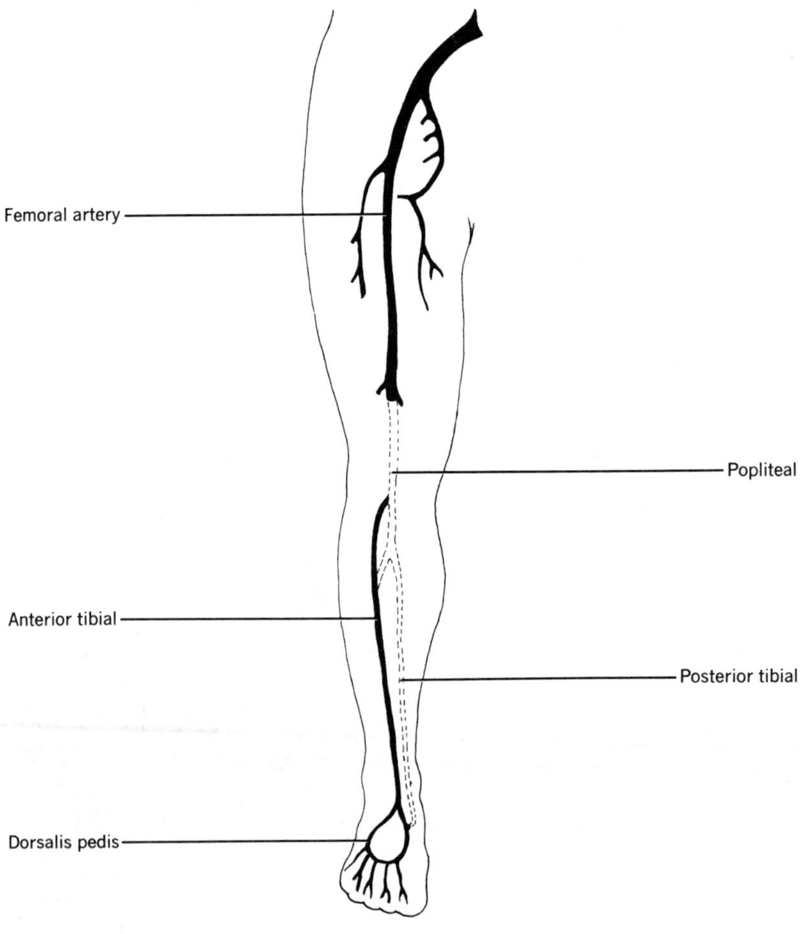

FIGURE 7-10
Blood supply to the lower extremity.

Metatarsal and digital arteries branch from this arch (see Figure 7–10).

CAPILLARIES

The arteries branch to smaller and smaller arterioles and finally to microscopic capillaries. It is in the capillaries that the work of the blood is done. Into the tissue space the blood pressure filters water and crystalline substances such as glucose, amino acids, and electrolytes needed by the cells for their metabolism. Oxygen, which is

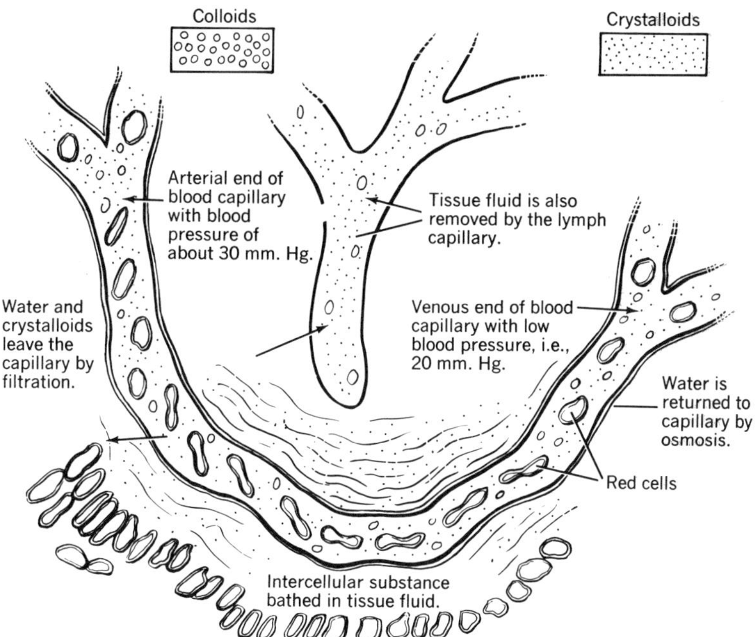

FIGURE 7-11

Tissue fluid is formed by a process of filtration at the arterial end of the blood capillary where the blood pressure exceeds the colloidal osmotic pressure. The fluid is absorbed by the blood capillaries and lymphatics. It will be returned to the venous end of the capillary when the colloidal osmotic pressure exceeds the blood pressure. The fluid is absorbed into the lymphatic capillary when the interstitial fluid pressure is greater than the pressure within the lymphatic capillary. Normally, only very little colloid escapes from the blood capillary. The escaped colloid is returned to the blood circulation by the lymphatics. [From S. R. Burke, *Composition and Function of Body Fluids,* 1st ed. (St. Louis, Mo.: The C. V. Mosby Co., 1972).]

in greater concentrations in the capillary bloodstream than it is in the intercellular spaces, diffuses from the blood to the cells. Carbon dioxide, which is in greater concentrations in the tissue spaces than it is in the blood, diffuses into the bloodstream to be returned to the pulmonary circuit, where it can be eliminated in the expired air. Excess water in the tissues can be returned to the blood capillary by osmosis, or to the lymph capillary (see Figure 7–11).

VENOUS RETURN

As we consider the venous circulation, we shall begin in the periphery, where the veins are the smallest, and shall trace the

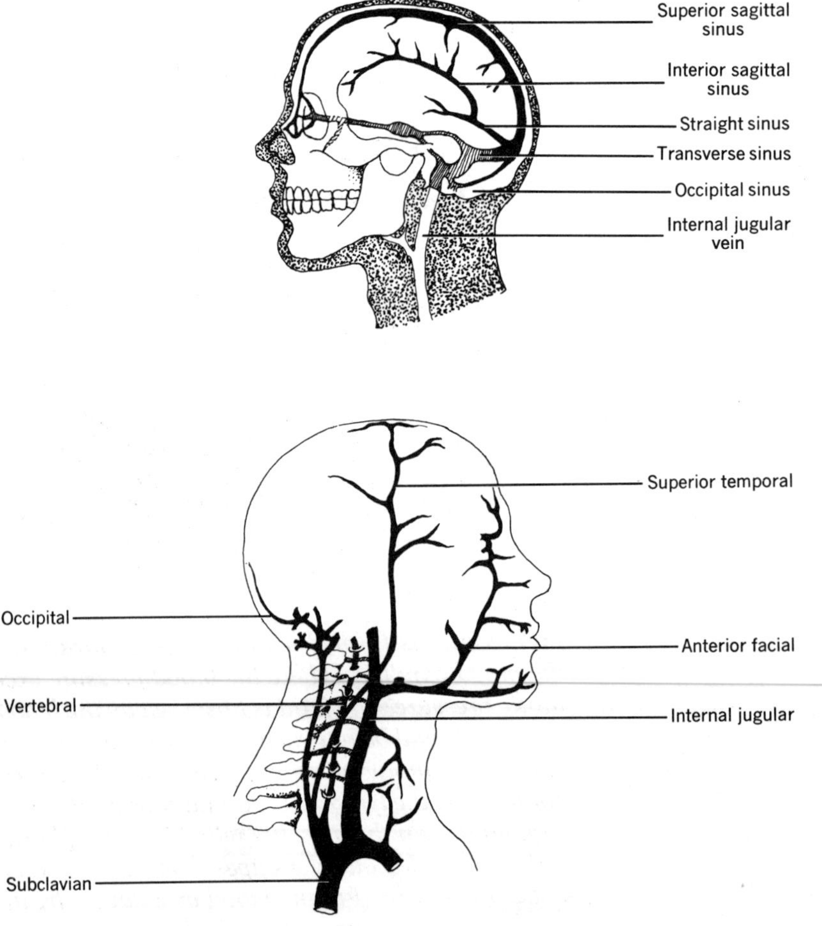

FIGURE 7-12
Venous return from the head.

blood as it moves into larger and larger veins until finally it is returned to the right atrium.

The external and internal jugular veins are the chief veins returning blood from the head. They receive their blood from the cranial venous sinuses and return it to the subclavians and then into the right and left innominates. The innominates empty into the superior vena cava, which enters the right atrium.

Both deep and superficial veins are found in the upper extremity. The deep veins have the same names as their corresponding arteries. However, the superficial ones have different names. They begin in the dorsal network of the hand. The cephalic goes up the lateral side of the forearm and empties into the axillary. The basilic, which is on the ulnar side of the forearm, also empties into the axillary. The prominent superficial vein at the elbow is the median cubital.

Veins of the thorax include the innominates, which receive

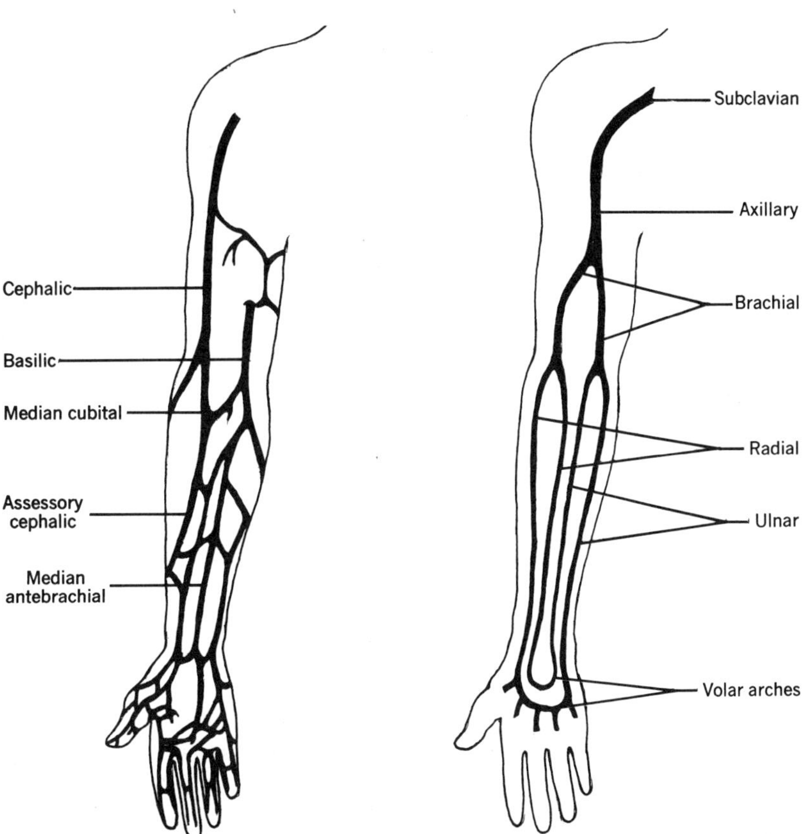

FIGURE 7-13
Venous return from the upper extremity.

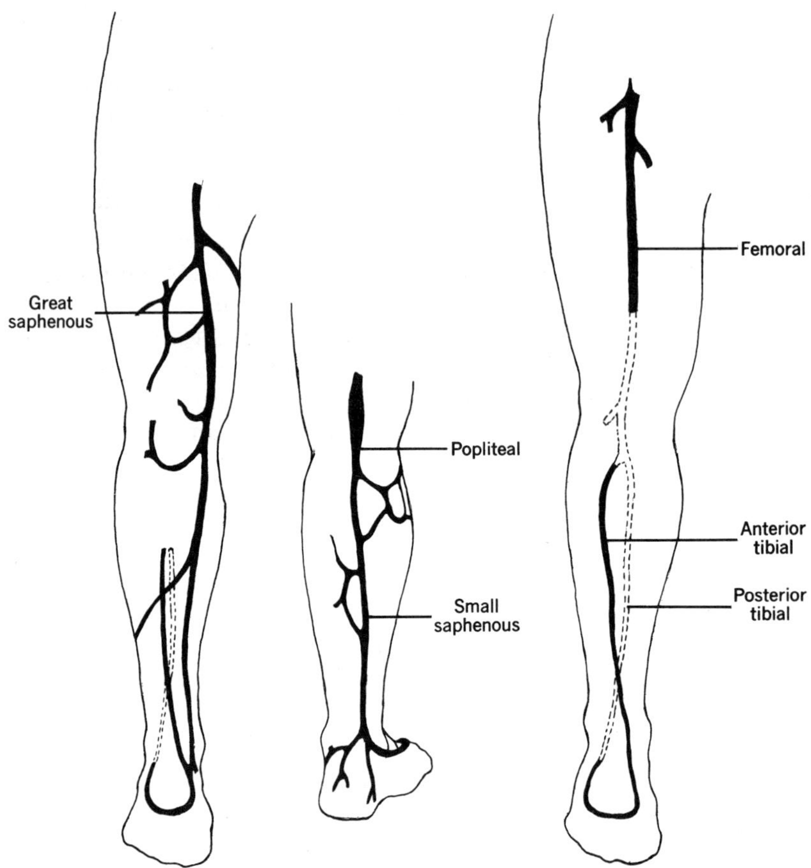

FIGURE 7-14
Venous return from the lower extremity.

from the jugulars, subclavians, and others. Blood flows into the superior vena cava from these vessels and is returned to the right atrium. Small azygos veins, which also empty into the superior vena cava, bring venous blood from the lower parts of the body.

The veins of the lower extremities have the most valves in them. The deep veins have the same names as their corresponding arteries. The superficial veins, called the saphenous veins, are those that are most likely to become varicosed.

Veins of the abdomen and pelvis include the external iliac, which is a continuation of the femoral; and the internal iliac, which drains the pelvis and flows into the common iliac. The inferior vena cava receives blood from four lumbars—the spermatics or ovarians; the renal, which drains the kidney; the suprarenals from the adrenal glands; and the hepatic, which drains the liver.

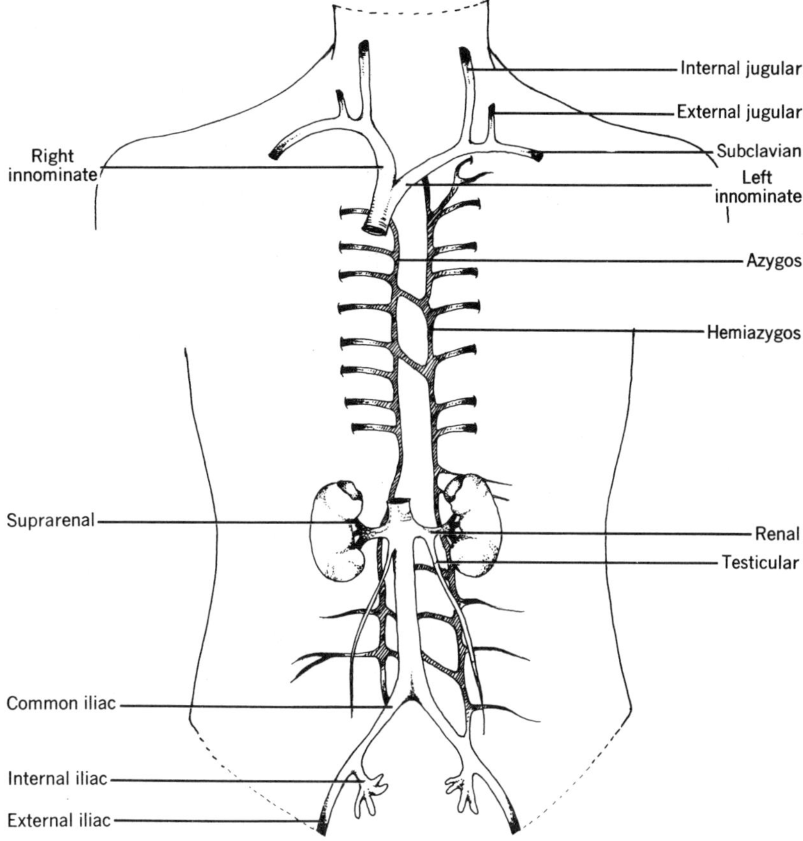

FIGURE 7-15
Veins of the pelvis, abdomen, and thorax.

The portal system is very important: It receives blood from the organs of digestion and the spleen and takes it to the liver so that the liver can perform many important functions on the substances contained in this blood (see Figure 7–16). The superior mesenteric and inferior mesenteric venous blood flows into the portal vein, which once it enters the liver branches to smaller and smaller vessels until the blood finally comes into contact with the liver cells in the sinusoids. Here blood glucose can be converted to glycogen and stored for future use. Blood proteins are made in the liver from amino acids in this blood. Also flowing into the sinusoids is the blood from the hepatic artery, which is nutrient to the cells of the liver. From the sinusoids the blood flows into the central vein and then into the hepatic vein (see Figure 7–17). The hepatic vein empties into the inferior vena cava.

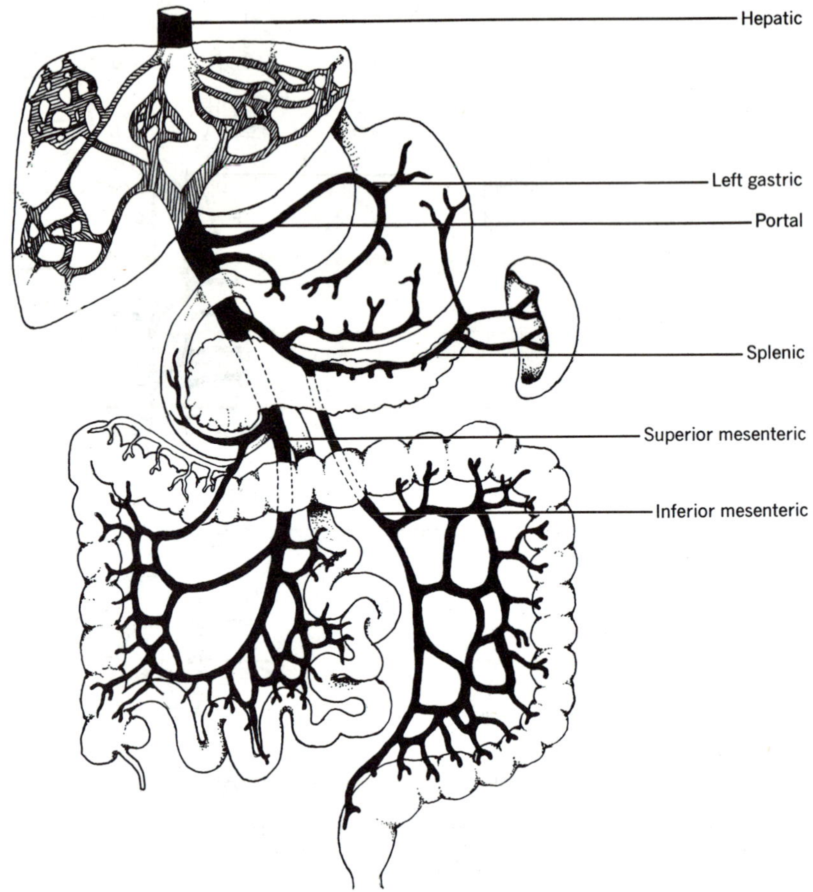

FIGURE 7-16
Venous drainage of the organs of digestion.

You should note that the portal vein differs from other veins in that it does not have valves to prevent the back flow of blood. Thus, in certain liver diseases in which some of the sinusoids become blocked, there may be large accumulations of blood in the mesenteric veins and a great increase in venous pressure there.

THE COMPOSITION OF BLOOD

The liquid portion of the circulating blood is called plasma. The formed elements of the blood are the platelets and blood cells. Red blood cells, which are very small and have no nuclei, are manufac-

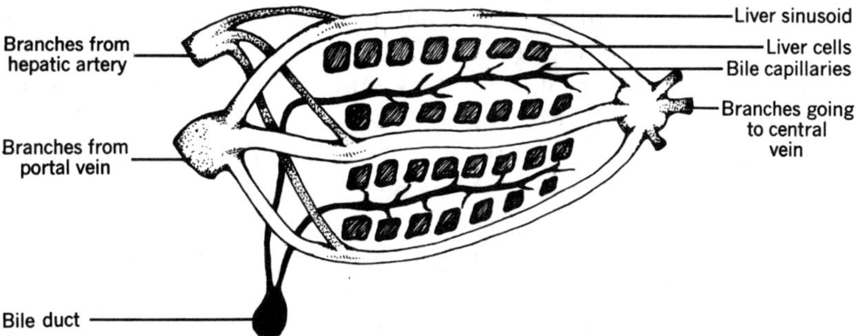

FIGURE 7-17
A schematic diagram of the liver sinusoids and bile capillaries. Blood from the hepatic artery flows into these sinusoids. This arterial blood is nutrient to the cells of the liver. Venous blood from the portal vein also flows into the sinusoids. This venous blood is rich in the products of digestion and is brought to the sinusoids in order that the liver cells can process these products.

tured in the red bone marrow and have a life span of about 120 days. They are destroyed in the reticuloendothelial system (the liver, the spleen, and bone marrow). Red blood cells contain hemoglobin for oxygen transport. The amount of hemoglobin can be measured to estimate the degree of anemia. Normal hemoglobin is about 15 G./100 ml. It is normally a little higher in men than in women. The number of red blood cells is about 4.5 to 5 million cells/cu. mm. The hematocrit is the fraction of the total volume of blood which consists of cells, normally about 45 percent. Patients who are hemorrhaging have low hematocrits; patients who are losing the fluid portion of their blood (for example, in burns with a great deal of edema) have high hematocrits and their blood has a much greater clotting tendency.

The leukocytes, or the white blood cells, are formed in both the red bone marrow and lymphatic tissue. The normal white blood cell count—called the absolute count—is about 5,000 to 10,000 cells/cu. mm. Monocytes make up about 5 percent of the total white count. Their numbers increase in chronic infections. Lymphocytes, which account for about 27.5 percent of the total, are important in helping to develop immunity. The neutrophils or polys make up about 65 percent of the total WBC. These increase in acute infections. Eosinophils and basophils together comprise about 2.5 percent of the total white cell count.

The thrombocytes or platelets, which are smaller than the red blood cells, play a very important role in blood clotting. Because they are sticky and will adhere to a damaged vessel wall, they help to plug the leak and also allow other cells to stick to them and to form a clot.

Clotting time is normally about 7 to 10 minutes. If this time is prolonged, the patient will have a bleeding tendency. If the clotting time is less than normal, he will have a tendency toward intravascular clots.

An intravascular clot is called a thombus; if the clot moves in the venous system it is called an embolus. Intravascular clotting is promoted by injury to the vessel wall, prolonged bed rest, foreign material in the blood, or trauma to the components of the blood itself. It is hindered by early ambulation, stable platelets, the presence of heparin or other anticoagulants, and a vitamin K deficiency. Vitamin K deficiency is not a common clinical problem except in newborn babies who are normally vitamin K deficient. Normally, the bacteria in the large bowel form vitamin K, and because the bowel of a newborn is sterile the infant has no vitamin K unless his mother has had an injection of vitamin K shortly before delivery or the infant has received an injection.

Prothrombin time, the time it takes plasma to clot, is normally about 12 to 13 seconds. If it is much longer, a person will have a bleeding tendency. People who are on anticoagulants because they have thrombosis or have had a heart attack must have frequent prothrombin times done, in order that their prothrombin time does not become too prolonged.

Extravascular clotting is hastened by contact with a rough surface such as gauze and is hindered by oxalates and citrates, which remove the calcium from the blood.

For a more complete picture of the composition of blood and the types of clinical studies done on blood, see Tables 7–1 and 7–2.

Blood typing is done to assure that a person who needs a blood transfusion will receive the right type of blood. Typing helps but does not assure that he will not have a transfusion reaction. The ABO typing system is most commonly used. Type O is called the universal donor. In theory, a person with type O blood can give blood to anyone but can receive only from type O donors. Type O blood will not agglutinate with either anti-A or anti-B sera. Type AB is the universal recipient. People with type AB blood can receive blood from anyone but give only to type AB recipients. This blood agglutinates with both anti-B and anti-A sera. Type A blood agglutinates with anti-A serum. People with type A blood can receive from donors of either type A or type O and can give to recipients of either type A or AB. Type B blood agglutinates with anti-B serum. People with type B blood can give to people with either

TABLE 7-1

Hematology Findings in Health and in Disease

Normal Range	Conditions in Which Variations from Normal May Occur
Volume 7–9% of body weight (4000–6000 ml.)	Decreased: hemorrhage, surgical shock, burns
Erythrocytes 4.5–5 million/mm.^{3}	Increased: polycythemia, anoxia, chronic pulmonary disease, high altitudes, renal disease with increased secretion of erythropoietin, Cushing's Syndrome Decreased: anemia, hemorrhage, leukemia
Reticulocytes 0.8–1% of rbc (red blood cells)	Increased: hemolytic jaundice, anemia with increased bone marrow activity
Leukocytes 5,000–10,000 mm.^{3}	Increased: infections and tissue destruction, leukemia, metabolic disorders Decreased: irradiation, bone marrow aplasia
Neutrophils (PMN) 60–70% of wbc (white blood cells)	Increased: acute infections, gout, uremia, neoplastic diseases of the bone marrow, uremia, diabetic ketosis, massive necrosis, poisoning by mercury, lead, or digitalis Decreased: agranulocytosis, acute leukemia, measles, malaria, overwhelming bacterial infections
Lymphocytes 25–33% of wbc.	Increased: whooping cough, chronic infections, infectious mononucleosis, chronic lymphatic leukemia, thyrotoxicosis Decreased: Hodgkin's Disease, in response to adrenal cortical steroids, whole-body irradiation
Monocytes 2–6% of wbc.	Increased: chronic bacterial diseases, tetrachlorethane poisoning, monocytic leukemia
Eosinophils 1–3% of wbc.	Increased: hypersensitivity (hay fever, asthma, chronic skin diseases), helminthic infestations, leukemia Decreased: steroid therapy
Basophils 0.25–0.5% of PMN	Increased: acute severe infections, leukemia
Metamyelocytes 5% of PMN	Increased: acute severe infections, leukemia

Platelets	250,000–350,000	Increased: after trauma or surgery, after massive hemorrhage, polycythemia Decreased: thrombocytopenic purpura, lupus erythematosis, following massive blood transfusions
Hemoglobin ml.	14–16 G./100	Increased: conditions in which there is an increase in erythrocytes Decreased: anemia, hemorrhage, leukemia
Hematocrit	42–47%	Increased: dehydration, plasma loss, burns, conditions in which there is an increase in erythrocytes Decreased: hemorrhage, anemia
Sedimentation rate Men 0–12 mm./hr. Women 0–20 mm./hr.		Increased: infection, coronary thrombosis, leukemia, anemia, hemorrhage, malignancy, hyperthyroidism, kidney disease Decreased: severe liver disease, malaria, erythremia, sickle-cell anemia
Bleeding time 1–3 min.		Increased: thrombocytopenia, acute leukemia, Hodgkin's Disease, hemorrhagic disease of the newborn, hemophilia, thrombocytopenia
Coagulation time 6–12 min.		Increased: hemophilia, anticoagulant therapy
Clot retraction time Begins in 1 hr. Completes in 24 hr.		Increased: thrombocytopenia, acute leukemia, pernicious anemia, multiple myeloma, malignant granuloma, hemorrhagic disease of the newborn
Prothrombin time 10–15 sec.		Increased: treatment with anticoagulants, hemorrhagic disease of the newborn, liver disease, hemophilia

From Shirley R. Burke, *The Composition and Function of Body Fluids* (St. Louis, Mo.: The C.V. Mosby Co., 1972).

type B or AB blood and can receive from either type B or type O donors.

About 85 percent of humans are Rh positive. The remaining 15 percent, who are Rh negative, are susceptible to trouble if they receive an Rh positive blood transfusion. This factor is probably overrated as a troublemaker in pregnancy: An Rh negative mother must be carrying an Rh positive baby and a placental leak must

TABLE 7-2

Blood Chemistry Findings in Health and in Disease

Normal Range*	Conditions in Which Variations from Normal May Occur
Sodium 133–143 mEq/liter	Increased: dehydration, brain injury, steroid therapy Decreased: gastrointestinal loss, sweating, renal tubular damage, water intoxication
Potassium 3.9–5.0 mEq/liter	Increased: shock, crush syndrome, anuria, Addison's disease, renal failure, diabetic ketosis Decreased: severe diarrhea, bowel fistula, diuretic therapy, Cushing's Syndrome
Calcium 4.5–5.7 mEq/liter	Increased: hyperparathyroidism, hypervitaminosis D Decreased: hypoparathyroidism, acute pancreatitis, Vitamin D deficiency, steatorrhea, nephrosis
Chloride 95–105 mEq/liter	Increased: dehydration, hyperchloremic acidosis, brain injury, steroid therapy, respiratory alkalosis, hyperparathyroidism Decreased: gastrointestinal loss, potassium depletion associated with alkalosis, diabetic ketosis, Addison's disease, respiratory acidosis, mercurial diuretic therapy
Magnesium 1.5–2.4 mEq/liter	Increased: administration of magnesium compounds in the presence of renal failure Decreased: severe malabsorption
Phosphate 1.8–2.6 mEq/liter	Increased: Vitamin D excess, healing fractures, renal failure, hypoparathyroidism, diabetic ketosis
pO_2 95–100 mm. Hg (arterial blood)	Increased: administration of high concentrations of oxygen Decreased: hypovolemia, decreased cardiac output, chronic lung diseases
pCO_2 36–44 mm. Hg (arterial blood)	Increased: respiratory acidosis, metabolic alkalosis Decreased: respiratory alkalosis, metabolic acidosis, hypothermia
CO_2 combining power 45–70 vol./100 ml. 21–28 mEq/liter	Increased: emphysema, metabolic alkalosis Decreased: metabolic acidosis
CO_2 25–35 mEq/liter (measured as HCO'_3)	Increased: respiratory acidosis, metabolic alkalosis Decreased: respiratory alkalosis, metabolic acidosis
Cholesterol 150–250 mg.%	Increased: obstructive jaundice, renal disease, pancreatic disease, hypothyroidism Decreased: severe liver disease, starvation, terminal uremia, hyperthyroidism, cortisone therapy
Bilirubin Total 0.3–1.5 mg.% Direct 0.1–0.3 mg.%	Increased: biliary obstruction, impaired liver function
Protein total 6–7.8 G.%	Increased: dehydration Decreased: renal disease, malnutrition, liver disease, severe burns
Albumin 3.5–5.5 G.%	Increased: dehydration Decreased: renal disease, liver disease, malnutrition

Globulin 2.5–3.0 G.%	Increased: chronic infectious diseases Decreased: sarcoidosis, cirrhosis
BUN (blood urea nitrogen) 20–40 mg.% (adult)	Increased: fever, excess body protein catabolism, renal failure Decreased: growing infant
Creatinine 0.9–1.7 mg.%	Increased: acromegaly, renal failure
Uric acid 1.5–4.5 mg.%	Increased: gout, gross tissue destruction, renal failure, hypoparathyroidism Decreased: administration of uriosuric drugs (cortisone, salicylates)
Glucose 70–120 mg.% (fasting)	Increased: diabetes mellitus, severe thyrotoxicosis, pheochromocytoma (during attack), burns, shock, after adrenalin injection Decreased: insulin overdosage, hyperplasia of islet cells, hypothalmic lesions, postgastrectomy dumping syndrome
Amylase 80–180 Somogyi units/100 ml.	Increased: acute and chronic pancreatis, postgastrectomy, cholecystitis, salivary gland disease Decreased: hepatitis, thyrotoxicosis, severe burns, toxemia of pregnancy
Alkaline phosphatase 0.8–3.1 Bessy units	Increased: bone diseases, liver disease, obstructive jaundice
Acid phosphatase 0.3–0.7 Bessy units	Increased: prostatic malignancy
SGOT less than 40 units/ml. (glutamic pyruvic transaminase)	Increased: myocardial infarction, acute rheumatic carditis, cardiac surgery, hepatitis, pulmonary infarction, acute pancreatitis, trauma
SGPT less than 30 units/ml. (glutamic pyruvic transaminase)	Increased: acute hepatitis, cirrhosis of the liver, myocardial infarction, infectious mononucleosis
CPK 0–4 units/ml. (creatine phosphokinase)	Increased: myocardial infarction, crush injury, hypothyroidism, tissue transplant rejection, cerebral vascular accident
ICD 60–290 units/ml. (isocitric dehydrogenase)	Increased: early hepatitis, pancreatic malignancy, pre-eclamptic toxemia, carcinomatosis of the liver
LDH 200–425 units/ml. (lactic dehydrogenase)	Increased: myocardial infarction (LDH in heart muscle is heat stable—when the specimen is incubated, the level remains elevated), hepatitis, skeletal muscle damage. (LDH in liver and muscle is heat labile so elevated levels return to normal after incubation)
Glutathione reductase 10–70 units/ml.	Increased: cancer, hepatitis
Aldolase 3–8 units/ml.	Increased: muscular atrophy, cancer, acute or chronic diseases
Lipase 0.2–1.5 units/ml.	Increased: pancreatitis, fat embolism following trauma

From Shirley R. Burke, *The Composition and Function of Body Fluids* (St. Louis, Mo.: The C.V. Mosby Co., 1972).

*Values depend to a certain extent on the technique used for the determination. Therefore one can expect to find occasional discrepancies when consulting different normal value charts.

take place in order for the mother to form antibodies against the Rh factor. No problems occur with a first baby. In subsequent pregnancies, if the baby is Rh positive these antibodies may cross the placenta and destroy the baby's red blood cells. The baby must have a complete exchange transfusion. Rhogram, a drug that can be given to an Rh negative mother after the birth of an Rh positive baby will destroy the mother's antibodies and prevent damage to the blood cells of babies in future pregnancies.

Blood transfusion reactions may be extremely dangerous. Because the symptoms of a reaction usually occur within the first 10 minutes of the transfusion, all transfusions should be started at a very slow flow rate until it is reasonably certain that a reaction is not going to occur. Symptoms of a transfusion reaction include chills and fever. There may be a rash and itching; shortness of breath; nausea; and pain usually in the kidney region, chest, or legs. Hematuria and shock are present if the reaction is severe. Because such a reaction is an emergency situation, the transfusion should be stopped immediately. However, the needle should not be removed as the transfusion equipment should be examined by blood bank personnel in order to help determine the cause of the reaction. Urine specimens should be collected and examined for hemoglobin.

THE LYMPHATIC SYSTEM

The general functions of the lymphatic system are to return tissue fluid to the bloodstream and to filter the tissue fluid. In contrast to the blood system, these vessels begin in the tissues and end in the veins of the chest. There is no pumping mechanism. The movement of lymph is slow; it depends upon the contraction of skeletal muscles, peristaltic contractions, and pulsations of nearby arteries. The many valves in the lymphatic vessels prevent the backflow of the fluid.

Lymph capillaries open into the tissue spaces and lead to larger lymph vessels. These vessels penetrate lymph nodes as they proceed toward the innominate vein where they return the tissue fluid to the blood. Lymph nodes function as filters to trap bacteria or foreign substances so that they will not be poured into the bloodstream. If you have a severe infection, the lymph nodes proximal to the site may become enlarged, because they are filled with the bacteria that are being destroyed by the lymphocytes there.

The lymphatic collecting vessels that receive lymph from the capillaries eventually empty into one of two terminal ducts. The

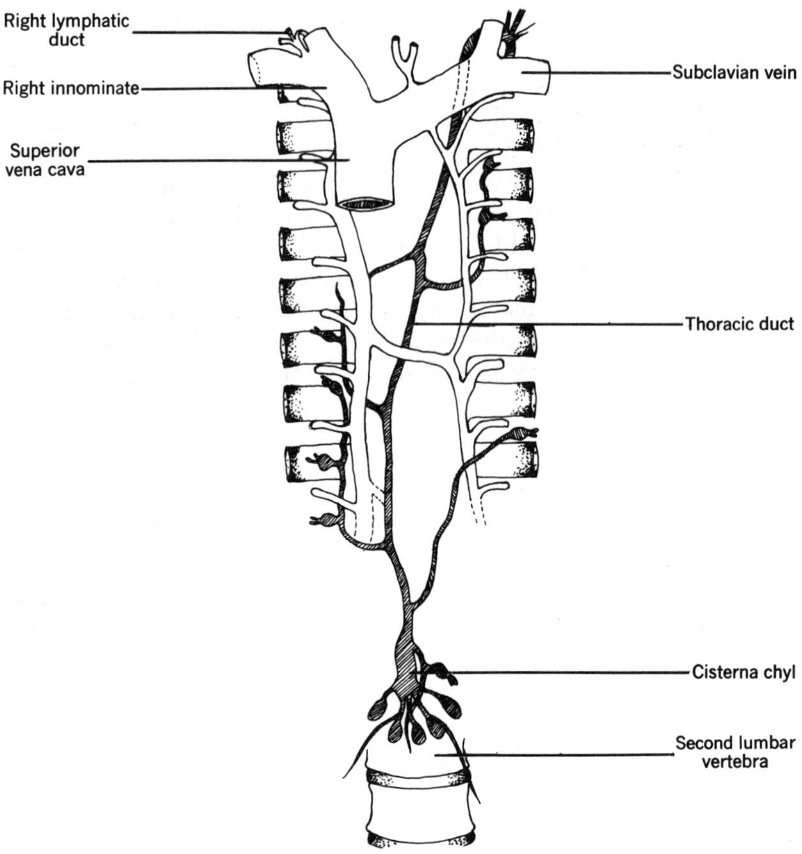

FIGURE 7-18
A schematic drawing of some of the major lymphatic vessels.

thoracic duct is terminal for the upper left part of the body and all of the lower parts of the body. It begins in a dilatation known as the cisterna chyli, which is located about the level of the second lumbar vertebra. It ascends with the aorta through the chest and receives lymph from intercostal lymphatic trunks and subclavian trunks. It ends in the left innominate vein. The right lymphatic duct drains the rest of the body and empties into the right innominate vein.

The thymus, spleen, and tonsils are also a part of the lymphatic system. Sometimes a person's tonsils must be removed if he has had repeated upper respiratory infections, because the tonsils can become so loaded with bacteria that they themselves become a source of infection.

SUMMARY QUESTIONS

1. Trace a drop of blood from the time it enters the right atrium until it enters the aorta.
2. What is the function of the valves of the heart?
3. What is the name of the valve that is located between the right atrium and right ventricle? What is the name of the valve between the left atrium and left ventricle?
4. How does the heart receive its blood supply?
5. What causes the heart sounds, lupp and dupp?
6. What is the contraction of the ventricles called?
7. What is the relaxation of the ventricles called?
8. Describe the intrinisc nerve supply of the heart.
9. Why must a weak, failing heart beat faster than the heart of a well-trained athlete?
10. What causes the closure of the semilunar valves?
11. Why must the blood pressure in the pulmonary circuit be lower than the blood pressure in the peripheral arteries?
12. Where is the work of the blood done?
13. What are the functions of the lymphatic system?
14. What is the normal value for hemoglobin?
15. What is the hematocrit?
16. Under what circumstances would you expect a patient to have an increased hematocrit?
17. Trace a drop of blood from the time it leaves the aorta until it gets to the capillaries of the small intestines and then returns to the right atrium.

SUGGESTED READINGS

Brady, R. J. *A Programmed Approach to Anatomy and Physiology.* 2d ed. Washington D. C.: Robert J. Brady Co., 1968.

Burke, S. R. *The Composition and Function of Body Fluids.* 1st ed. St. Louis Mo.: The C. V. Mosby Co., 1972.

Chaffee, E., and E. M. Greisheimer. *Basic Physiology and Anatomy.* 2d ed. Philadelphia, Pa.: J. B. Lippincott, 1971.

Jacob, W. W., and C. A. Francone. *Structure and Function in Man.* 2d ed. Philadelphia, Pa.: W. B. Saunders Co., 1970.

Simmons, A., *et al.* "Factors Affecting White Cell Differential Counts," *The American Journal of Medical Technology,* September 1973.

8 Diseases of the Circulatory System

VOCABULARY

Cyanosis
Dyspnea
Diuretic
Hemoglobin
Hypovolemia
Hypertrophy
Ischemia
Infarct
Lipid
Necrosis
Neoplasm
Phlebotomy
Petechiae
Radiopaque
Steroid
Toxicity
Vasopressor

OVERVIEW

I DIAGNOSTIC PROCEDURES
- **A.** Fluoroscopy and X-ray
- **B.** Angiocardiogram
- **C.** Aortagram
- **D.** Cardiac catheterization
- **E.** Electrocardiogram

II DISEASES OF THE BLOOD CELLS
- **A.** Leukemia
- **B.** Anemias
- **C.** Hodgkins Disease
- **D.** Infectious mononucleosis

III HEART DISEASES
- **A.** Coronary artery disease
  - **1.** Myocardial infarction
  - **2.** Angina pectoris
- **B.** Heart block
- **C.** Rheumatic heart disease
- **D.** Bacterial endocarditis

IV CONGENITAL HEART DEFECTS
- **A.** Patent foramen ovale
- **B.** Patent ductus arteriosis

V VASCULAR DISEASES

VI HYPERTENSION

VII HEMORRHAGE AND SHOCK

DIAGNOSTIC PROCEDURES

In addition to hematology and blood chemistry studies, several other diagnostic procedures are helpful in determining the presence of diseases of the circulatory system and the extent of the disease process.

Fluoroscopy is a radiological study in which a physician can observe the action of the heart. In addition to being able to observe the contractions of the heart, the physician can also see any hypertrophy of the myocardium and the extent of the enlargement.

An angiocardiogram is a procedure in which a radiopaque dye is injected into a vein and a rapid series of X-rays is taken as the dye travels through the heart and pulmonary circulation.

In an aortogram, dye is injected into the aorta and then X-rays are taken which will show the outline of the major arteries.

In a cardiac catheterization, a flexible catheter is passed into a vein and then into the heart. With this catheter pressures can be measured and blood can be withdrawn and analyzed for oxygen and carbon dioxide content.

The electrocardiogram, (ECG) is probably the most common procedure used to help determine the efficiency of the myocardium. In this procedure, the electrical changes that occur throughout the cardiac cycle are recorded, and the tracings made can be saved for comparison with future tracings. (You should review the discussion of the ECG on pages 91–92.)

As with other disease processes, careful observation and reporting of these observations are very important. The physician is greatly dependent upon an accurate and ongoing record of the patient's temperature, pulse, respirations, and blood pressure. The patient's own report of his subjective symptoms is also important. Thus, you should learn to be observant and accurate in reporting your observations. They may seem relatively unimportant in comparison to some of the other, more sophisticated diagnostic techniques, but they sometimes make the difference between "ease and disease" in a patient.

DISEASES OF THE BLOOD CELLS

Leukemia is a serious disease in which there is a tremendous increase in the number of white cells; however, these white cells are immature cells and are not capable of fighting infection. For this reason, what otherwise might be a relatively mild infection may prove fatal for someone with leukemia. In acute leukemia, the white count may be 30,000 to 50,000 cells per cubic millimeter, or higher. These patients usually also have anemia, which accounts for their pallor, fatigue, and weakness. At the present time there is no cure for most leukemias. However, for temporary relief patients are given antibiotics for their secondary infections and steroids. They may also benefit from antimetabolites — drugs that interfere with the multiplication of cells, particularly cells undergoing rapid proliferation such as the leukocytes of patients with leukemia. Blood transfusions may be used to increase the patient's hemoglobin and to relieve the affects of the anemia.

Anemia is a condition in which there is a reduction in the amount of hemoglobin. Anemia can be caused by loss, destruction, or faulty production of red blood cells and hemoglobin. Symptoms of anemia are similar regardless of the cause and are mainly the result of the inability of the blood to transport sufficient oxygen to the tissues. Patients may feel faint, tire easily, and have pallor. They are particularly sensitive to chilling and usually complain of being cold when others find the temperature comfortable. Their pulse is rapid, and exertional dyspnea occurs when the anemia is severe.

The treatment of anemia is directed toward correcting the cause of the disease. If it is due to blood loss, the bleeding must be controlled and blood transfusions may be necessary. If it is pernicious anemia, caused by faulty production of red blood cells, treatment is dietary, with supplements of vitamin B_{12}. Iron-deficiency anemia may also respond to dietary treatment, which should include extra citrus fruits because vitamin C is necessary for the proper utilization of dietary iron. It may also be necessary to give ferrous sulfate, 0.3 Gm., three times a day.

Sickle-cell anemia is a hereditary disease in which the red cells sometimes take on abnormal shape and are less able to carry oxygen. Recent evidence suggests that the administration of urea helps to prevent the sickling of the red cells and increases their ability to transport oxygen.

Hodgkin's Disease is characterized by painless enlargment of the lymph nodes. The cervical nodes usually enlarge first; the inguinal and axillary nodes are affected later. Patients experience marked weight loss, anorexia, fatigue, and weakness. Frequently they have bleeding tendencies and anemia. Most patients survive only 4 or 5 years. Treatment usually includes radiation, steroids, and antineoplastic drugs, such as nitrogen mustard.

Infectious mononucleosis is a relatively common disease, particularly of teenagers and young adults. Fortunately, the disease is usually self-limiting. However, it may last for 2 or 3 weeks, after which time the patient usually has a fairly prolonged period of weakness and fatigue. Although the disease is infectious, it is not highly contagious and reasonable precautions will prevent its spread.

HEART DISEASES

Before discussing heart disease we shall consider for a moment some common phrases and their meaning to you. "Let's have a heart-to-heart talk." "I had my heart set on it." "Let's get to the heart of the matter." "Deep in your heart your know . . ."

Clearly if something is wrong with your heart the psychological implications are great indeed. You should think also about the meaning of words such as "failure" and "attack," which are used in connection with heart disease. Heart disease patients need supportive and understanding attention as much as they do medicine, surgery, and rest. Fear and anger are mediated by the sympathetic nervous system. (You should refer to page 182 for a discussion of the influence of the sympathetic nervous system on the heart and circulation.) No patient with heart problems needs the added problems of fear and anger, yet he has good reason to be troubled by both.

Arteriosclerosis is a fairly common disease. It is the result of hardening and narrowing of the arteries. Atherosclerosis is a form of arteriosclerosis in which fatty deposits form within the walls of the arteries. For all practical purposes, the two are essentially the same. One of the most serious consequences of this condition is coronary artery disease. With the narrowing of the lumen of the coronary arteries, not as much blood can be transported to the myocardium as is necessary to meet the requirements of the muscle. Arteriosclerosis usually occurs in people over 50 years of age; however, it may be evident in younger people, particularly men. It tends to be familial, especially when it occurs at an early age. Obesity, smoking, emotional strain, and lack of exercise may all be contributing factors.

Despite controversy, much evidence supports the theory that diets high in cholesterol contribute to coronary artery disease. Whole milk, cheese, butter, and other fatty foods are high in cholesterol. Although it probably does not help a great deal to reduce cholesterol intake after a person has reached middle age, it is wise to do so in the younger years and to continue the low cholesterol diet into adulthood. Children obviously need milk; however, 2 percent butter fat milk will provide for their needs.

Severe coronary artery disease can result in myocardial infarction, that is, the death of a portion of the cardiac muscle due to ischemia. Symptoms include sudden, severe pain in the chest, usually substernal and sometimes radiating to the shoulder and arm. This pain is not relieved by rest or nitroglycerine (a vasodilator). It may last several hours or as long as a day or two. There is usually some degree of shock, pallor, sweating, a severe drop in the blood pressure, and a rapid weak pulse, in addition to nausea and vomiting. Patients are usually quite restless and fearful. Tissue necrosis may cause a low-grade fever for 4 or 5 days. The patient should have complete bed rest; if there are symptoms of shock, he should be kept flat in bed, unless he experiences respiratory distress. A narcotic—demerol or morphine—may be needed for both the pain and the restlessness. Anticoagulant drugs such as heparin or coumadin may also be necessary to increase the prothrombin time and to prevent further clot formation. If the shock is severe, a vasopressor drug such as leophed or hypertensin may be used to maintain blood pressure.

Oxygen is used to reduce the work of breathing and to increase the oxygen saturation of the blood, so that cells near the infarcted area are well supplied with oxygen. Recovery is slow: Following convalescence, the patient will probably have to change many of his well-established life patterns in order to avoid a recurrence.

Angina pectoris is also the result of myocardial ischemia. However, pain is fleeting and usually occurs when the patient experiences unusual stress. This pain is relieved by rest and by nitroglycerine taken sublingually. A person who has had one attack of angina should have nitroglycerin tablets with him at all times; he should also not smoke and should avoid strenuous activities.

Heart block is a condition in which interference occurs with the nerve impulses being transmitted through the bundle of His. It may be a partial or a complete block. In the event of a complete block, the pulse will be very slow, about 30 to 40 beats per minute. This condition may be associated with atherosclerosis, coronary artery disease, rheumatic heart disease, or congenital malformations. It may also be brought on by digitalis toxicity. It may be necessary to implant an artificial pacemaker (see Figure 8–1).

Rheumatic heart disease results in valvular damage, usually to the mitral valve. Vegetations grow on the valve and leave hardened stiff valve flaps that either will not open wide enough (stenosis) or will not close completely (insufficiency). The valve may be both stenosed and insufficient. It may be necessary to replace the valve (see Figure 8–2).

If the mitral valve is affected, the end consequence will be congestive heart failure if the process is untreated. Failure will cause severe respiratory difficulties. Too much blood will accumulate in the pulmonary circuit and will cause pulmonary hypertension.

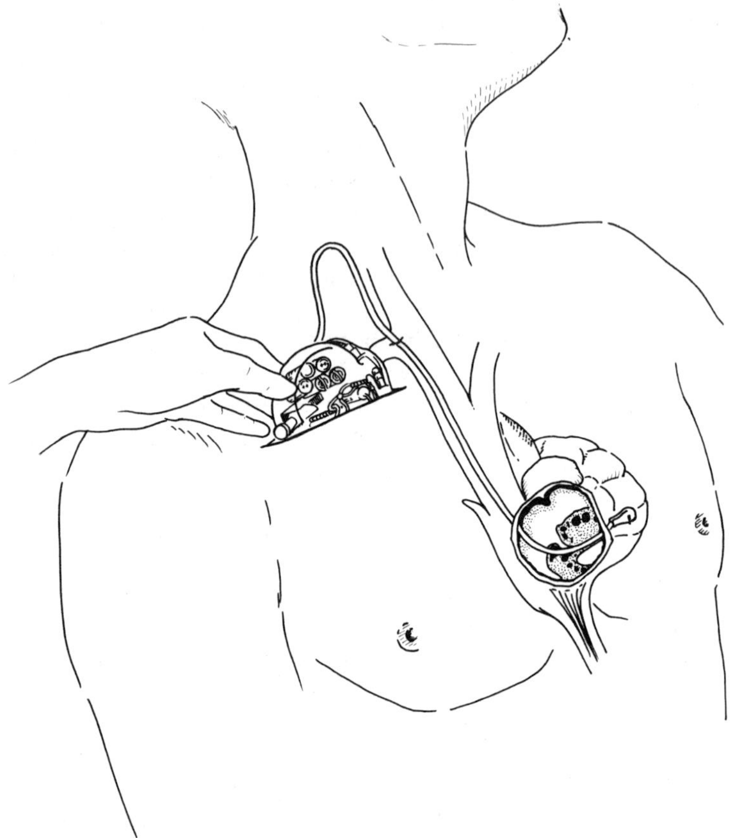

FIGURE 8-1
An isotopic demand pulse generator. The battery power for this pacemaker is derived from a 150-mg. metallic pellet of plutonium 238 which has an 87-year half-life. This device has an anticipated useful life of at least 10 years. The first nuclear-powered pacemaker was implanted in a patient on April 27, 1970. Within the following 4 years, more than 600 were implanted in patients. The pacemaker is designed to operate at a rate of 72 pulses per minute; however, the demand feature of the pacemaker permits a more physiologic heart response to the patient's varying needs by allowing variations in spontaneous rate and by limiting ventricular overdriving. (Courtesy of Medtronic, Inc., Minneapolis, Minnesota.)

One method of treating this life-threatening condition is the technique of rotating tourniquets. Tourniquets, ideally blood pressure cuffs, are applied high on three extremities. The cuffs are inflated to a level that will just permit the pulse to be palpated. All cuffs remain in place for a period of 10 to 15 minutes; then one cuff

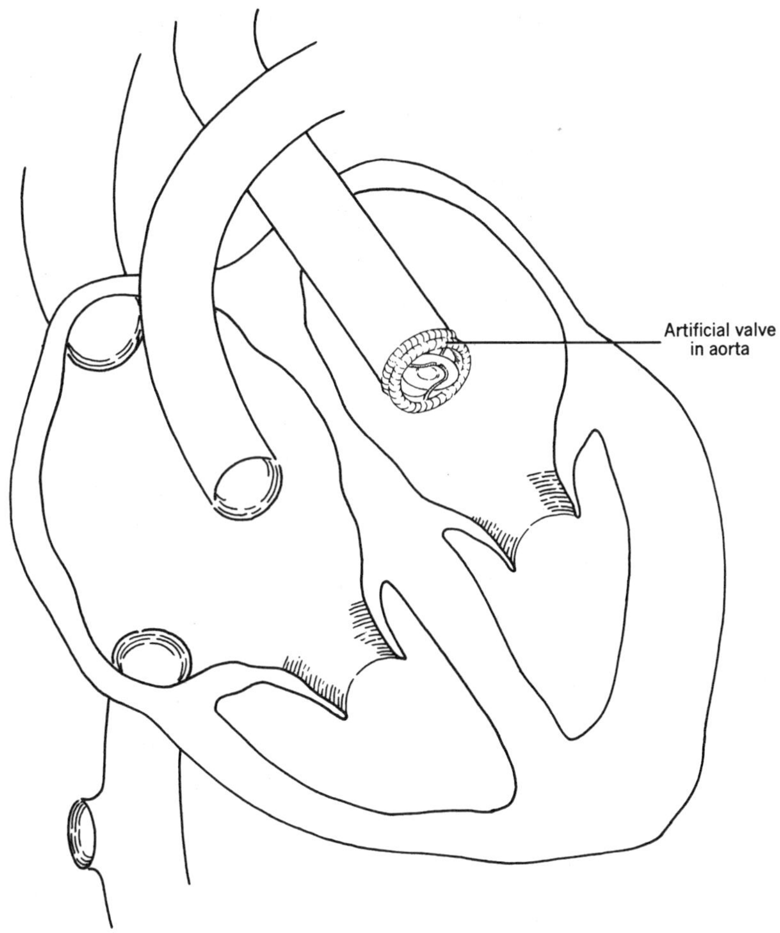

FIGURE 8-2
Bjork-Shiley cardiac valve prosthesis. For use in those cases when replacement of the natural aortic, mitral, or tricuspid valves is indicated. The disc opens to an angle of 60° to permit forward flow of blood during systole of the heart chamber and closes during diastole of that chamber. (Courtesy of Shiley Laboratories, Inc., Santa Ana, California.)

is moved to the free extremity. After another period of 10 to 15 minutes, the next cuff is moved to the free extremity. Finally, the third cuff is moved. Thus, the last extremity to have the cuff moved has had the veins compressed for 30 to 45 minutes.

Obviously, this procedure must be done systematically, and ideally by the same person for the entire time of the procedure, which may last for several hours. During the procedure, about 700 ml. of blood is trapped in the extremities; because it is not being returned to the pulmonary circuit, congestion there is relieved.

When the procedure is discontinued, only one tourniquet must be removed at a time; the other tourniquets are removed as they would be in the sequence of the rotation. If this is not done properly, it will be equivalent to a sudden infusion of 700 ml. of blood into the circulation, a tremendous circulatory overload.

Another treatment for pulmonary edema is the rapid withdrawal of about 500 ml. of venous blood (phlebotomy). Both the phlebotomy and the tourniquet procedures prevent blood from entering the pulmonary circuit and thereby minimize the pulmonary hypertension.

If the aortic valve is involved in rheumatic heart disease, the left ventricle has a tendency to hypertrophy. Although for some time it can compensate for the additional strain, after a prolonged period of hypertrophy of the myocardium the left ventricle may begin to dilate. Its walls become thin and unable to contract with sufficient force adequately to circulate the blood. Like the mitral valve, the aortic valve can be replaced with a similar prosthetic valve.

Bacterial endocarditis is an inflammation of the heart valves and the lining of the heart. Prior to the use of penicillin, the disease was almost always fatal. The patient has a slight fever, malaise, and fatigue. Later his complexion is sallow, his fever becomes more marked, and he may suffer chills and sweats, pronounced weakness, anorexia, and weight loss. These patients develop petechiae (small spots caused by bleeding under the skin) on the skin and mucous membranes. They are also prone to develop emboli, which may be life-threatening. Clubbing of the fingers may appear late in the course of the illness.

CONGENITAL HEART DEFECTS

Of the many types of congenital defects of the heart probably the most common is a patent foramen ovale or a patent ductus arteriosis. Each of these is surgically correctible. In the patent foramen ovale, some of the blood flows directly from the right atrium to the left atrium—bypassing the pulmonary circuit and failing to get oxygenated. If the defect is severe, cyanosis will result because the peripheral tissues will not receive adequate amounts of oxygen. Patent foramen ovale produces what is commonly called a "blue baby," so called because of the blue tinge to the skin.

In patent ductus arteriosis, some of the oxygenated blood leaving the aorta enters the ductus and returns to the pulmonary circuit. As a result, although the peripheral circulation does not receive as much blood as it should, what blood it does receive is well oxygenated. The pulmonary circuit, on the other hand, receives more blood than is required.

VASCULAR DISEASES

There are a variety of types of peripheral vascular disease, and each one has some particularly distinctive features. They all have in common a decreased blood flow, particularly to the extremities. Decreased blood flow causes pain on walking, particularly when walking upstairs or up a hill. This condition is called intermittent claudication.

With poor blood supply in the legs, patients often develop leg ulcers, which are very difficult to treat because wounds are dependent upon a good blood supply in order to heal. For the most part, these diseases are treated palliatively (sympamatically). Surgeons may do a sympathectomy (the removal of some of the sympathetic chain). However, the results are often disappointing, particularly if the disease is advanced.

HYPERTENSION

Hypertension is a major cause of death and disability in young adults. It may lead to serious problems such as congestive heart failure or cerebrovascular accidents (brain damage due to diseased vessels in the brain). Primary hypertension, which is very common, is high blood pressure due to some unknown cause. Secondary hypertension is high blood pressure that results from a known disease.

Pheochromocytoma (an adrenal tumor) causes intermittent severe hypertension. If the tumor is removed, the disease is cured. Hypertension may also be secondary to kidney disease, arteriosclerosis, or other types of peripheral vascular diseases. Malignant hypertension is a severe type of high blood pressure. The prognosis of this disease is very grave: It progresses rapidly and most of its patients live only for about 1 year.

Many patients with hypertension have no symptoms at all, in which case (although these patients have a very good prognosis) it is sometimes difficult to convince them that treatment is necessary. Diural is a diuretic that is helpful in lowering the blood pressure. Aldomet, reserpine, and some tranquilizers may also be effective in lowering pressure. Patients should not have excessive salt in their diets and sometimes must be placed on a low sodium diet. Overweight patients should be put on a reducing diet.

The symptoms of hypertension are varied and cannot always be correlated with the height of the blood pressure. Some of the more common symptoms are headache, dizziness, fatigue, and nervousness. Frequently, patients also have dependent edema. Figure 8–3 shows the capillary dynamics associated with the edema of hypertension. You should compare this drawing with the drawing of the normal capillary in Figure 7–11 on page 101.

SHOCK AND HEMORRHAGE

Before we discuss in detail the consequences of shock and hemorrhage, it might be well for you to review the initial introduction of the topic of shock on pages 31 and 32.

Hypovolemia resulting from hemorrhage can result in shock because blood pressure, among other things, is dependent upon a fairly constant volume of vascular fluid.

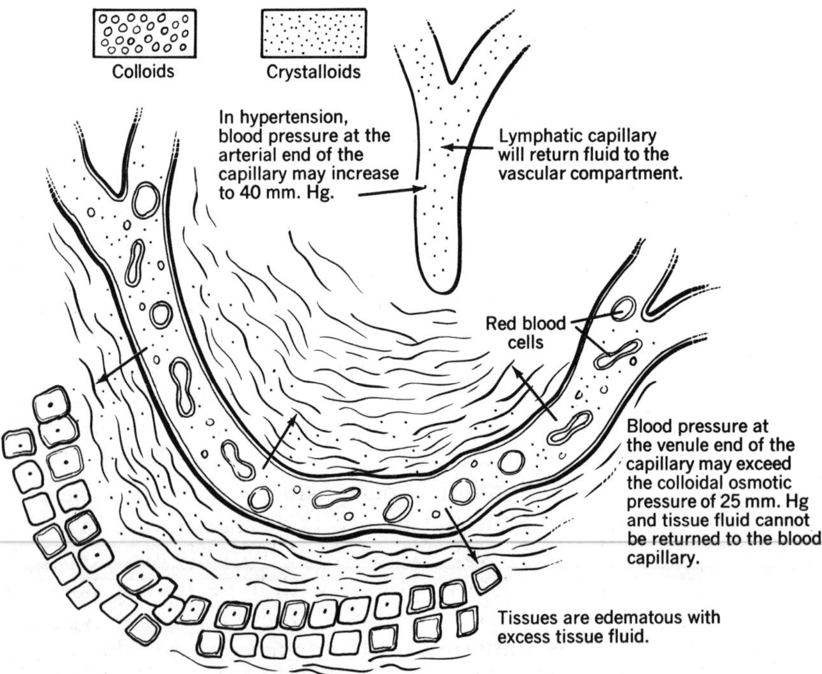

FIGURE 8-3
Capillary dynamics in hypertension that will result in edema. [From S. R. Burke: *The Composition and Function of Body Fluids,* 1st ed. (St. Louis, Mo.: The C. V. Mosby Co., 1972.)]

Blood pressure on the capillary level is the mechanism by which the peripheral body cells receive their supplies of nutrients for their metabolic processes. Without a sufficiently high capillary blood pressure to cause the filtration of these nutrients into the intercellular spaces, cellular metabolism decreases and patients in shock will feel cold. They are pale because the peripheral tissues are not well perfused with blood. The pulse is rapid and weak. They are thirsty and dehydrated because more intercellular fluid is being returned to the vascular compartment to help increase vascular volume. All of these signs of shock, although alarming, are compensatory mechanisms of the body as it attempts to confine the remaining vascular volume to a smaller vascular bed in order that functions of the vital centers (heart and brain) can be maintained.

Urine production also decreases, as the kidney will conserve fluid for the vascular compartment. For a detailed discussion of renal compensatory mechanism in shock, you should see pages 31 and 32. You should also study Figure 8–4, which shows the capillary and intercellular dynamics in the presence of hypovolemia and compare this with the normal capillary dynamics shown in Figure 7–11 on page 101.

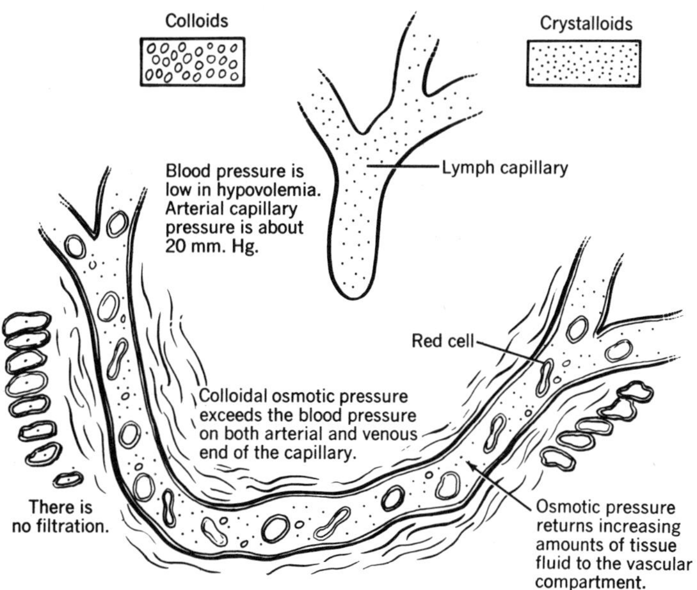

FIGURE 8-4

Capillary compensation in hypovolemia. [From S. R. Burke: *The Composition and Function of Body Fluids,* 1st ed. (St. Louis, Mo.: The C. V. Mosby Co., 1972).]

A patient in shock should be placed flat in bed with his legs elevated. Elevation helps return venous blood to the heart so that cardiac output can be increased. Although at one time physicians thought that a patient in shock should have extra blankets and hot water bottles because he is cold, today they believe that these measures probably counteract the normal compensatory mechanisms of the body in as much as heat increases metabolism and the patient in shock needs blood flow to his vital centers more than he does to the periphery. A light blanket will probably be sufficient for his comfort and still not increase peripheral metabolism excessively. The patient will probably need vasopressor drugs and intravenous fluid therapy. Recently some doctors have started to question the advisability of using vasopressor drugs to treat shock. They believe such drugs may be contraindicated in the presence of shock because doses adequate to produce vasoconstriction intensify the shock by diminishing blood flow to organs through increasing vascular resistance. Dextran, a plasma expander, is effective in increasing vascular volume; however, it will not help in replacing the formed elements of the blood (cells). If the hemorrhage is severe, it may be necessary to give whole blood transfusions.

SUMMARY QUESTIONS

1. Name the major symptoms of leukemia.
2. Name the causes and symptoms of anemia.
3. What disease is characterized by painless enlargement of the lymph nodes and is usually fatal within about 5 years?
4. How does the pain of a myocardial infarction differ from the pain of angina pectoris?
5. What is the difference between valvular stenosis and insufficiency?
6. What type of heart defect will result in a "blue baby"?
7. What is the difference between primary and secondary hypertension?
8. Name some of the causes of heart block.
9. Name the symptoms of shock.
10. What are some appropriate emergency measures used to treat shock?

SUGGESTED READINGS

Bruhn, J. C., and S. Wolf. "Studies Reporting Low Risk of Ischemic Heart Disease: a Critical Review," *American Journal of Public Health*, August 1970.

Butler, H. H. "How to Read an ECG," *R. N. Magazine,* January–March, 1973.

Gertler, H. E., *et al.* "Carbohydrates, Insulin and Lipid Interrelationship in Ischemic Vascular Disease," *Geriatrics,* April 1970.

Gros, G., *et al.* "Relationship of Sodium and Potassium Intake to Blood Pressure," *The American Journal of Clinical Nutrition,* June 1971.

Isler, C. "How to Insert an IV Step by Step," *R. N. Magazine,* October 1973.

Jernigan, A. K. "Diet for the Stroke Patient," *Hospitals,* June 1970.

Keller, M. D., *et al.* "A Study of the Primary Prevention of Coronary Heart Disease," *American Journal of Public Health,* August 1970.

Miller, G. D. "An Epidemiological View of Hypertensive Disease," *The Journal of Urban Health,* April 1973.

Moyer, J. H., III. "Optimum Therapy for Essential Hypertension," *American Family Physician,* June 1972.

Netter, F. H. *Heart, The Ciba Collection of Medical Illustrations,* Summit, N.J.: Ciba Pharmaceutical Products, Inc.

Rodman, M. J. "Drugs Used in Cardiovascular Disease," *R. N. Magazine,* March–May 1973.

9 The Respiratory System

VOCABULARY

Carbaminohemoglobin
Cholecystectomy
Cilia
Enzyme
Esophagus
Hemoglobin
Hyperventilation
Lacrimal glands
Nasal conchae
Olfactory
Otitis media
Peripheral

OVERVIEW

I STRUCTURES OF THE RESPIRATORY SYSTEM
- **A.** Nasal cavities, including the paranasal sinuses and nasolacrimal ducts
- **B.** Pharynx and tonsils
- **C.** Larynx
- **D.** Trachea
- **E.** Primary bronchi
- **F.** Secondary bronchi and bronchioles
- **G.** Alveoli
- **H.** Lungs and pleura

II PHYSIOLOGY OF RESPIRATIONS
- **A.** Pressures involved
- **B.** Internal and external respirations
- **C.** Factors facilitating the combining of oxygen with hemoglobin
- **D.** Gas transport
- **E.** Regulation of breathing

III AIR VOLUMES
- **A.** Tidal air
- **B.** Inspiratory and expiratory reserves
- **C.** Residual air
- **D.** Minimal air
- **E.** Vital capacity

IV GAS LAWS

Although Hippocrates, the father of medicine, thought the main purpose of breathing was to cool the heart, we have known for hundreds of years that this is not the case. Let us consider more recent ideas concerning the necessity of breathing.

Oxygen is essential for most of the chemical reactions taking place in our body. Carbon dioxide is a major waste product resulting from metabolism. It is the role of the respiratory system to help the circulatory system in the exchange of these gases. Because the respiratory system and circulatory system are so closely related, patients whose disease is actually respiratory in nature may have complaints that suggest circulatory problems, whereas those who have pathology of the circulatory system may complain mostly of symptoms such as shortness of breath. We shall examine the structure of the respiratory system and consider the functions of these vital structures.

All of the respiratory structures leading to the microscopic alveoli where the actual exchange of gases takes place are lined with mucous membrane. The mucous membrane is highly vascular. The inhaled air is warmed by the blood flowing through the vessels of this membrane. Microscopic cilia on the outer surface of the membrane in most of the prealveolar structures of the respiratory tract help to wave inhaled particles of dust backward into the throat where they may be swallowed and eliminated through the gastrointestinal tract. As air passes over the moist membrane, dust may also be collected by the mucus. The mucous membrane functions to warm and to moisten the air before it reaches the delicate alveoli.

Air enters the tract through the nose. The internal surface of the nose is greatly increased by the nasal conchae, which project from the lateral wall of each nasal cavity. The nasal septum, which separates the nasal cavities, is mostly cartilage covered by mucous membrane. However, the posterior part of the septum is bone, the vomer and the perpendicular plate of the ethmoid. The olfactory region of the nasal mucosa is located on the superior part of the nasal septum and the superior conchae. The sense of smell is discussed more completely in Chapter 13.

The paranasal sinuses are air spaces in some of the bones of the skull that open into the nose. These sinuses are lined with mucous membrane that is continuous with that of the respiratory tract. The sinuses give resonance to our voice. If they are infected and filled with mucous, some of the resonance is lost. The paranasal sinuses take the names of the bones in which they lie: maxillary, frontal, ethmoid, and sphenoid sinuses (see Figure 9–1).

The nasolacrimal ducts, which open into the nasal cavity, are also lined with mucous membrane. These ducts convey into the nose the tears, which are constantly being produced by the lacrimal glands to cleanse the surface of the eye. The release of tears serves as an added source of moisture to humidify the air. Tears, as well as the secretions of the mucous membrane, contain lysozyme, an enzyme that helps to destroy bacteria that we inhale.

From the nose the air passes into the nasopharynx, a tubular passageway posterior to the nasal cavities and the mouth. The walls of the pharynx are skeletal muscle and the lining is mucous membrane. The pharynx is subdivided into three parts: nasal, oral, and laryngeal.

The eustachian tube from the middle ear opens into the nasopharynx. This tube helps to equalize the pressure in the middle ear with that of the atmosphere. Upper respiratory infections—particularly those in children—can spread through the tube to the ear and cause otitis media.

The pharyngeal tonsil is lymphatic tissue located in the posterior part of the nasopharynx. This lymphatic tissue, as well as others elsewhere, plays an important role in protecting the body from bacterial invasion. If enlargement of the pharyngeal tonsil obstructs the upper air passage to such an extent that the individual is obliged to breathe through his mouth, air is not properly moistened, warmed, and filtered before it reaches the lungs.

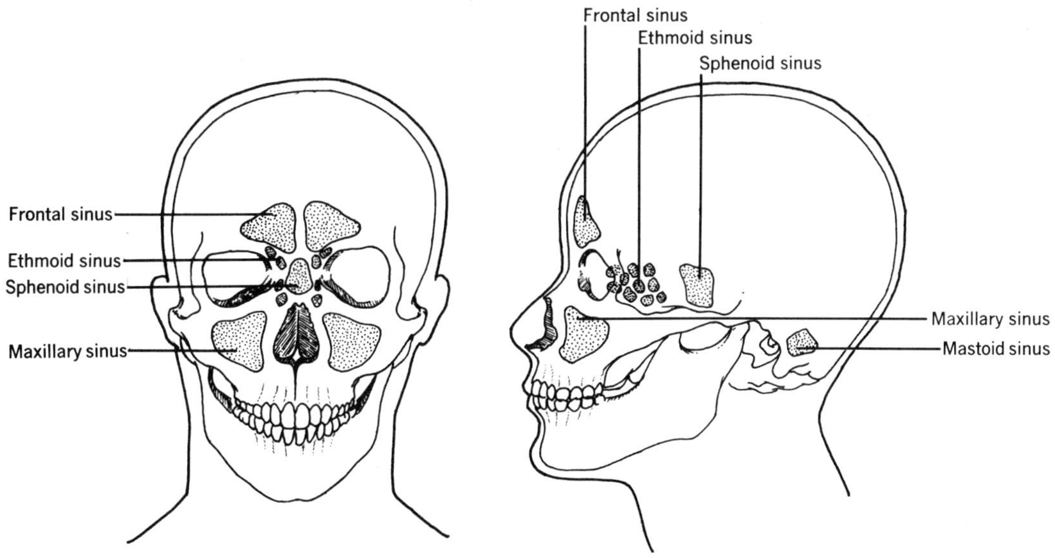

FIGURE 9-1

A frontal and lateral view of the air spaces or sinuses in the bones of the skull. The sinuses that drain into the nose, called the paranasal sinuses, include the frontal, ethmoid, and maxillary sinuses, and the sphenoid sinus. The mastoid sinuses drain into the middle ear.

The oropharynx, which is inferior to the nasopharynx, serves as a passageway not only for air but also for food. Lymphatic tissue in the oropharynx includes the palatine and lingual tonsils. From the oropharynx the air passes into the laryngopharynx. The laryngopharynx opens posteriorly into the esophagus and anteriorly into the larynx.

The larynx or voice box is made of nine cartilages joined together by ligaments and lined with mucous membrane. The larynx is controlled by skeletal muscle. The thyroid cartilage is the largest of the laryngeal cartilages. It forms the laryngeal prominence, or "Adam's apple." The epiglottis is a leaf-shaped cartilage that closes the opening to the remainder of the respiratory tract during the act of swallowing (see Figures 9–2 and 9–3).

At the upper end of the larynx are the vocal folds, cordlike structures that can vibrate as expired air passes over them and produces sound. The loudness of the sound depends upon the force of the vibration, and the pitch depends upon the frequency of the vibrations. Because children and women usually have shorter vocal cords than do men, their voices tend to be higher pitched. The glottis, the space between these vocal folds, is the narrowest part of the laryngeal cavity. Any obstruction, such as a foreign body (for example, a piece of food), can cause suffocation if it is not promptly removed.

The trachea, which is a cylindrical tube, extends from the larynx to the bronchi. The anterior and lateral aspects of the tube are composed of fibrous membrane and C-shaped cartilages. The cartilage rings keep the trachea open for the passage of air to and from the lungs. Smooth muscle layers and connective tissue fill in the posterior interval of the tube. During the act of swallowing, the

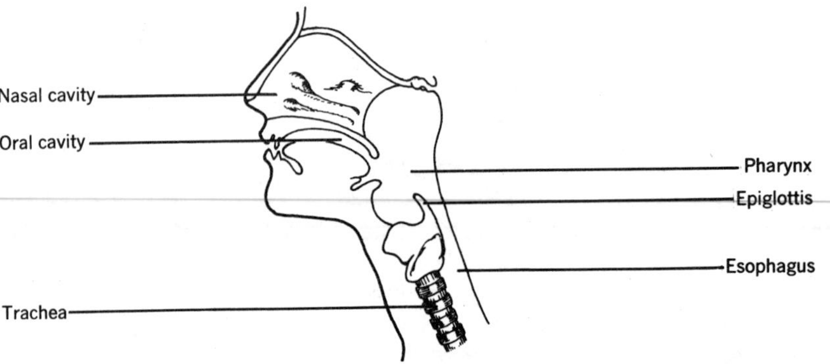

FIGURE 9-2

A lateral view of the larynx, trachea, and esophagus. Note the position of the epiglottis, which closes the respiratory passageway during the act of swallowing.

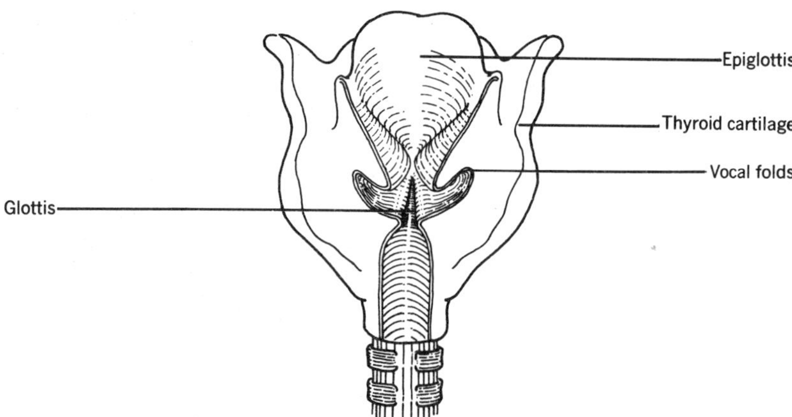

FIGURE 9-3
A posterior view of the larynx, cartilages, vocal folds, and the glottis.

esophagus, which is posterior to the trachea, can bulge into this region. The trachea is lined with ciliated mucous membrane containing goblet cells for the production of mucus.

From the trachea the air enters the primary bronchi. There are two primary bronchi, one entering each lung. The right bronchus is shorter, wider, and more vertical than the left; for this reason, any foreign body that has been aspirated is likely to enter the right bronchus.

The bronchi are similar in structure to the trachea. They branch into secondary bronchi, each of which enters a lobe of the lungs, three on the right and two on the left. The secondary bronchi continue to branch and form smaller and smaller bronchioles. As the branches get smaller, the cartilagenous rings begin to disappear and the structures are mostly smooth muscle. The terminal bronchioles and respiratory bronchioles are entirely smooth muscle passageways. The respiratory bronchioles lead to the alveoli, where the actual exchange of gases between the bloodstream and the respiratory system takes place. All of the prealveolar structures comprise what is known as the anatomical dead space because air contained in these structures following inspiration does not reach the alveoli and will be exhaled. During a normal quiet inspiration, about 500 ml. of air is inhaled. About 150 ml. of this air remains in the anatomical dead space until expiration.

Each microscopic alveolus is surrounded with a rich capillary network arising from the pulmonary artery. The blood pressure in the pulmonary capillaries is very low, so that normally no fluid filters into the alveoli. Because the alveoli must be fluid-free in order to accommodate adequate amounts of air, low pulmonary blood pressure is very important. The chief reason that the blood pressure in these pulmonary capillaries is so much lower than it is in most other capillaries is the enormous size of the capillary bed and the small amount of blood contained within the capillaries at any one time. The bed covers an area of about 100 sq. m. and contains only about 60 ml. of blood at any one time.

The blood in the arteriole end of the pulmonary capillary is high in carbon dioxide and low in oxygen. Most of the carbon dioxide in this blood diffuses into the alveoli, and oxygen from the alveoli diffuses into the blood capillary.

The two lungs are located in the thoracic cavity, one on either side of the heart (see Figure 9–4). The apex of the lung is about 1.5 inches above the first rib, and the base of the lung rests on the diaphragm. The right lung has three lobes, and the left has two. The lungs have a dual blood supply. Bronchial arteries, which are nutrient to the lungs, supply these tissues with food and oxygen. The pulmonary circuit takes care of the exchange of gases.

The pleura is a serous membrane covering the lungs. The visceral layer of this membrane closely covers the lungs; the parietal layer lies on the inner surface of the chest wall, the diaphragm, and the lateral aspect of the mediastinum (the space between the two lungs containing the heart, blood vessels, trachea, esophagus, lymphatic tissue, and vessels). A small amount of serous fluid in the space between the visceral and parietal pleura prevents friction.

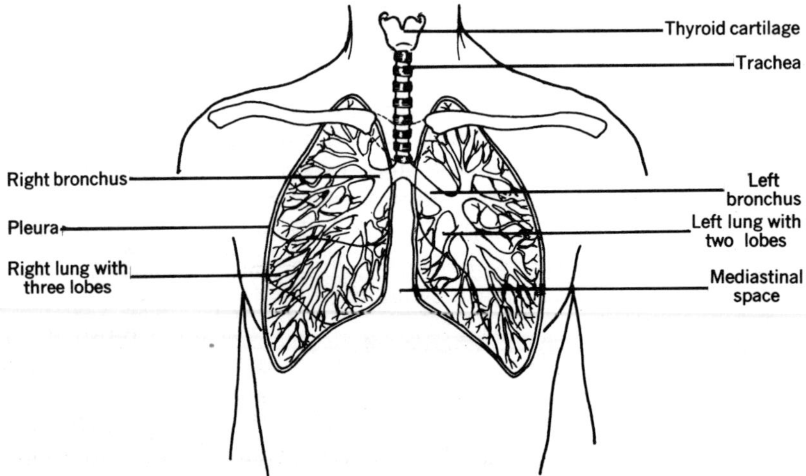

FIGURE 9-4
The lungs in the thoracic cavity and the distribution of bronchi within the lungs.

PHYSIOLOGY OF RESPIRATIONS

Air flows from an area of higher pressure or concentration to an area of lower pressure or concentration. Involved in the mechanics of respiration are the atmospheric pressure, intrapulmonic pressure, and intrapleural or intrathoracic pressure.

The atmospheric pressure is that of the air around us. At sea level, this pressure is 760 mm. Hg. The intrapulmonic pressure is the pressure of the air within the bronchi and bronchioles. This pressure varies above and below 760 mm. Hg, depending upon the size of the thorax. The intrapleural or intrathoracic pressure is the pressure in the pleural space. It is normally always less than atmospheric, about 751 to 754 mm. Hg. This pressure, however, may exceed atmospheric pressure when a person coughs or strains at stool.

During inspiration, the size of the thorax increases chiefly because of the contraction and descent of the diaphragm. The external intercostal muscles also contract to increase the size of the thorax by raising the ribs, and the sternum pushes forward to increase the anterior–posterior diameter of the thoracic cavity. As the volume of the thorax increases, the intrapulmonic and intrapleural pressures decrease. Thus, air enters the lungs until the intrapulmonic pressure is equal to atmospheric pressure.

Expiration is mainly a passive action—the relaxation of the diaphragm and external intercostal muscles, which decreases the size of the thorax. As these muscles return to their resting state, the elastic lungs recoil. This recoil increases the intrapulmonic pressure slightly above the atmospheric pressure and the air is forced out of the lungs.

External respiration involves the exchange of gases between the circulating blood in the alveolar capillaries and the air in the alveoli. Internal respiration involves the exchange of gases between the circulating blood in the peripheral capillaries and the tissue cells as they use oxygen and produce waste carbon dioxide.

FACTORS FACILITATING THE COMBINING OF OXYGEN WITH HEMOGLOBIN

The partial pressure* of oxygen in the lungs is greater than that in the bloodstream, and therefore oxygen goes from the lungs to the blood. The partial pressure of oxygen in the blood is higher than

*In a mixture of gases, the combination of the pressures exerted by all of the gases is called the total pressure, and the pressure exerted by a single gas is called the partial pressure.

that in the peripheral tissues and oxygen diffuses from the blood into the tissues.

A pH toward the alkaline side favors the combining of oxygen with hemoglobin, whereas a pH toward the acid side favors the dissociation of oxygen from hemoglobin. In the pulmonary circuit, because carbon dioxide and water are being expelled, the blood is slightly more alkaline; here, oxygen readily combines with hemoglobin. In the active peripheral tissues, carbon dioxide is being produced. Therefore, because carbon dioxide combines with water to produce carbonic acid, blood in these capillaries is slightly more acid. This acidity favors the release of oxygen from hemoglobin.

Temperature also influences the association and dissociation of oxygen and hemoglobin. The increased metabolism produces slightly higher temperatures in the active peripheral tissues than in the pulmonary circuit, which favor the dissociation of oxygen with hemoglobin in the periphery and the association of oxygen with hemoglobin in the pulmonary circuit.

GAS TRANSPORT

Oxygen is transported mainly as potassium oxyhemoglobin in the red blood cell; however, a small amount is dissolved in the plasma. Carbon dioxide is mainly carried in the plasma in the form of bicarbonate. Some of the carbon dioxide in the plasma is carried as carbonic acid. In the red blood cells, carbon dioxide is carried in the form of carbaminohemoglobin and as bicarbonate.

REGULATION OF BREATHING

The respiratory center in the medulla (of the brain) controls the rate and depth of respiration according to the needs of the body. Normal respiratory rates are about 14 to 20 respirations per minute. Obviously, vigorous exercise can greatly increase the rate of respiration, as active skeletal muscles require more oxygen than do resting muscles.

When the respiratory center is stimulated, it sends efferent (motor) nerve impulses via the phrenic nerve to the diaphragm to increase the rate of contraction. Various stimuli affect this respiratory center, the most powerful and important of which is an increased level of carbon dioxide. Excess oxygen or decreased amounts of carbon dioxide will depress the respiratory center and respirations will become slower.

Pressoreceptors in the alveoli are stimulated by the expansion of the lung during inspiration. They then send afferent (sensory) messages to the medulla to inhibit respiration and to relax the diaphragm. As the lung deflates, the pressoreceptors cease sending their impulses and the diaphragm contracts. This is called the Hering–Breuer reflex.

In the aorta and carotid arteries, chemoreceptors respond to oxygen lack and to increased levels of carbon dioxide. As the oxygen level in the blood falls, these chemoreceptors are stimulated and nerve impulses are sent to the respiratory center to increase the rate and depth of respirations. Although these chemoreceptors are not of major importance in regulation of respirations in the normal healthy individual, they may be very important in regard to patients with serious respiratory diseases (see Chapter 10).

Other factors such as fever and pain can also modify the rate and depth of respiration. Fever increases the body metabolism, which in turn increases respiration. Pain may also cause an increase in the rate of respiration. However, if the pain is due to thoracic surgery or to a cholecystectomy or other high abdominal surgery, the patient is likely to have slow, shallow respirations because his respiratory movements will be painful. As the result of shallow respirations following surgery, the patient may develop pneumonia. To help prevent this, the patient should be encouraged to cough and to breathe deeply in spite of the discomfort.

AIR VOLUMES

Tidal air is the air that is moved during quiet inspiration and expiration. Normally, the volume of this air is about 500 ml. Of the 500 ml., about 150 ml. stays in the anatomical dead space, the bronchi, bronchioles, and other prealveolar structures until the following expiration, when it is exhaled.

The inspiratory reserve is the amount of air that can forceably be inspired after a normal inspiration. Its volume is about 2000 to 3000 ml. The expiratory reserve is the amount of air that can be forceably expired after a normal expiration. The normal expiratory reserve volume, somewhat less than the inspiratory reserve, is about 1200 ml.

Residual air, which is about 1000 to 1200 ml., is the air that remains in the lung as long as the thorax is airtight. Minimal air is the air that remains in the alveoli after the lung collapses because of trauma or thoracic surgery.

Vital capacity is the quantity of air moved on deepest inspiration and expiration. On the average, vital capacity is about 3000 to 5000 ml., although in a healthy, young adult it may be much greater.

GAS LAWS

Boyle's law concerning gases states that at a constant temperature the volume of a gas is inversely proportional to the pressure of the gas. Dalton's law states that each gas exerts its pressure independently of other gases. All of the chemical and physiological activities of gases are determined by the pressure under which the gas is maintained. These pressures, which can be measured, yield valuable diagnostic information.

In a mixture of gases such as oxygen and carbon dioxide in the bloodstream, the combination of the pressures of each of the gases is called the total pressure. Clinically, we are concerned with the pressure of each gas, which is called the partial pressure. The partial pressure of oxygen, or the pO_2, in arterial blood is about 104 mm. Hg. It is obviously influenced by the oxygen content of the inspired air and by blood volume. The partial pressure of carbon dioxide, or the pCO_2, of arterial blood is 40 mm. Hg. The pCO_2 decreases with hyperventilation as more and more carbon dioxide is blown off. The pressure is increased with shallow respirations and diseases of the respiratory system that decrease the ventilatory space. The activity of carbon dioxide of particular clinical concern is its tendency to combine with water to form carbonic acid. The formation of too much carbonic acid due to decreased lung volume in a patient with diseased lungs will disturb the acid–base balance of the blood and the patient may develop acidosis.

SUMMARY QUESTIONS

1. Discuss internal and external respirations.
2. List in sequence all of the structures through which inhaled air passes from the time it enters the nose until it reaches the alveoli.
3. What are the functions of the nose?
4. List the paranasal sinuses.
5. What are the functions of the paranasal sinuses?
6. Where are the eustachian tubes located and what is the function of these tubes?
7. What is lysozyme?
8. Where is the glottis?
9. What is meant by the anatomical dead space?
10. What is the name of the membrane that covers the lungs?
11. Discuss the pressures involved in the process of external respiration.
12. What muscle action is involved in normal inspiration?
13. What muscle action is involved in normal expiration?

14. Discuss the factors that facilitate the combining of oxygen with hemoglobin.
15. Where is the respiratory center located and how is this center stimulated?
16. How is oxygen transported in the bloodstream?
17. How is carbon dioxide transported in the bloodstream?
18. Discuss the Hering–Breuer reflex.
19. What is tidal air?
20. Discuss Boyle's and Dalton's laws concerning gases.

SUGGESTED READINGS

Burke, S.R. *The Composition and Function of Body Fluids.* 1st ed. St. Louis, Mo.: The C. V. Mosby Co., 1972.

Guyton, A. C. *Function of the Human Body.* 3rd ed. Philadelphia, Pa.: W. B. Saunders Co., 1969.

Jacob, W. W., and C. A. Francone. *Structure and Function in Man.* 2d ed. Philadelphia, Pa.: W. B. Saunders Co., 1970.

10 Diseases of the Respiratory System

VOCABULARY

Amnionic fluid
Analgesic
Anorexia
Bacteremia
Calcification
Cyanosis
Cytology
Dyspnea
Exuate
Fibrosis
Hemoptysis
Induration
Mastoiditis
Otitis Media
Pleura
Pneumococcus
Pneumothorax
Purulent
Staphylococcus
Streptococcus
Thoracentesis
Tracheostomy
Thoracotomy
Viable

OVERVIEW

I COMMON DIAGNOSTIC PROCEDURES
- **A.** Roentgenogram
- **B.** Fluoroscopy
- **C.** Sputum Specimens
- **D.** Bronchoscopy
- **E.** Bronchogram
- **F.** Vital capacity
- **G.** Bronchospirometry

II DISEASES OF THE UPPER RESPIRATORY TRACT
- **A.** Sinusitis
- **B.** Epistaxis
- **C.** Tonsilitis
- **D.** Common Cold
- **E.** Influenza

III DISEASES OF THE LUNGS AND PLEURA
- **A.** Pneumonia
- **B.** Pleurisy
- **C.** Tuberculosis
- **D.** Chronic obstructive lung diseases
- **E.** Bronchial asthma

IV TUMORS OF THE RESPIRATORY SYSTEM

V HYALINE MEMBRANE SYNDROME

VI CHEST INJURIES

COMMON DIAGNOSTIC PROCEDURES

Roentgenography, or X-rays, of the chest help the physician to diagnose many types of chest diseases. Usually, both anteroposterior (AP) and lateral views are taken, as the view of some lesions in a single X-ray may be obstructed by surrounding structures. Although it is common practice to use the term "AP chest X-ray," the film is actually taken from the posterior to the anterior position (see Figure 10–1).

Fluoroscopy enables the physician to view the thoracic cavity in motion, but unlike an X-ray examination it gives no permanent record. No special preparation is required for either a fluoroscopic examination or a chest X-ray other than the removal of any object—such as a religious medal—that obstructs a view of the chest area.

Sputum specimens help the physician to diagnose infections of the respiratory system. The causative organism can be cultured out of the sputum and be tested for its sensitivity to a variety of antibiotics. If tuberculosis is suspected, three 24-hour sputum specimens are usually required for examination. If any one of these specimens is positive, the diagnosis is confirmed. However, that all three specimens are negative does not necessarily negate the possibility of active tuberculosis. It is entirely possible that the bacillus was simply not present in these particular specimens.

Sputum specimens may also be used for cytological examination. Sometimes abnormal cells from tumors are present in the sputum.

Bronchoscopy can be used to visualize the bronchi directly and also to obtain a small specimen of tissue for microscopic examination if a tumor is suspected. Bronchoscopy can also be used to remove foreign bodies that have been aspirated.

The bronchoscope, the hollow instrument used for the bronchoscopy examination, is passed into the trachea and on to the bronchi. To be prepared for this examination, the patient should have nothing to eat or to drink for 8 to 12 hours before the bronchoscopy to avoid the danger of aspirating vomitus. A local anesthetic, such as cocaine or pontocain, is usually administered. Because the patient's gag reflex will be absent for several hours following the examination, the period of fasting must be continued until the return of the reflex.

For a bronchogram, a radiopaque substance is injected into the trachea of the patient and he is then tilted in various positions so that the dye flows throughout the bronchial tree. X-ray films will show the outline of the bronchi and bronchioles (see Figure 10–2). Before this procedure, the patient should have postural drainage (see page 53) to drain as much mucus as possible from

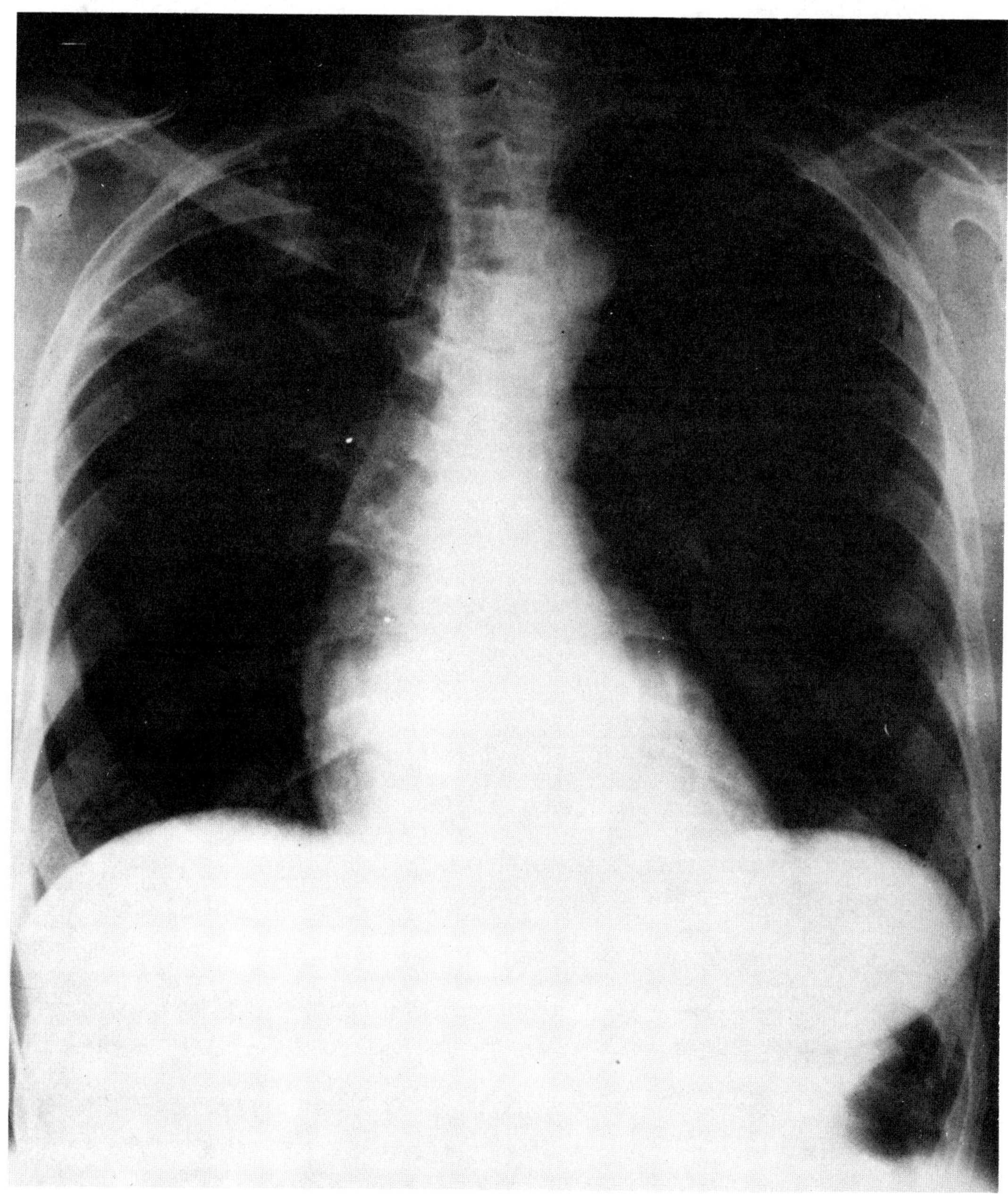

FIGURE 10-1
Normal chest X-ray. (Courtesy of James W. Lecky, M.D., Professor of Radiology, University of Pittsburgh.)

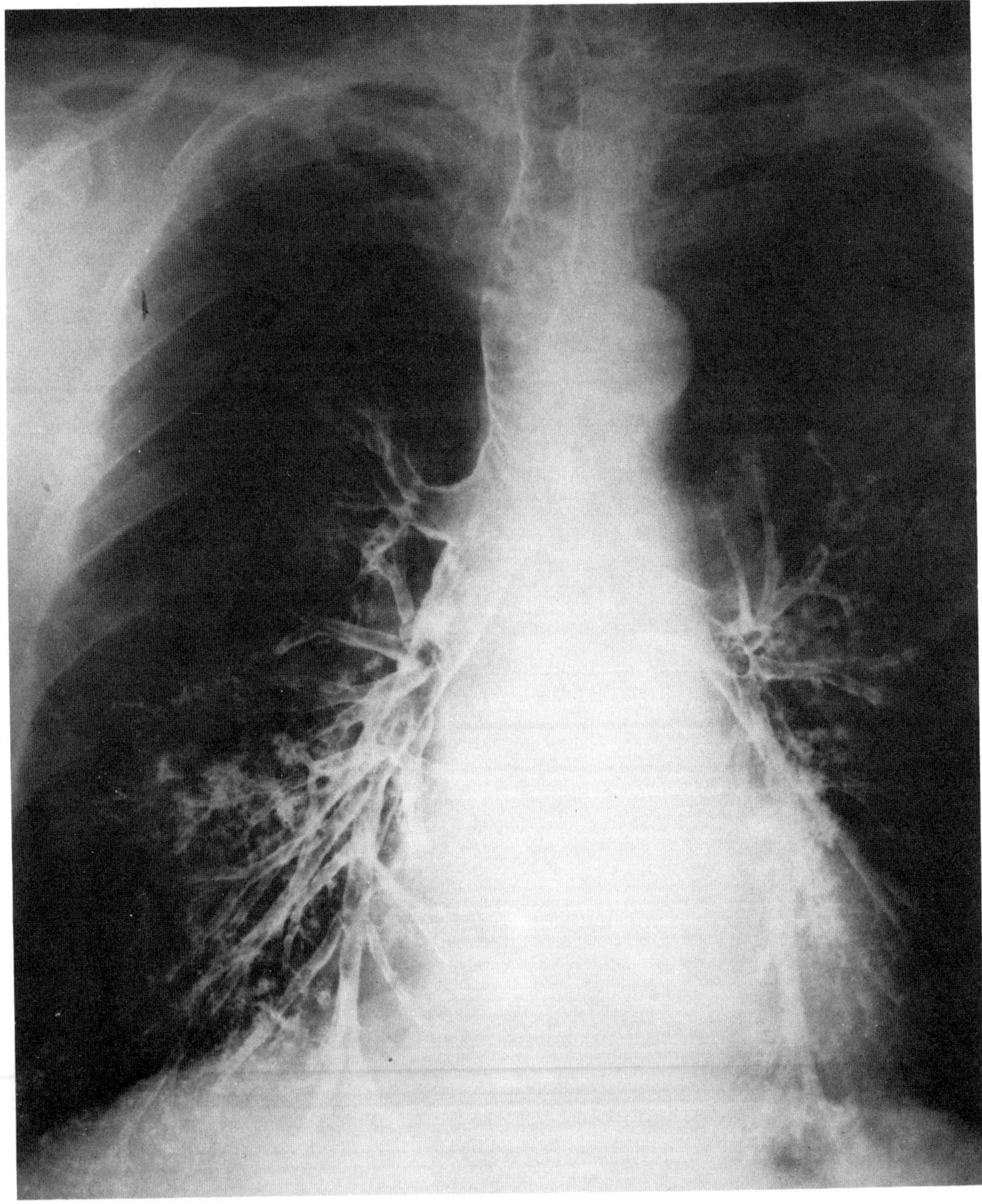

FIGURE 10-2
Normal bronchogram. The air passageways are visible because a radiopaque substance has been instilled. (Courtesy of James W. Lecky, M.D., Professor of Radiology, University of Pittsburgh.)

the tract so that the dye is able to reach the smaller bronchioles. The pharynx and larynx are sprayed with a local anesthetic, and the patient is also given a sedative prior to the examination. Following the examination, the patient should again have postural drainage to help remove the dye, and, as with the bronchoscopy, he should fast until his gag reflex returns to normal.

The most common respiratory function tests are the vital capacity measurement and a procedure called a bronchospirometry. To measure vital capacity the patient takes as large an inspiration as he can and then expires as much of it as possible into a machine that records the amount. About 3000 to 5000 ml. is the normal vital capacity. It may be decreased considerably in most diseases of the lung.

For a bronchospirometry, the patient breathes through a double lumen tube placed in his trachea. The amounts of air inspired and expired are measured, and arterial blood is analyzed for oxygen and carbon dioxide content.

Several different types of tuberculin skin tests exist. All are based on the fact that after the body has been invaded by the tubercle bacillus the body develops an allergy to the organism in about 6 to 8 weeks. About 48 to 72 hours following an intradermal injection of tuberculin, a person whose body has been invaded by the tubercle bacillus will develop an area of induration (swelling and redness) at the site of the injection. A positive reaction does not necessarily mean that a person has tuberculosis, but it does mean that at some time bacilli have invaded his body and created a primary lesion that has probably healed. A person who has had a recent conversion from a negative to a positive tuberculin reaction should, however, have a chest X-ray and may take isoniazid to help prevent the development of clinical tuberculosis.

RESPIRATORY DISEASES

SINUSITIS

Sinusitis is an infection of the paranasal sinuses. Because the mucous membrane that lines these sinuses is continuous with that of the nose and throat, sinusitis is a fairly common complication of any upper respiratory infection. It can be particularly painful if the mucosa lining the ducts leading from these sinuses into the nose is so swollen that there is little or no drainage of the mucus produced by the inflammed tissues into the nose. Nose drops such as epinephrine may help to decrease the swelling and to facilitate drainage.

EPISTAXIS

Epistaxis is a nosebleed that may be caused by trauma or by ulcerations of the lining of the nose. Small tumors or polyps can also cause epistaxis. Sometimes an abnormally high blood pressure causes the vessels in the nasal mucosa to break and results in varying degrees of hemorrhage. In this instance, the nosebleed is of course a more fortunate hemorrhage than if the hypertension had caused a rupture of vessels in the brain.

Epistaxis usually can be controlled effectively if the patient remains quiet with his head elevated. Because cold causes vasoconstriction, cold compresses are also helpful. In some cases, the nasal cavity may have to be packed with gauze in order to apply pressure on the bleeding vessels and to encourage blood clotting.

TONSILLITIS

Tonsillitis may involve any or all of the tonsils. Repeated upper respiratory infections often result in tonsillitis. This disease is much more common in children that in adults. Its onset is usually sudden and includes chills and fever as high as 105° or 106°F. Malaise and headache are frequently present. Patients suffer severe pain in the tonsillar area, especially during swallowing.

Although tonsillitis is usually self-limiting, serious complications, including otitis media, mastoiditis, scarlet fever, bacteremia, and rheumatic fever, may occur. Sometimes, repeated attacks of acute tonsillitis will lead to chronic tonsillitis, characterized by a low-grade fever, lassitude, and failure of growing children to gain weight. Such children may have to have a tonsillectomy (surgical removal of the tonsils).

THE COMMON COLD

The common cold, the most widespread of all respiratory diseases, is caused by a virus that is very easily spread from one person to another. One characteristic of the cold virus is that it usually fails to produce an immunity so that some persons suffer from one cold after another. Aspirin, bed rest, and an increased fluid intake are helpful in relieving the symptoms of the cold. Laryngitis or other upper respiratory infections often accompany colds. Laryngitis can also be caused by excessive smoking or voice strain.

INFLUENZA

Although there are many different types of influenza, all are characterized by an inflammatory condition of the upper respiratory tract and varying degrees of malaise. From time to time, many people have died during widespread epidemics of influenza. The deaths result because the influenza has developed into a severe form of pneumonia. Several vaccines have been developed which provide some immunity to particular types of influenza; however, the immunity is of short duration.

PNEUMONIA

Pneumonia is an acute inflammation of the lungs which may be caused by a variety of organisms. The pneumococcus is the most frequent causative organism. Pneumococci can frequently be cultured from the nose and throats of healthy individuals. Under circumstances in which they develop a decreased resistance, these individuals are likely to get pneumonia. For this reason, pneumonia is a fairly common complication of other disease processes.

Streptococci, staphylococci, and many viruses can also cause pneumonia. These pathogens, too, may be carried by a healthy person in the mucosa of the upper respiratory tract.

There are two main kinds of pneumonia: In lobar pneumonia an entire lobe of a lung is involved, and in bronchopneumonia the disease process is scattered throughout the lung. Both types involve copious sputum, chest pain, dyspnea, and fever. Specific antibiotic therapy and bed rest are indicated. If the dyspnea is severe enough, oxygen therapy may be required.

Although it is not very common, oil aspiration pneumonia is a very serious disease. It can result from aspirating a peanut, popcorn, or other oily substances. Nose drops that have an oily base should never be given to a young child: Because it is frequently difficult to get such a child to cooperate, he may aspirate the medication.

PLEURISY

Pleurisy, which usually is secondary to some other respiratory disease, is an inflammation of the pleura. It takes two forms: Pleurisy with effusion, in which large amounts of fluid collect in the pleural cavity and may have to be removed by means of a thoracentesis (see Figure 10–3); and dry pleurisy, in which very little exudate is produced and sharp pain occurs as the two pleural surfaces rub together during respirations. A tight chest binder may provide some relief. In both types of pleurisy bed rest is indicated. If pleurisy is a complication of another disease, it usually heals spontaneously when the primary disease is successfully treated.

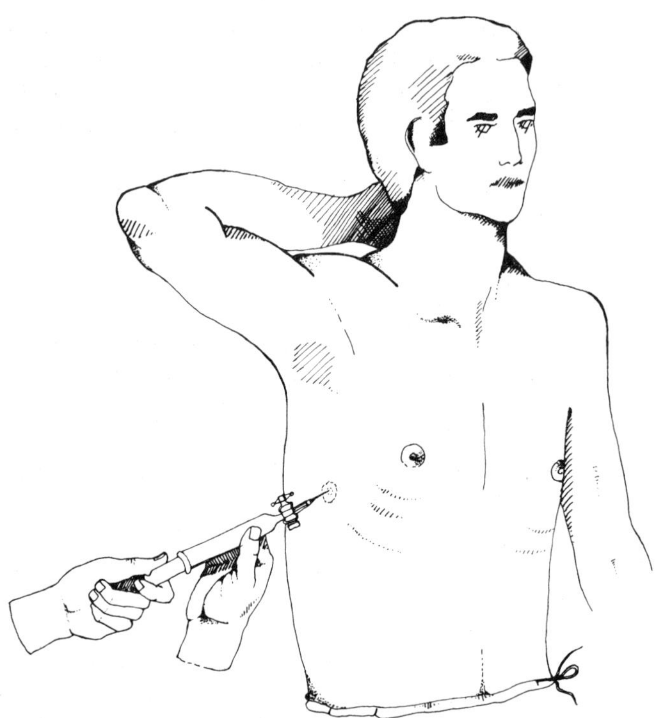

FIGURE 10-3
Fluid being removed from the pleural cavity by means of a thoracentesis. If there is a large amount of fluid to be removed, tubing will be attached to the three-way stopcock between the needle and syringe. This tubing will allow the fluid to drain into a container.

TUBERCULOSIS

Tuberculosis is a chronic disease caused by the bacillus Mycobacterium tuberculosis. The organism can live for months in dried sputum and can also withstand exposure to many disinfectants. It is, however, vulnerable to sunlight.

Tuberculosis is usually transmitted by droplet infection. When a person with active tuberculosis sneezes or coughs, the organisms spread though the air and may be inhaled by someone else. For this reason, tubercular patients must be taught to use proper tissue technique and to cover their nose and mouth when they cough. Gastrointestinal tuberculosis may be a result of eating contaminated foods or using contaminated tableware. Children are more likely to develop gastrointestinal tuberculosis because they are exposed to mouth-contaminated articles more often than are adults.

Pulmonary tuberculosis develops directly following the first implantation of the tubercle bacillus in the lung, which is known as the primary phase of the disease. If the infected person's resistance is high, the lesion may heal by fibrosis and calcification. The remnants of the healed primary focus are usually visible on X-ray. They may contain viable bacilli that are walled off and cannot spread to other lung tissue unless a marked decrease in resistance occurs at some subsequent time. The patient will have a positive tuberculin test and some degree of immunity to subsequent exposure to the organism.

At times, the primary complex does not heal but goes on to a progressive form of the disease. The focus of the infection may enlarge and undergo caseation, a form of necrosis having cheeselike appearance. The area of caseation may slough away and leave a cavity in the lung.

Because the onset of tuberculosis is insidious, for a long time the patient may have no symptoms at all. If there are early symptoms, they may be readily dismissed. Fatigue, anorexia, weight loss, and a slight cough are all symptoms that can be attributed to overwork, excessive smoking, or poor eating habits; however, they are also early symptoms of tuberculosis. A low-grade fever in the late afternoon and evening, night sweats, and increased fatigue occur as the disease progresses. The cough produces purulent sputum that may be blood-streaked. Hemoptysis, the coughing up of blood, may also occur. Later stages produce marked weakness, wasting, and dyspnea. Chest pain may be present if the infection has spread to the pleura.

Chemotherapy is one of the most important aspects of the treatment of tuberculosis. Streptomycin was the first drug to be used for tuberculosis; however, there are now several other effective drugs that have fewer side affects than streptomycin. Isonicoinic acid—hydrazide or isoniazid (INH), *para*-aminosalicylic acid (PAS), kanamycin, ethionamide, and cycloserine are all used in the treatment of tuberculosis. Frequently, a patient is prescribed a combination of several drugs.

Surgical treatment of tuberculosis may be required for patients with advanced disease or for those who do not respond to medical treatment. Radical surgery, such as a pneumonectomy (removal of a lung), is done less frequently today than it was in the past. Today, the surgery is more likely to involve the removal of only a part of the lung or, if the diseased area is larger, the removal of a lobe of the lung (lobectomy).

Collapse of a diseased portion of the lung is sometimes done to facilitate healing. If a cavity exists, collapse therapy puts that portion of the lung at rest and brings the edges of the cavity closer together. A pneumothorax, the injection of air into the pleural cavity, may be used to collapse the lung temporarily until healing can take place. In a thoracoplasty, the space into which the lung can expand is permanently reduced by the removal of some of the ribs on the affected side.

Regardless of whether the prescribed treatment for tuberculosis is medical or surgical, rest—both physical and mental—is an important aspect of the care of a patient with tuberculosis. The patient's attitude toward the disease will greatly affect his ability to rest. If he is anxious and worried about financial problems and his prognosis, he will be unable to rest even though he physically complies with the recommended rest periods.

Optimum diet and other general good hygienic regimen are equally important in increasing the body's defenses against tuberculosis. For many patients the prescribed chemotherapy causes gastric disturbances that interfere with their appetite. A carefully planned menu, attractively prepared meals, and pleasant dining facilities are of no avail unless the patient eats his meals. Gastric disturbances resulting from the drugs can sometimes be avoided if pills are taken with milk.

CHRONIC OBSTRUCTIVE PULMONARY DISEASE

The incidence of chronic obstructive pulmonary diseases has greatly increased in recent years, probably because our population is made up of greater numbers of elderly individuals who, in general, have decreased resistance. Increased air pollution may also have contributed to the increase in chronic obstructive lung diseases.

Patients usually have a combination of three pathological changes in their respiratory systems: bronchitis, emphysema, and bronchiectasis. All of these are seriously disabling.

Chronic bronchitis causes scarring of the mucosa lining the bronchi and bronchioles which greatly interferes with the normal bronchial eliminating mechanism. Normally, longitudinal ridges in the mucous membrane of the bronchi and cilia help move particles of dust and mucus toward the outside of the body. After repeated bronchial infections, many of these longitudinal ridges and cilia are destroyed. Abnormal circumferencial ridges develop and make it very difficult for patients to cough up the thick tenacious mucus characteristic of chronic bronchitis. The thick mucus interferes with the free flow of air through the respiratory tract, and the scarred thickened mucosa of the terminal and respiratory bronchioles may trap in the alveoli air that is high in carbon dioxide and low in oxygen.

A second aspect of chronic obstructive pulmonary disease is emphysema. In emphysema, the alveoli are greatly distended and a loss of alveolar structure and elasticity takes place. The alveoli do not normally empty with expiration. Exertional dyspnea is usually the first symptom of emphysema. The dyspnea progresses slowly until it is present even at rest. A chronic cough may be productive of mucopurulent sputum. Expiration is prolonged and difficult. As the disease progresses, the patient is able to exhale only by contracting accessory muscles of respiration. You should recall that expiration is normally accomplished by the relaxation of the diaphragm and external intercostal muscles. Because he can exhale only by contracting his abdominal muscles, the work of breathing for the patient with emphysema is doubled. In advanced emphysema, respiratory function may be so impaired that the amount of carbon dioxide in the blood is greatly increased and the respiratory center in the medulla of the brain may become insensitive to the normal stimulus of carbon dioxide. The patient's respirations can become entirely dependent upon the chemoreceptors that respond to oxygen lack. This condition, called carbon dioxide narcosis, is usually characterized by increasing lethargy and even stupor. On the other hand, some patients who have carbon dioxide narcosis are hyperactive.

In bronchiectasis, the pathological change occurring is a widening of the secondary bronchi and bronchioles. The widening of the air passageways greatly increases the anatomic dead space so that, whereas only about 150 ml. of an inspiration of air remains in the anatomic dead space of a healthy person, as much as 300 ml. of inspired air may never reach the alveoli of a patient with bronchiectasis. Thus, the blood is not properly oxygenated. The patient will be cyanotic and dyspneic; produce large quantities of foul, greenish-yellow sputum; and suffer fatigue, weight loss, and anorexia.

Compare Figures 10–4 and 10–5. Figure 10–4 illustrates the normal exchange of gases between the alveoli and the blood stream. Figure 10–5 shows some of the pathologic changes that occur in chronic obstructive pulmonary disease. All the pathologic changes that occur in chronic obstructive pulmonary disease are for the most part irreversible. Much can be done, however, to slow the progressive deterioration of the respiratory structures and to relieve the symptoms of this crippling disease.

Postural drainage (see Figure 10–6) helps to remove the secretions by gravity. The exact position of the patient during this treatment depends upon the location of the area to be drained. However, the general principle involves positioning the patient so that the area to be drained is higher than are the bronchi and other

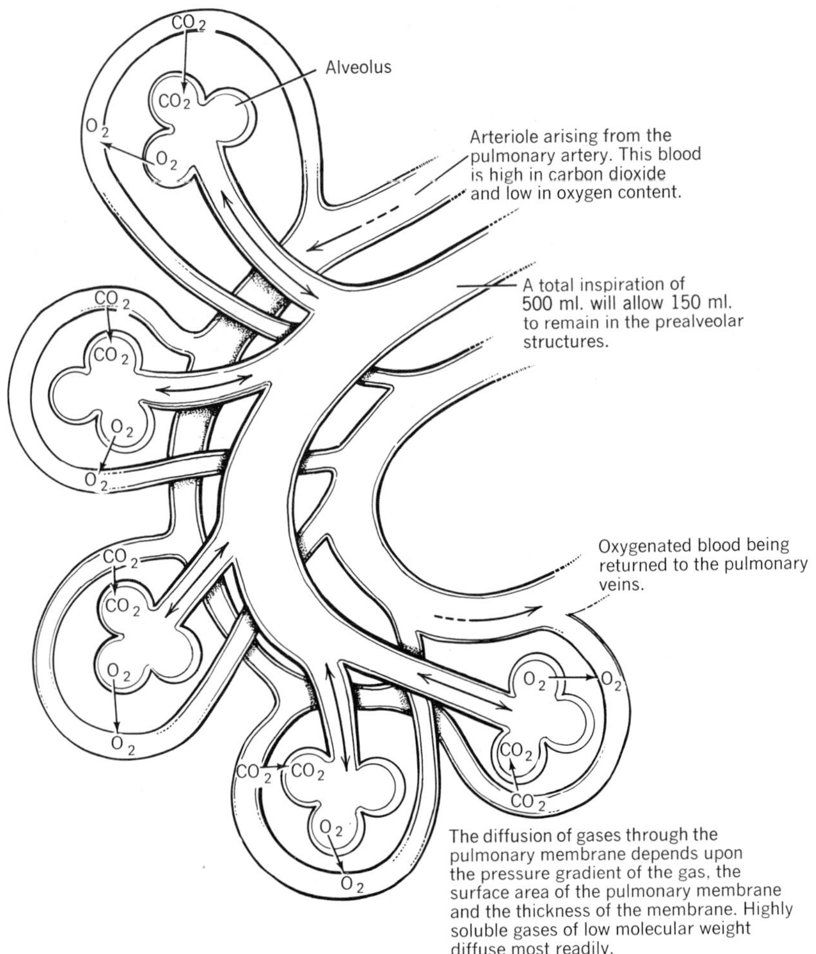

FIGURE 10-4

Normal exchange of oxygen and carbon dioxide between the alveoli and the pulmonary capillaries. [From S. R. Burke: *The Composition and Function of Body Fluids,* 1st ed. (St. Louis, Mo.: The C. V. Mosby Co., 1972.)]

passageways through which the sputum must pass to be expectorated. While the patient is in the postural drainage position, he should be encouraged to cough and to breathe deeply in order to help raise the secretions. Percussion of the chest also may help dislodge the tenacious sputum so that it can more easily be expelled. Mouth care is important following treatment because the sputum leaves an unpleasant taste and odor in the mouth.

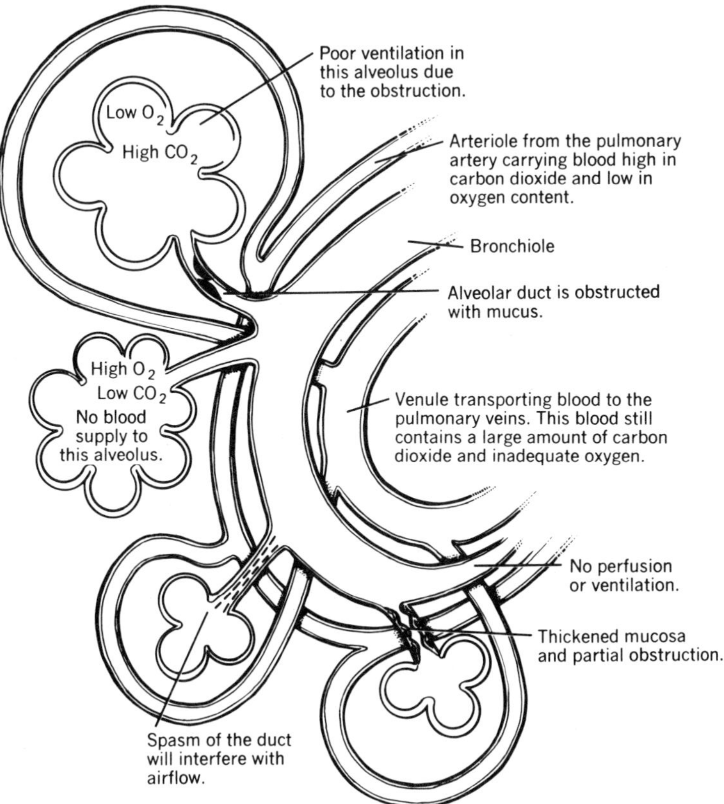

FIGURE 10-5

A diagrammatic presentation of some pathologic changes that will interfere with the exchange of gases between the alveoli and pulmonary capillaries in chronic obstructive pulmonary disease. [From S. R. Burke: *The Composition and Function of Body Fluids,* 1st ed. (St. Louis, Mo.: The C. V. Mosby Co., 1972.)]

Antibiotics are used to control infection. Patients must be careful to avoid exposure to possible respiratory infections. If a superimposed infection does occur, it must be treated as soon as possible.

Expectorants, such as potassium iodide, are helpful in clearing the respiratory passages of mucus. Other drugs that help to liquefy sputum come in the form of aerosol sprays. Warm, humidified air may also be useful.

Intermittent positive pressure with compressed air or oxygen may provide a more adequate aeration of the lungs. A variety of

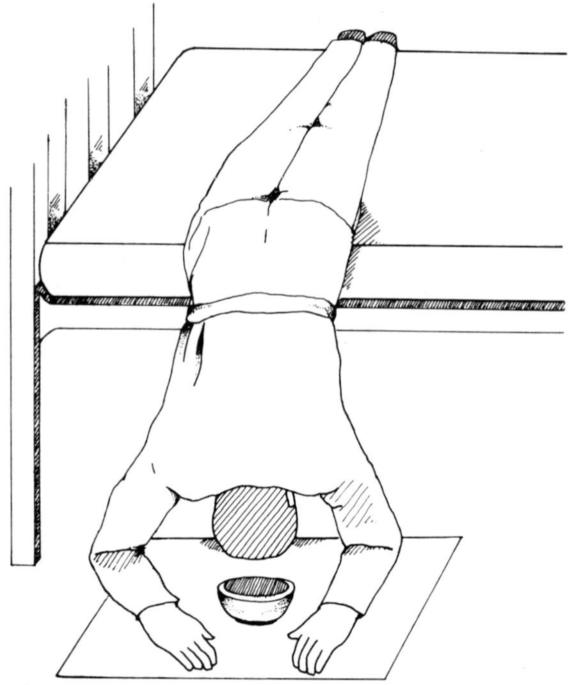

FIGURE 10-6
A common position used for postural drainage.

types of equipment are available for this kind of treatment, but the general principle is that positive pressure is used during inspiration. Usually, equipment is designed so that it operates according to the patient's own rate of respiration; however, with some of the older models the patient must adapt his breathing to the rate set by the apparatus (see Figure 10–7).

As with any chronic disease, patients with chronic obstructive pulmonary disease should adapt a regimen of good health and dietary habits. Obviously, smoking is forbidden or certainly discouraged. Frequently, for patients who have been heavy smokers, abandoning a long-standing habit, particularly under anxiety-producing circumstances such as discovering that he has a serious disease, may be a difficult adjustment.

BRONCHIAL ASTHMA

Asthma may be caused by emotional stress, infection, or allergy. The cause is often a combination of these factors. During an asthmatic attack there is spasm of the bronchioles, edema of the

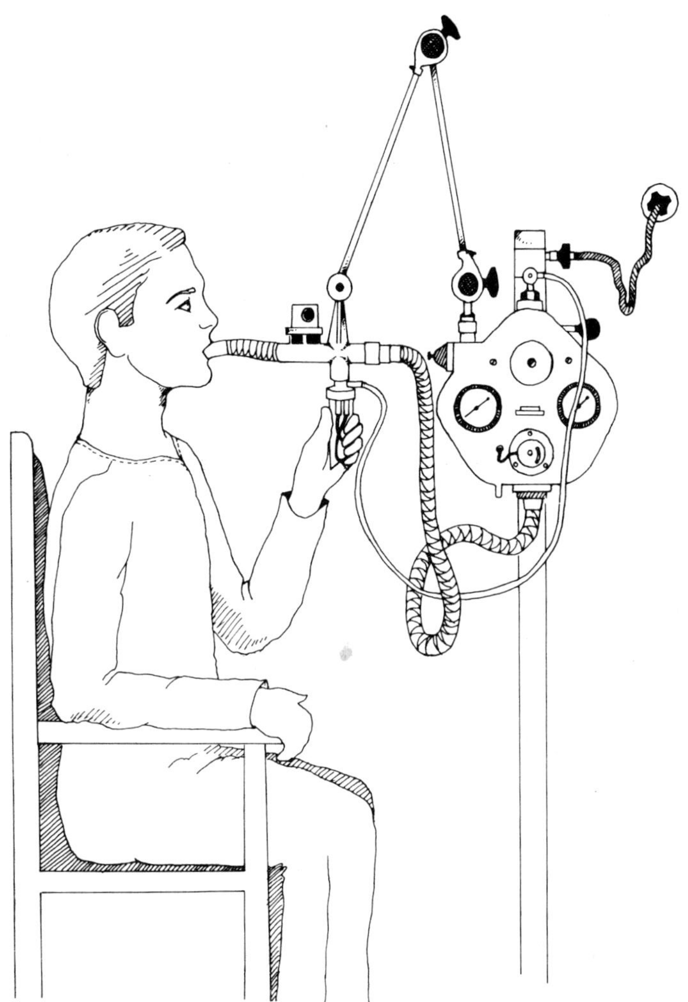

FIGURE 10-7
A patient receiving an intermittent positive pressure treatment.

mucosa of the respiratory tract, and thick tenacious mucus. Air is trapped in the alveoli and every breath is an effort. Breathing is easier if the patient is sitting up and uses accessory muscles of respiration. Wheezing is characteristic and is usually more pronounced on expiration. Perspiration is usually profuse and the patient's skin is pale. A severe attack may produce cyanosis.

Treatment during an asthmatic attack is symptomatic. Bronchial dilator drugs such as aminophylline, Isuprel, and epinephrine may be used. Epinephrine, and to a lesser extent ephedrine, although effective bronchial dialators, have unpleasant side effects such as tachycardia, tremors, and anxiety. Because the attack itself is anxiety producing, it may be well to give a sedative, such as phenobarbital, simultaneously with the bronchial dialator drugs in order to help counteract such unpleasant side effects.

Steroid preparations such as dexamethasone also help to produce bronchial dilation and are anti-inflammatory. An added advantage of the steroid preparations is that they do not cause anxiety. Unlike some other steroid preparations, the side effects of dexamethasone are minimal.

Expectorants such as ammonium chloride or potassium iodide help the patient to raise the thick secretions that are obstructing his respiratory tract. Encouraging fluid intake helps to replace the fluids lost.

It may be necessary to administer oxygen if the attack is particularly severe and cyanosis is present. Oxygen administered by mask may increase the patient's anxiety and fear of suffocation, in which case it may be more desirable to administer the oxygen by a nasal catheter or to place the patient in an oxygen tent.

Long-term measures used to treat the asthmatic patient are important in preventing future attacks and lessening the possibility of complications such as emphysema. If the asthma is associated with allergy, it may be possible to desensitize the patient to the allergen. If this is not possible, measures must be taken to avoid exposure to dust, pollens, and animal danders, the most common inhalants that cause asthma.

The asthma patient is particularly susceptible to respiratory infections. He needs to avoid factors that predispose him to respiratory infections; for example, fatigue and contact with persons who have colds.

Emotional stress may precipitate an asthmatic attack. Every effort should be made to assist a patient in adjusting to or avoiding anxiety-producing situations. If the patient's attacks seem closely related to emotional factors, psychotherapy may be recommended.

TUMORS OF THE RESPIRATORY SYSTEM

Cancer of the larynx generally has a better prognosis than does cancer of other organs. It usually does not metastasize as early as cancers in some other parts of the body and also produces fairly early symptoms. As with any disease process, the earlier it is recognized and treated, the better the prognosis. Early symptoms are persistent hoarseness and difficulty in swallowing. Although radiotherapy may be used, the common treatment is surgery. If the tumor is large, it may be necessary to do a laryngectomy. If the tumor is discovered while it is still relatively small, a surgeon sometimes can remove it without removing the entire larynx; this procedure, called a laryngofissure, does not result in the patient's losing his voice.

If laryngectomy is necessary, the patient will have a permanent tracheostomy and will not be able to speak normally. Many U.S. cities have Lost Chord Clubs, groups of people who have had this surgery and who help one another to cope with the disability. In addition to providing emotional support, club members are often able to teach the recent patient esophageal speech. The patient learns to swallow and to regurgitate air and gradually is able to produce speech. As he practices, his speech becomes less jerky and he makes himself understood.

Also in use are artificial battery-operated devices that the laryngectomy patient holds against his throat as he speaks. The instrument transmits vibrations to the throat, and the patient learns to use these vibrations to produce speech.

Incidence of carcinoma of the lung has increased markedly in recent years. It is more common in men than in women, and most patients are over 40 when the disease is discovered. Tumors usually produce no symptoms until the growth is fairly advanced. At this point, the patient may have a cough producing mucopurulent or blood-streaked sputum. The patient then begins to experience fatigue, anorexia, and weight loss. Dyspnea, chest pain, and hemoptysis are later symptoms of lung cancer.

As with other malignant growths, treatment usually involves surgery, radiation, and chemotherapy. Unless the disease is diagnosed early, prognosis is poor. Metastasis occurs to the mediastinal and cervical lymph nodes, the opposite lung, and the esophagus. Frequently, lung cancer metastasizes to the brain.

HYALINE MEMBRANE SYNDROME

Hyaline membrane syndrome takes the lives of several thousand newborn infants every year. It is rarely found in infants who have survived 5 days or more; however, hyaline membrane formation has been found in persons who have died of kerosene poisoning.

In the newborn, the symptoms are usually evident within a few hours after birth. Respirations are rapid and irregular. Later there is sternal retraction, cyanosis, and an expiratory grunt. At autopsy, a hyaline albuminoid membrane is found within the bronchioles and alveoli.

Aspiration of large amounts of amniotic fluid, particularly by infants who are delivered by Caesarean section, is a major cause of hyaline membrane disease. Recently, physicians have found that there is a decrease amount of lecithin (a fatty substance) in the amnionic fluid that surrounds the fetus who will be born with hyaline membrane disease.

Amniocentesis is a procedure in which a sample of the amnionic fluid is withdrawn from the amnionic sac that surrounds the fetus. This fluid can be analyzed for the lecithin content; if it is less than normal, measures can be taken to prevent the infant from developing this fatal disease.

CHEST INJURIES

Fractured ribs, although painful, are not usually serious unless there is injury to other structures as well; for example, if the sharp ends of the broken bones penetrate the lung and cause it to collapse. The usual treatment for uncomplicated fractures of ribs includes supporting the chest with an elastic bandage and administering analgesics for pain.

Penetrating wounds of the chest are very serious. An open wound of the chest will cause a pneumothorax (air in the pleural cavity which can cause the collapse of the lung). It must be treated promptly with an airtight dressing. When medical aid is obtained, the air in the pleural cavity will be aspirated by means of a thoracentesis. Later, a thoracotomy may be necessary to repair or to remove injured tissues.

SUMMARY QUESTIONS

1. Differentiate between a roentgenographic examination and a fluoroscopic examination of the chest.
2. For what purpose might sputum specimens be required?
3. How is a patient prepared for a bronchogram?
4. What is vital capacity and how is it measured?
5. Why is sinusitis a common complication of any upper respiratory tract infection?
6. What are the symptoms of acute tonsillitis?
7. What organisms commonly cause pneumonia?
8. Differentiate between bronchopneumonia and lobar pneumonia.
9. Discuss both the medical and surgical treatment of tuberculosis.
10. Discuss the pathological changes that occur in chronic obstructive lung disease.

11. What measures are used to treat a patient with chronic obstructive lung disease?
12. Discuss some possible causes of asthma.
13. Why does cancer of the larynx generally have a better prognosis than does lung cancer?
14. How are penetrating wounds of the chest treated?

SUGGESTED READINGS

Burke, S. R. *The Composition and Function of Body Fluids.* 1st ed. St. Louis, Mo.: The C. V. Mosby Co., 1972.

Carr, D. "Care of the Patient with Inoperable Lung Cancer," *Clinical Note on Respiratory Diseases,* Summer 1973.

Cohen, E. P. "New Asthma Treatment Enables Sufferers to Ward Off Attacks," *Today's Health,* September 1973.

Eagan, D. "Postural Drainage," *American Journal of Nursing,* April 1973.

Green, G. M. "Defenses of the Lung," *Clinical Notes on Respiratory Diseases,* Winter 1972.

Mantripp, A. J. "Care of Chest and Heart Patients," *Nutrition,* Summer 1970.

Moody, L. "Asthma: Physiology and Patient Care," *American Journal of Nursing,* July 1973.

Nett, L., and L. Petty. "Oxygen Toxicity," *American Journal of Nursing,* September 1973.

Traver, G. A. "Assessment of Thorax and Lungs," *American Journal of Nursing,* March 1973.

Wasserman, K. "Cardiovascular Manifestations of Respiratory Insufficiency," *Clinical Notes of Respiratory Diseases,* Fall 1973.

11 The Nervous System

VOCABULARY

Centriole
Cytoplasm
Dorsal
Foramen magnum
Golgi apparatus
Lipid
Lysosome
Midsaggital
Mitochondria
Ribosome
Vertebral canal
Viscera

OVERVIEW

I DIVISIONS OF THE NERVOUS SYSTEM
- **A.** Central nervous system
- **B.** Peripheral nervous system
- **C.** Autonomic nervous system

II NERVE TISSUE
- **A.** Properties of nerve tissue
- **B.** Cell processes
  - **1.** Dendrites
  - **2.** Axons
- **C.** Coverings of the cell processes
  - **1.** Myelin
  - **2.** Neurilemma

III CLASSIFICATIONS OF NEURONS
- **A.** Motor or efferent
- **B.** Sensory or afferent
- **C.** Connecting neurons

IV NERVE CONDUCTION
- **A.** Electrical transmission
- **B.** Chemical mediators

V REFLEXES

VI REACTION TIME

VII THE SPINAL CORD
- **A.** Location
- **B.** Protective coverings
  - **1.** Vertebral column
  - **2.** Meninges

VIII SPINAL NERVE ORIGINS AND DISTRIBUTION
- **A.** Nerve plexuses
  - **1.** Cervical
  - **2.** Brachial
  - **3.** Lumbar
  - **4.** Sacral
  - **5.** Pudendal
- **B.** Functions of spinal nerves

IX CORD BRAIN CONNECTIONS
- **A.** Ascending tracts
- **B.** Descending tracts

X THE BRAIN
- **A.** Brain stem
  - **1.** Medulla oblongata
  - **2.** Pons
  - **3.** Cerebellum
  - **4.** Midbrain
- **B.** The diencephalon
  - **1.** Hypothalamus
  - **2.** Thalamus
- **C.** Reticular formation
- **D.** Cerebrum
  - **1.** Cortex
  - **2.** White matter tracts
  - **3.** Basal ganglia
- **E.** Meninges

XI CEREBROSPINAL FLUID
- **A.** Formation
- **B.** Composition
- **C.** Pressure
- **D.** Circulation of cerebrospinal fluid

XII CRANIAL NERVES

XIII THE AUTOMATIC NERVOUS SYSTEM
- **A.** Sympathetic division
- **B.** Parasympathetic division

The nervous system aids in the control and coordination of the other systems of the body. It provides the tools by which we reason, learn, remember, and indulge in activities that are distinctly human. It assists us in making choices. The freedom to choose is a valuable freedom that carries with it tremendous responsibility. Frequently, the choices we make determine what future choices are available to us.

DIVISIONS OF THE NERVOUS SYSTEM

The central nervous system is made up of the brain and spinal cord. The brain is the largest and most complex mass of nerve tissue and is well protected by the bones of the skull. The spinal cord is also well protected by the bones of the vertebrae. This bony protection to the nerve tissue of the central nervous system is particularly important because nerve tissue in the central nervous system will not regenerate if it is injured.

The peripheral nervous system is composed of nerves that connect the central nervous system with the rest of the body. Parts of the peripheral nervous system include 12 pairs of cranial nerves; 31 pairs of spinal nerves; and autonomic nerves, which supply the viscera. The autonomic nerve fibers are located within the spinal nerves and some of the cranial nerves. Although the autonomics cannot be separated anatomically from the cranial and spinal nerves, they can be separated on the basis of function.

NERVE TISSUE

Nerve tissue, one of the primary tissues of the body, is composed of neurons, which are the nerve cells, and a variety of supporting cells called neuroglia. Neurons have special properties of irritability and conductivity. The cell body of the neuron has a spherical nucleus. Neurofibrils cross through the cytoplasm of the nerve cell body and extend into the processes that convey impulses to and from the cell body. Also in the cytoplasm are Nissl bodies, which are granular masses. There are many Nissl bodies in a resting neuron but few in the working neurons, a fact that suggests that these structures together with numerous free ribosomes are concerned with protein synthesis needed to replace proteins that are used by the neuron in various metabolic activities. Neurofibrils and Nissl bodies are unique to nerve cells. These cells have cytoplasmic structures that are common to other types of cells such as mitochondria, Golgi apparatus, and lysosomes. However, because there are no centrioles in the adult neurons, these cells do not multiply. The cell bodies of neurons are located only in the gray matter of the brain, cord, and ganglia. The cell body must be well protected because if it is destroyed the whole nerve dies.

Cell processes, called dendrites and axons, are extensions of the cell body. The dendrites carry impulses toward the cell body. Some dendrites are very long, whereas others are short. The axons convey impulses away from the cell body. Each neuron has only one axon; however, frequently it has one or more branches that come off the axon at right angles.

Some of the cell processes have myelin and neurilemma coverings. Myelin is a white, lipid-and-protein substance that surrounds many of the nerve fibers of the body. It is not a continuous sheath, as it is interrupted at intervals by constrictions called the nodes of Ranvier. At these nodes, where the nerve fiber is not insulated, there is an exchange of sodium and potassium into and out of the nerve fiber. This exchange of ions, according to the membrane theory of nerve transmission, assists in the transmission of the nerve impulse. The nerve fibers in the gray matter of the brain and cord do not have the myelin covering.

All peripheral nerves have neurilemma, a thin multinucleated covering essential to the repair of peripheral nerves. Nerve fibers in the central nervous system do not have neurilemma; thus, when these nerve cells are destroyed, they will not regenerate. Sometimes, however, other central nervous system cells take over the function of cells that have been destroyed. The neurilemma sheath probably also plays a protective role and aids in maintaining the integrity of normally functioning nerves. Figure 11–1 shows a typical neuron.

CLASSIFICATION OF NEURONS

Neurons are commonly classified according to their function. Afferent neurons are sensory and carry impulses from the periphery to

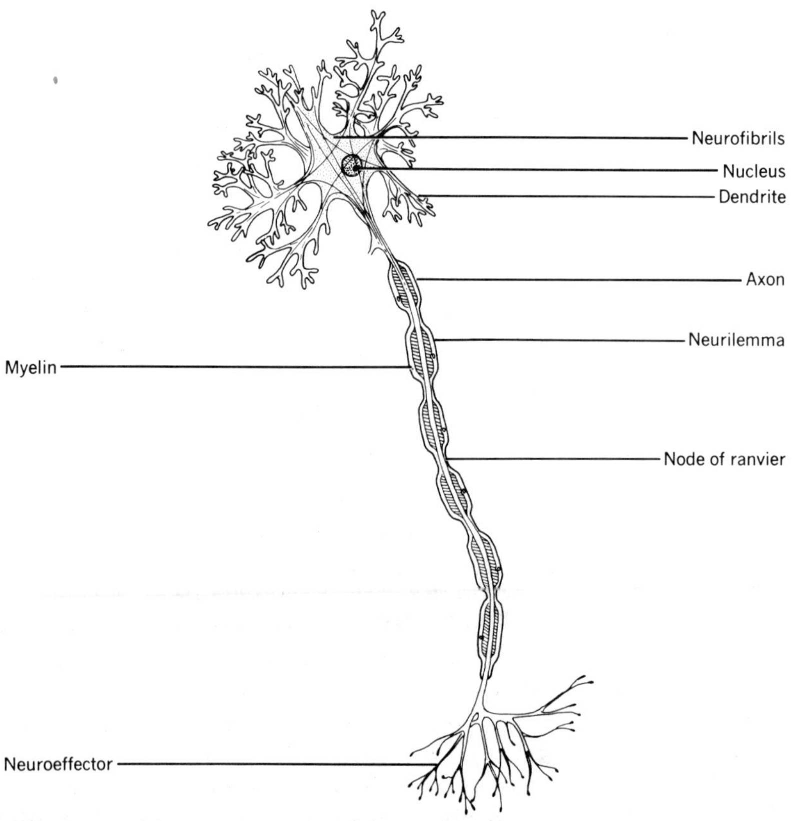

FIGURE 11-1
A typical neuron.

the central nervous system. Their dendrites have special receptor end organs called exteroceptors, interoceptors, and proprioceptors which convert stimuli into nerve impulses. The exteroceptors convey stimuli such as temperature, pain, or touch from the external environment. Interoceptors, which are found in the viscera, give rise to visceral sensation. Proprioceptors, which are located in muscles, joints, tendons, and the labyrinth of the inner ear, are concerned with muscle sense, position sense, and movement of the body in space. The interpretation of all sensations is done by the brain.

Efferent fibers are motor and secretory. They transmit impulses from the central nervous system to muscles and glands delivering orders for activity. Most nerves contain both motor and sensory fibers. The optic nerve from the eye is an example of a purely sensory nerve. There are no purely motor nerves because all must have afferent fibers for muscle sense.

Connecting (internuncial) neurons transmit impulses from one part of the brain or cord to another. These neurons do not leave the central nervous system.

PROPERTIES OF NERVE TISSUE

Properties of irritability or excitability enable the nerve tissue to respond to stimuli. This property is especially well developed in the receptor endings; however, a nerve fiber may be stimulated at any point along the course. Occasionally, receptors may lose their irritability temporarily because of prolonged stimulation. This property is called sensory adaptation. The olfactory nerves for smell become sensory adapted easily; for this reason we become unaware of unpleasant odors such as stale cigarette smoke. If we leave a smoky room, go into fresh air for awhile, and then return, we readily notice the unpleasant odor.

Conductivity is the ability of the nerve to transmit impulses. The velocity of the impulse is very rapid and self-propagating. The impulse slows down a bit at the synapse, for example, where a sensory neuron gives the impulse to a connecting neuron. Resistance at the synapses varies considerably. It is high when we are learning a new skill, but with practice the resistance is lowered and we are able to perform the skill more rapidly. According to the membrane theory of nerve conduction, the outside of the nerve fiber has a positive charge and the inside has a negative charge. The conducting nerve membrane becomes more permeable and allows sodium to enter the nerve and potassium ions to leak out. The result is a reversal of the electrical potential and a depolarization of the membrane. Immediately after repolarization takes place the membrane becomes relatively impermeable to sodium and the normal resting membrane potential is restored. The electrical transmission of the nerve impulse continues over each segment of the nerve until in the case of a sensory impulse it is interpreted by the brain, or in the case of a motor impulse muscle activity is accomplished.

During conduction of a nerve impulse, chemical and thermal changes take place. The chemical changes involve the use of glucose and oxygen by the nerve tissue. Heat is also produced during the conduction of a nerve impulse but not in the quantity that it is during muscle activity. You will get quite warm when playing a fast game of tennis, but you will not get overheated from nerve activity required for study!

Nerve, like muscle tissue, abides by the All or None Law. When stimulated, the nerve fiber will give its best response or none at all.

CHEMICAL MEDIATORS OF THE SOMATIC SYSTEM

Acetylcholine is liberated at the synapse and helps the nerve impulse over to the receiving dendrite. Cholinesterase immediately destroys the acetylcholine so that the impulse does not continue indefinitely. Because these chemical mediators are of clinical importance, in Chapter 12 we shall discuss in depth some drugs that act at the synapses by their action on these chemical mediators.

REFLEXES

Reflexes can be classified in several ways. Clinically they are frequently classified according to the part affected. A deep tendon reflex, such as a knee jerk when the patellar tendon is tapped, is an example of a reflex classified according to the structure involved. Other examples are the corneal reflexes, which cause us to blink when the cornea is touched, and the pupil reflex, which causes the pupil of the eye to constrict in bright light and to dilate in the dark.

Reflexes may also be classified according to the level of the nerve structures involved. During a first-level reflex a sensory impulse goes only as high as the spinal cord and a motor impulse is immediately sent from the cord to cause a muscle response. First-level reflexes are protective reflexes; the response is very rapid, easily predictable, and difficult to inhibit. An example is that of jerking your hand away from a hot stove. Quickly following the first-level reflex, second- and third-level reflexes will probably occur, but the initial protective action is the simple first-level reflex (see Figure 11–2).

Second-level reflexes travel as high as the brain stem. These also are protective in nature. Examples of a second-level reflex are gasping (as you might when you have burned your hand), cough, and vomiting.

Third-level reflexes, called learned or conditioned reflexes, involve the cerebral cortex. Control of bladder and bowel are third-level reflexes. The number of these reflexes is almost unlimited and examples vary considerably from one individual to another. Simple job skills that a specific worker has performed many times become reflex acts. Although others of us can accomplish these tasks, we must give the matter more conscious attention and we shall not be able to do the particular tasks as rapidly as the worker does.

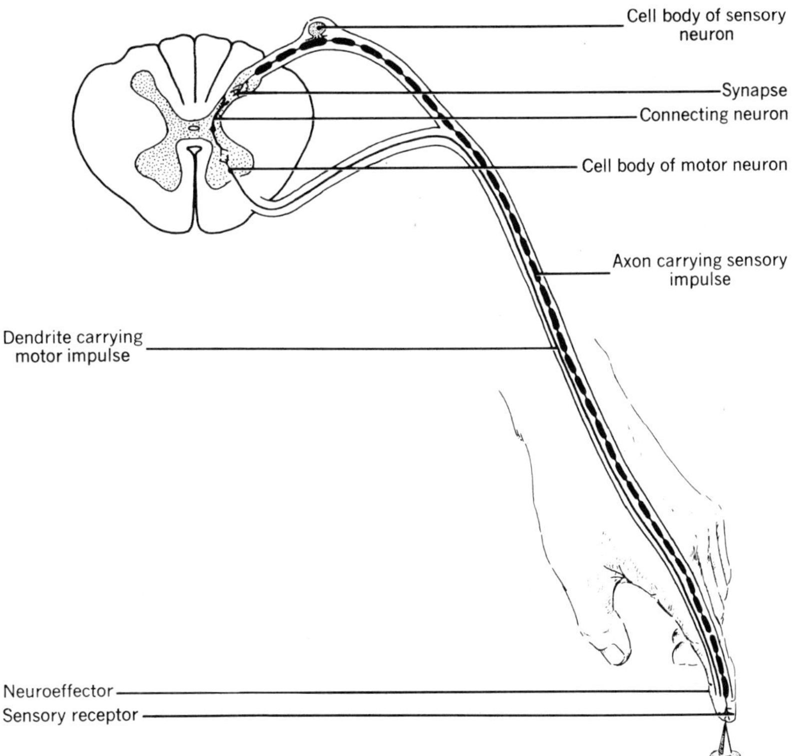

FIGURE 11-2
The pathway of a first-level reflex.

REACTION TIME

Reaction time is the time from the application of the stimulus to the start of the response. Reaction time is influenced by several factors, for example, the nature and strength of the stimulus, how the situation is perceived, and the number and condition of the synapses the impulses must cross as well as how many muscles must act. Removing your hand from the hot stove involves few synapses and relatively few muscles. On the other hand, if you are driving a car and are alerted to a traffic hazard, in order to avert an accident you may have to apply the brakes, alter the direction of the car, and honk the horn to warn other motorists of the danger. Age, fatigue, and drugs such as alcohol can also influence reaction time. A motorist who has consumed too much alcohol, which depresses the cerebral cortex, may not be able to react fast enough to avert an accident.

SPINAL CORD

The spinal cord, located in the vertebral canal, extends from the foramen magnum to the level of the disc between the first and second lumbar vertebrae. It is thicker in the cervical and lumbar regions, where the nerves supplying the extremities arise. Thirty-one pairs of spinal nerves arise from the cord. These nerves leave the bony canal by way of the intervertebral foramen. At the lower end of the cord the nerves droop and form the cauda equina (horse's tail) (see Figure 11–3).

The cord is protected not only by the bony vertebral column but also by meninges. The outermost meningeal covering is the dura mater, which is tough fibrous membrane. This covering extends down to the level of the second sacral vertebra. The arachnoid beneath the dura is a delicate spiderweb-like covering that also extends to the level of the second sacral vertebra. The pia mater closely covers the cord and extends only to the distal end of the cord. The pia mater is very vascular.

The spaces between the meninges are of clinical significance. Extradural space or epidural space lies above the dura. Subdural space lies between the dura and arachnoid. Beneath the arachnoid is the subarachnoid space, which contains the cerebrospinal fluid. Analysis of this fluid, samples of which can be obtained by a lumbar puncture, is a valuable diagnostic aid in many cases. We shall discuss this procedure in detail in Chapter 12.

If we examine the spinal cord in cross section, we note that there are two fissures or indentations. The deepest of these fissures

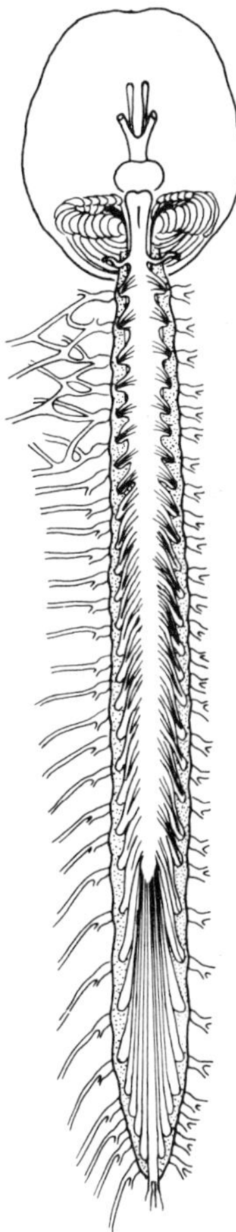

FIGURE 11-3
The spinal cord and spinal nerves.

is the anterior fissure. The gray matter of the cord contains unmyelinated nerve fibers, connecting neurons, the cell bodies of motor nerve fibers. White matter surrounds the gray matter, and the nerve fibers here are myelinated (see Figure 11–4).

SPINAL NERVE ORIGINS AND DISTRIBUTION

The dorsal or sensory root ganglion contains the cell bodies of afferent nerve fibers. The anterior root contains the motor fibers. The cell bodies of these fibers lie in the gray matter of the cord. These two roots unite to form a spinal nerve just as they leave the inter-

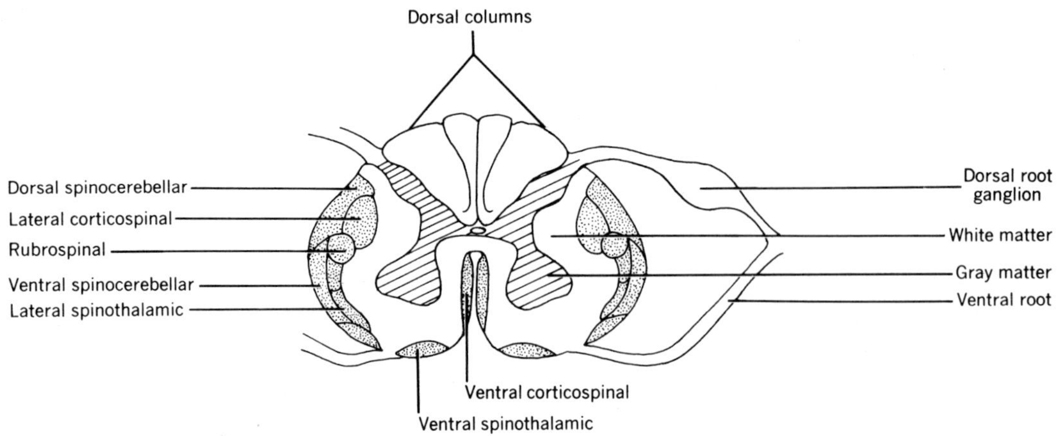

FIGURE 11-4
A section of the spinal cord showing the ascending and decending cord tracts.

vertebral foramen. Thus, the spinal nerve is mixed (both motor and sensory) and it is myelinated.

Each spinal nerve has three main branches after it leaves the canal. The dorsal branches go to the skin and muscles of the back; the visceral branches go to the internal organs; and the ventral branches, which are the largest, go to the front of the body and to the extremities.

Nerve plexuses are formed by the interlacing of the ventral branches of certain spinal nerves. The cervical plexus is formed by the first four cervical nerves. The most important nerve arising from this plexus is the phrenic nerve, which is motor to the diaphragm. Specifically, the phrenic nerve has nerve fibers of the third, fourth, and fifth cervical nerves. A transection of the cord above the level of the third cervical vertebra will obviously result in a cessation of respirations and death.

The brachial plexus involves cervical nerves five through eight and the first thoracic nerve. This plexus supplies the entire upper extremity. The radial nerve is the largest from the brachial plexus. It supplies the triceps and the posterior forearm. Injury to this nerve may result in wrist drop. The ulnar nerve supplies the anterior, medial forearm, and the fourth and fifth fingers. The median goes to the anterolateral forearm and to the rest of the fingers and thumb. Because the thumb is very important to the function of the hand, damage to this nerve can result in considerable disability. The musculocutaneous supplies the biceps and is both motor and sensory to the skin of the lateral forearm. Improper use of crutches can cause injury to the brachial plexus. For a detailed discussion of crutch palsy see page 80.

There is no plexus in the thoracic region. Twelve intercostal nerves arise from this level of the cord and supply the intercostal muscles, the skin in this region, and some of the abdominal muscles.

The lumbar plexus, which includes the first four lumbar nerves, supplies the lower abdominal muscles and the groin. The femoral nerve from this plexus supplies the quadriceps, and the obturator nerve supplies the adductor muscles of the thigh.

The sacral plexus arises from the fourth and fifth lumbar nerves and the first sacral nerve as well as some of the fibers from the second and third sacral. Nerves from this plexus supply the gluteal muscles, thigh, leg, and foot. The sciatic nerve supplies the hamstrings, leg, and foot. The peroneal nerve, which arises from the lateral aspect of the sciatic nerve, supplies the lateral and anterior muscles of the leg. Injury to this nerve may result in foot drop.

The pudendal plexus contains fibers from sacral nerves two through four. It is a small plexus; however, it is of clinical significance because sometimes a pudendal nerve block is done for local anesthesia during child birth.

The function of all of the spinal nerves is to carry impulses to and from the periphery and the spinal cord. We shall consider how these nerve impulses ascend to the brain for higher action and from the brain to control various motor activities.

CONNECTIONS OF THE CORD WITH THE BRAIN

Ascending tracts in the spinal cord carry sensory impulses to the brain. These tracts, which are located in the white matter of the cord, are sometimes named according to their origin and destination. The lateral spinothalamatic (from the spinal cord to the thalamus of the brain) carries impulses that are concerned with pain and temperature. The ventral spinothalamatic tracts transmit impulses concerned with pressure. The spinocerebellar tracts, which carry impulses to the cerebellum, play an important part in reflex adjustments of muscle tone and posture. These adjustments are not on the conscious level because the impulses do not reach the cerebral cortex and therefore you are unaware of them. The dorsal columns carry impulses that do reach higher centers in the brain. These impulses which originate in proprioceptors in skeletal muscle, tendons, and joints, involve conscious muscle sense.

Descending tracts in the cord carry motor impulses from the brain. The pyramidal or corticospinal tracts carry impulses for fine voluntary movement. The cell bodies of these nerve fibers are located in the motor areas of the frontal lobes of the brain; therefore, the muscle actions resulting from these nerve impulses are under the conscious control. The lateral corticospinal fibers cross in the medulla of the brain so that, for example, injury to motor areas in the right side of the brain will result in paralysis of the structures on the left side of the body that are controlled by the injured motor area. The smaller ventral corticospinal tracts do not cross in the medulla; some of these fibers cross at each segment of the spinal cord.

The extrapyramidal or rubrospinal descending cord tracts arise from lower levels of the brain and therefore are concerned primarily with unconscious voluntary movements. Muscle coordination and reflex control of equilibrium are examples of activities resulting from nerve impulses traveling down the extrapyramidal tracts.

Figure 11–4 shows the location of the ascending and decending cord tracts. Cord functions are by segments: In the event of cord transection, motor activities and sensation below the level of the transection will be absent; however, there will be no impairment of function or sensation above the level of the transection.

BRAIN OR ENCEPHALON

The brain is the largest and most complex mass of nerve tissue in the body. The brain is well protected by the bones of the skull and by the meninges. It has a rich blood supply, a fact that is discussed in detail in Chapter 7.

The brain attains full physical growth in about 18 to 20 years. Although physical growth may cease, an unlimited number of neural pathways can be developed well into advanced age. Whether these pathways will be developed, regardless of the person's age, depends upon many factors, the most important of which probably is the desire to learn. How learning takes place is a very complex subject, but there is no doubt that we learn by asking questions. Language was invented to ask questions. Answers can be given by grunts or gestures, but questions must be spoken. Humankind came of age when man asked the first question. Social and intellectual stagnation result not from the lack of answers but from the absence of the impulse to ask questions.

BRAIN STEM

The medulla oblongata, or "bulb" of the brain, is an expanded continuation of the spinal cord. This vital structure is only about 1 inch long. It contains centers for the regulation of respirations, heartbeat, and vasomotor activity. Second-level reflexes such as coughing, vomiting, and swallowing are mediated from the medulla. Many of the cord–brain pathways cross from one side to the other in the medulla.

The pons is a bridge that contains conduction pathways between the medulla and higher brain centers. It also connects the two halves of the cerebellum. Motor and sensory nuclei of cranial nerves five to eight are located in the pons.

The cerebellum is located in the lower back part of the cranial cavity. Its functions, although not on the conscious level, are essential for coordinating muscle activity for smooth, steady movements. Reflex centers for the regulation of muscle tone, equilibrium, and posture are located in the cerebellum. The cortex of the cerebellum is gray matter. Beneath the cortex are white matter tracts that resemble branches of a tree as they extend to all parts of the cortex. The cerebellar nuclei are masses of gray matter deep within the cerebellum. In the midsagittal section the arrangement of white and gray matter presents a treelike appearance that is called the arbor vitae, or tree of life.

The midbrain connects the hindbrain with the forebrain. It is a short, narrow segment that provides conduction pathways to and from higher and lower centers. Reflex centers in the midbrain include the righting, postural, and auditory visual reflexes. The righting reflexes are concerned with keeping your head right side up. Postural reflexes of the midbrain are concerned with the position of your head in relation to the trunk. The visual and auditory reflexes cause you to turn your head toward the direction in which a sound originates.

THE DIENCEPHALON

The structures of the diencephalon are located above the midbrain and are covered by the cerebral hemispheres. Although their functions are not on the conscious level, they are essential to life.

The hypothalamus is located in the floor of the third ventricle and part of the walls of this ventricle (see Figure 11–5). Some of the important functions of the hypothalamus include the manufacture of the hormones that are released from the posterior pituitary and causes the anterior pituitary to release its hormones. The

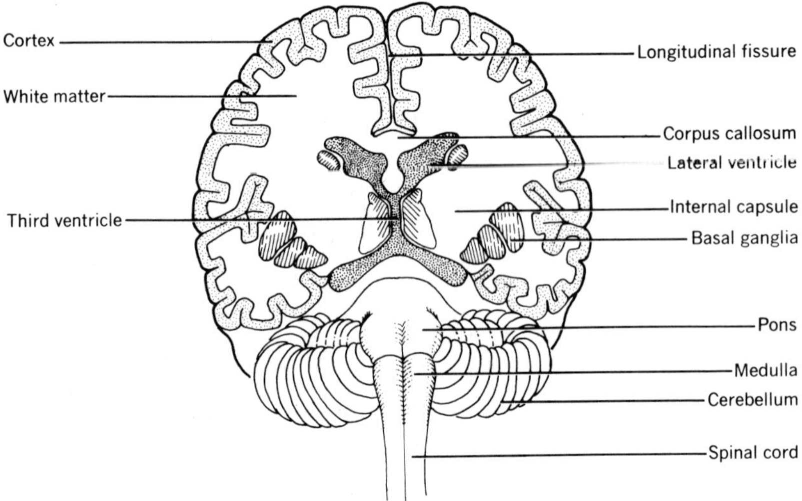

FIGURE 11-5
A coronal section of the brain.

hypothalamus contains centers for the regulation of visceral activity, water balance, temperature control, and sugar and fat metabolism. Our normal sleeping-waking mechanisms are also controlled by the hypothalamus. There are many connections between the hypothalamus and the frontal lobe of the brain, which is involved with such complex thought processes as judgment, reason, and creative thinking. It is not surprising, therefore, that our emotional state can affect visceral activity and influence our normal sleep patterns.

The thalamus is located in the walls of the third ventricle. All afferent (sensory) impulses go first to the thalamus, where they are sorted and grouped; they are then sent on to the proper area of the cerebral cortex, where they are interpreted. The Law of Specific Nerve Energies holds that the place where a stimulus ends in the thalamus determines what sensation will be experienced. If the nerve impulse arrives in the heat area of the thalamus, the sensation will be that of heat even though the stimulus may have been something quite different.

THE RETICULAR FORMATION

The reticular formation is composed of nerve fibers that spread through the upper portion of the spinal cord, medulla, pons, midbrain, and thalamus. This structure is concerned with inhibitory and facilitatory influences. When a stimulus needing no response reaches the reticular formation, it is prevented from going on to the cortex. This stoppage is called inhibitation. Even a weak stimulus that is important will be magnified or facilitated. When one portion of the reticular formation is stimulated, a generalized increase in cerebral activity and an increase in muscle tone throughout the body take place. When you are asleep, the reticular activating system is in an almost totally dormant state. Sensory impulses, such as pain or the sound of the alarm clock, can activate the system and you will wake up. This impulse is called the arousal reaction. Signals from the cerebral cortex can also stimulate the reticular activating system and thereby increase its activity. Damage to the brain stem that involves parts of the reticular formation may result in prolonged coma.

THE CEREBRUM

The cerebrum, the largest part of the brain, is separated into right and left hemispheres. Each hemisphere is subdivided into four lobes, which are named according to the covering cranial bones: frontal, parietal, temporal, and occipital lobes.

The cerebral cortex is a relatively thin layer of gray matter that is arranged in folds called convolutions. Between the convolutions are indentations, or fissures. Figure 11–5 shows how these convolutions and fissures increase the total surface area of gray matter.

A few of the fissures are important anatomical landmarks with which you should be familiar. The longitudinal fissure is located between the two hemispheres. This fissure extends downward to the corpus callosum, which is a white matter tract connecting the two hemispheres of the cerebrum. The transverse fissure lies between the cerebrum and cerebellum; the central fissures, or fissures of Rolando, lie between the frontal and parietal lobes; and the lateral fissures, or fissures of Sylvius, lie between the temporal and frontoparietal lobes.

The cerebral cortex contains billions of neurons. The function of the cortex is to retain, to modify, and to reuse information. These qualities are the bases of associative memory and the foundation of knowledge. Abstract thinking, reasoning, judgment, and moral sense use all parts of the cortex; however, scientists believe the prefrontal cortex to be particularly concerned with these complex intellectual functions.

We shall now consider certain regions of the cortex that are primarily concerned with the expressive or with the receptive phases of cortical activity. The general motor area, or expressive area, of the cortex is located just anterior to the central fissure. Specific areas of this general motor area are concerned with movement of specific parts of the body. For example, the motor speech area, located at the base of the general motor area, is concerned with the ability to form words both in speaking and in writing. This area is in the left hemisphere of right-handed people, and in the right hemisphere of left-handed people.

The general sensory area (receptive phase) is posterior to the central fissure. In this area, true discrimination of sensations is experienced; for example, you know if your hand is touching a soft, velvety surface, or that you have stepped on a tack. Other sensory areas such as those for sight and hearing are located elsewhere in the cortex. The visual area lies in the occipital lobe; the auditory area lies in the superior part of the temporal lobe; and the olfactory and gustatory areas lie in the medial aspect of the temporal lobe.

Beneath the cortex are white matter tracts. These tracts are myelinated fibers traveling in three principle directions. Commissural fibers connect the right and left hemispheres of the cerebrum. Projection fibers are ascending and descending fibers. An important example of projection tracts is the internal capsule, which is located lateral to the thalamus (see Figure 11–5). Association fibers transmit impulses from front to back on the same side of the brain.

The basal ganglia are four masses of gray matter lying deep within the white matter of each hemisphere. They are concerned with associative movements such as swinging your arms as you walk and unconscious facial expressions. Although you can obviously, consciously alter your facial expression and determine the degree to which you will move your arms as you walk, most of the time your acts are not on the conscious level but are controlled primarily by the basal ganglia rather than the cerebral cortex.

MENINGES

The meninges are the coverings of the brain beneath the cranial bones. The outermost covering is the dura mater. It has two layers. In some places these layers are in contact with each other and in other places they are separated. The internal dura covers the brain and sends extensions into some of the fissures. The falx cerebri, a fold of dura in the longitudinal fissure, forms the superior sagittal sinus in the space above the lower layer of dura and the inferior sagittal sinus in the space below. The tentorium cerebelli which lies in the transverse fissure, forms the straight sinuses in its folds. These sinuses contain venous blood that ultimately will flow into the internal jugular veins. For a detailed discussion of the cranial venous sinuses see Chapter 7.

The arachnoid is a delicate covering beneath the dura. The arachnoid villi are tiny projections of the arachnoid into the venous sinuses. It is through the villi that the cerebrospinal fluid is returned to the blood stream.

The innermost covering of the brain is the pia mater. The pia, which is very vascular, dips down into the fissures of the cortex.

All of these meningeal coverings of the brain are continuous with the meningeal coverings of the cord. The spaces between the coverings are called the subdural space and the subarachnoid space. Cerebrospinal fluid is found in the subarachnoid space. Because this fluid is of considerable clinical importance, we shall discuss it in detail here.

CEREBROSPINAL FLUID

The function of the cerebrospinal fluid is to provide a protective fluid cushion for the brain and cord and to aid in the exchange of nutrients and wastes between the central nervous system and bloodstream.

The blood vessels of the choroid plexus in the ventricles produce cerebrospinal fluid by the process of filtration. The healthy adult has about 100 to 150 ml. of cerebrospinal fluid. This fluid, which is normally clear, contains about 40 to 60 mg./100 ml. of glucose; however, the glucose content varies with the blood sugar level. In addition to glucose, the fluid also contains traces of protein and other nitrogenous substances as well as electrolytes such as sodium, potassium, chloride, calcium, and magnesium.

The cerebrospinal fluid is under pressure. This pressure is measured with a water manometer (see Figure 11–6). Two pressure readings are usually recorded: The initial pressure and then the pressure after a small amount of fluid has been removed for chemical analysis or microscopic study. The closing pressure will be lower than the initial pressure.

Cerebrospinal fluid is constantly being formed and circulates until it is returned to the bloodstream. The pathway for the circulation of the cerebrospinal fluid is as follows: lateral ventricles, foramen of Monroe, third ventricle, aqueduct of Sylvius, fourth ventricle, and then the subarachnoid space via the foramens of Magendie and Luschka. Ultimately, it is returned by the arachnoid

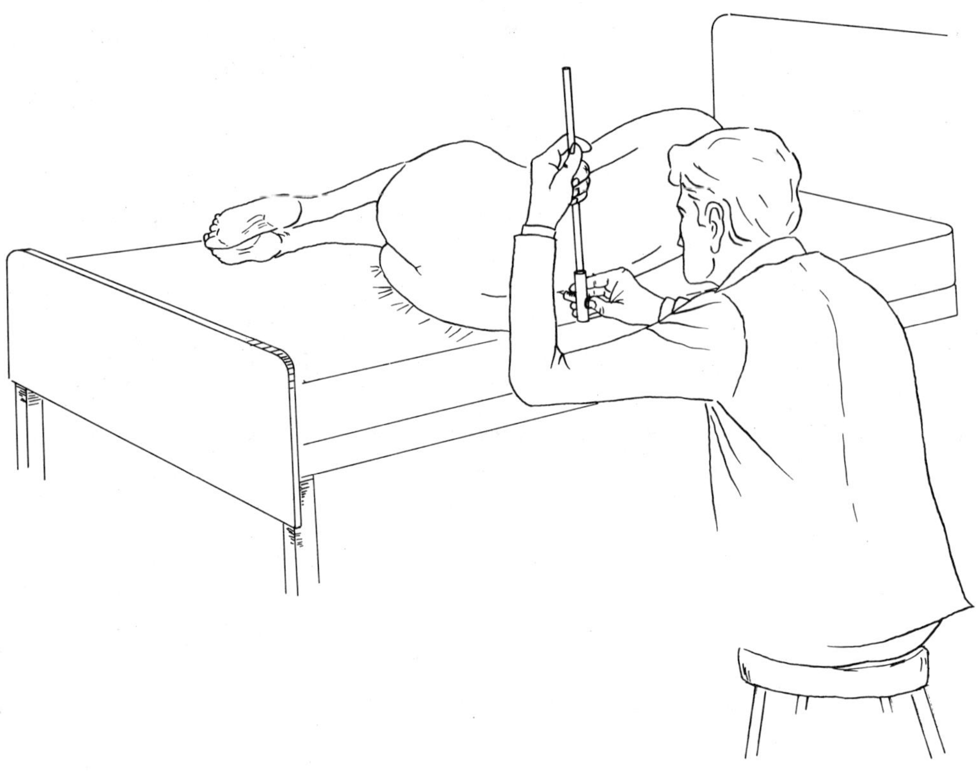

FIGURE 11-6
A patient on whom a lumbar puncture is being performed. By this procedure cerebrospinal fluid is obtained for examination and the pressure of the cerebrospinal fluid can be measured.

villi to the cranial venous sinuses. Some of these structures are shown in Figure 11–7.

CRANIAL NERVES

Twelve pairs of cranial nerves emerge from the undersurface of the brain. These nerves are numbered from front to back. The first, second, and eighth nerves are purely sensory; the others are both motor and sensory. Table 11–1 lists the cranial nerves and the principle functions of each.

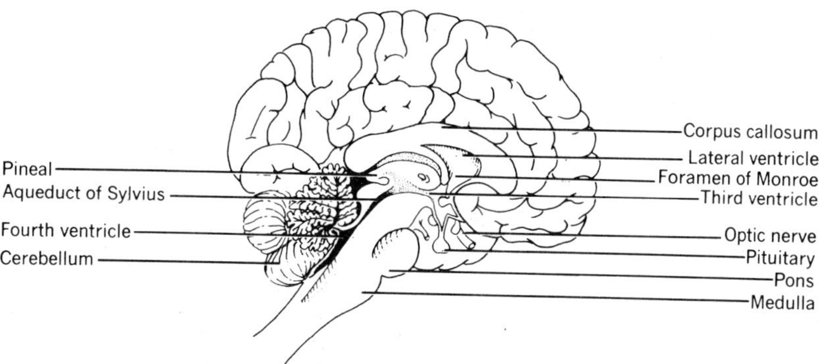

FIGURE 11-7
A midsagittal section of the brain.

TABLE 11-1

Cranial Nerves

Nerve	Function
I Olfactory	Smell
II Optic	Vision
III Oculomotor	Eye movements; regulates pupil size
IV Trochlear	Eye movements
V Trigeminal	Mastication; sensory to face and head
VI Abducens	Abduction of eye
VII Facial	Facial expression; taste; salivary secretion
VIII Acoustic	
A. Auditory branch	Hearing
B. Vestibular branch	Equilibrium
IX Glossopharyngeal	Swallowing; secretion of saliva; taste
X Vagus	Parasympathetic fibers to major viscera
XI Accessory	Motor and sensory fibers to shoulder and head
XII Hypoglossal	Tongue movements

THE AUTONOMIC NERVOUS SYSTEM

The autonomic nervous system is composed of visceral efferent nerve fibers that transmit impulses to smooth muscle, cardiac muscle, and glands. These nerve fibers lie within the cranial and spinal nerves; however, their impulses are not under conscious control.

The hypothalamus integrates these impulses and makes the

necessary routine adjustments in controlling such vital activities as digestion, and regulation of the blood pressure and heartbeat.

There are two major divisions of the autonomic system: the sympathetic and parasympathetic divisions. In general, these two divisions are antagonistic. For example, stimulation of the parasympathetic division increases the flow of digestive juices and the motility of the gastrointestinal tract, whereas stimulation of the sympathetic division inhibits peristalsis and the secretion of digestive juices.

The sympathetic division is concerned with preparing the individual to meet emergency situations. Stimulation of this division results in the "fight or flight" reactions that occur when you are angry or frightened. This division has a much more extensive distribution than does the parasympathetic division. When it is stimulated, the responses happen very rapidly: Specifically, the pupils of the eyes dilate; the bronchioles dilate; and respirations, heart rate, and blood pressure increase. The vessels in the skeletal muscles dilate, whereas those in the skin and viscera constrict. Inhibition of the digestive processes takes place.

Sympathetic nerve fibers originate in the thoracic and lumbar segments of the cord. For this reason, the division is sometimes called the thoracolumbar division. In addition to the action of acetylcholine and cholinesterase at the sympathetic synapses, sympathin is liberated at the neuroeffector junction (the point at which the sympathetic nerve reaches the organ). Because sympathin is similar to adrenalin, the sympathetic nerve fibers are sometimes also called adrenergic fibers. Sympathin is inactivated by monamine oxidase.

Because its fibers are contained within sacral spinal nerves and some of the cranial nerves, the parasympathetic division is also called the craniosacral division. The chemical mediators of the division are acetylcholine and cholinesterase; therefore, these fibers are sometimes referred to as cholinergic fibers.

The general function of the parasympathetic division is to conserve energy and to reverse the actions of the sympathetic division. Because the distribution of the parasympathetic fibers is not so great as that of the sympathetic division, parasympathetic actions do not occur as rapidly as do sympathetic actions.

SUMMARY QUESTIONS

1. Differentiate between the central nervous system and the peripheral nervous system.
2. Where are the cell bodies of neurons located?

3. What are the names of the cell processes of neurons?
4. Which nerve fibers have neurilemma and which do not?
5. Differentiate between an afferent neuron and an efferent neuron.
6. Where are connecting neurons located?
7. Give an example of sensory adaptation.
8. With respect to nerve tissue, what is the All or None Law?
9. Give examples of first-, second-, and third-level reflexes.
10. List several factors that influence reaction time.
11. What structures protect the spinal cord?
12. What type of nerve fibers are contained in the dorsal root, and in the ventral root?
13. What is the most important nerve arising from the cervical plexus?
14. List four nerves that arise from the brachial plexus and tell what structures are supplied by each.
15. From what nerve plexus does the femoral nerve arise?
16. From what nerve plexus does the sciatic nerve arise and what structures does this nerve supply?
17. What type of nerve impulses is carried in the ascending cord tracts? Give specific examples of these impulses.
18. What is the nature of the impulses carried in the corticospinal cord tracts?
19. Where is the medulla of the brain located and what is its function?
20. What type of reflex centers are located in the cerebellum?
21. Where is the center for righting reflexes located?
22. List several functions of the hypothalamus.
23. Where is the thalamus located and what is its function?
24. What is the function of the reticular formation?
25. Where in the brain are the longitudinal, transverse, lateral, and central fissures located?
26. Where in the brain are the general motor and general sensory areas located?
27. Where is the area for motor speech located?
28. Where are the basal ganglia located and what is the function of the basal ganglia?
29. Trace the pathway of the cerebrospinal fluid from the place of its formation until it is returned to the bloodstream.
30. Which cranial nerves are purely sensory?
31. List physiological changes that occur when the sympathetic nervous system is stimulated.
32. What is the general function of the parasympathetic nervous system?

SUGGESTED READINGS

Brady, R. J. *A Programmed Approach to Anatomy and Physiology.* 2d ed. Washington, D. C.: Robert J. Brady Co., 1968.

Guyton, A. C. *Function of the Human Body.* 3rd ed. Philadelphia, Pa.: W. B. Saunders Co., 1969.

Jacob, W. W., and C. A. Francone. *Structure and Function in Man.* 2d ed. Philadelphia, Pa.: W. B. Saunders Co., 1970.

Netter, F. H. *The Nervous System, The Ciba Collection of Medical Illustrations,* Summit, N. J.: Ciba Pharmaceutical Products, Inc.

Schifferes, L. J. *Healthier Living.* 3rd ed. New York, N.Y.: John Wiley & Sons, Inc., 1970.

12 Diseases of the Nervous System

VOCABULARY

Aneurysm
Arteriosclerosis
Basal ganglia
Flaccid
Hematoma
Hemiplegia
Hypertension
Internal capsule
Myelin
Subarachnoid
Supraorbital
Thrombosis

OVERVIEW

I DIAGNOSTIC TESTS
- **A.** Neurological examination
- **B.** Lumbar puncture
- **C.** Cisternal puncture
- **D.** Pneumoencephalogram
- **E.** Cerebral angiogram
- **F.** Electroencephalogram

II ALTERATIONS IN LEVELS OF CONSCIOUSNESS
- **A.** Anxiety
- **B.** Apprehension
- **C.** Convulsions
- **D.** Delirium
- **E.** Coma

III SPECIFIC DISEASES OF THE NERVOUS SYSTEM
- **A.** Meningitis
- **B.** Encephalitis
- **C.** Poliomyelitis
- **D.** Parkinson's Disease
- **E.** Cerebral vascular accidents
- **F.** Multiple sclerosis
- **G.** Myasthenia gravis
- **H.** Myotonia
- **I.** Cord injuries

DIAGNOSTIC TESTS

NEUROLOGICAL EXAMINATION

During a neurological examination the doctor notes carefully any disturbances in motor and sensory functions. He tests the patient's reflexes, such as the knee jerk or the reaction of the pupils to light. He also notes the patient's coordination and balance.

To test vibratory sense, the doctor uses a tuning fork. He also needs pins, a bit of cotton, and test tubes of hot and cold water. The patient must identify whether he is being touched with a sharp object, a warm or cold object, or the cotton. In assisting the doctor, the objects should be handed to him in such a way as to assure that the patient does not see them.

Aside from the neurological physical examination, the patient should be required to sign a statement of consent for other neurological tests. The doctor should explain these tests to the patient to assure that he understands the nature of the procedure.

LUMBAR PUNCTURE

In order to obtain cerebrospinal fluid for chemical analysis and to measure the pressure of the cerebrospinal fluid, the medical assistant usually positions the patient as shown in Figure 11–6. Necessary equipment includes materials to cleanse the skin for the injection, a syringe with hypodermic needle, some local anesthetic such as novocain, a lumbar puncture needle, a three-way stopcock, a manometer for measuring the pressure, and test tubes (usually three) for the specimens. All equipment must be sterile and, the doctor will need sterile gloves and sterile drapes.

It is well that the patient be advised that, although a local anesthetic is being used, he is likely to experience some discomfort during the procedure, but the pain does not indicate nerve damage. Knowing this possibility helps the patient avoid sudden movements that could result in injury.

The doctor then cleans the patient's skin in the lumbar region with an antiseptic solution and injects a local anesthetic into the region of the third and fourth lumbar vertebrae. The lumbar needle is usually inserted between the third and fourth lumbar vertebrae into the subarachnoid space. The pressure of the cerebrospinal fluid is measured and specimens of the fluid are taken. The assistant numbers and labels these specimens. After all of the specimens are collected, a second pressure reading is made (see Table 12–1).

If the doctor suspects that there may be some blockage in the pathway of the cerebrospinal fluid, he will do a Queckenstedt test.

To do this his assistant must make pressure over the jugular veins. Normally, when this is done the cerebrospinal fluid pressure rises markedly.

Following the lumbar puncture the assistant should place a Bandaid over the puncture site. Some doctors require that their patients stay flat in bed for several hours after the procedure as they believe that this position minimizes the likelihood of headache. Other doctors make no restrictions on their patients' activities.

CISTERNAL PUNCTURE

A cisternal puncture is similar to the lumbar puncture except that the cerebrospinal fluid is withdrawn from the cisterna magna (just below the occipital bone). The same equipment and assistance are needed as are for the lumbar puncture.

PNEUMOENCEPHALOGRAM

A pneumoencephalogram enables a doctor by the use of X-rays to visualize lesions of the brain. A lumbar puncture is performed and cerebrospinal fluid is withdrawn. Air is injected into the subarachnoid space through the spinal needle. The air rises and fills the ventricles of the brain so that X-ray films will show the size, position, and shape of the ventricles.

As preparation for a pneumoencephalogram, a patient should be given a sedative the evening before the procedure and usually again just prior to the examination. He should have nothing by mouth for at least 6 hours prior to the examination.

Following the pneumoencephalogram the patient should be placed flat in bed and have complete bed rest for 2 or 3 days. Patients frequently have severe headaches, nausea, and vomiting. Because they may also be in shock and experience convulsions and respiratory distress, they must be observed very closely. These symptoms gradually subside over a period of about 2 days.

CEREBRAL ANGIOGRAM

In a cerebral angiogram a radiopaque substance is injected into the left and right carotid arteries and sometimes into the brachial arteries. X-ray films are taken of the blood vessels of the brain. The test might reveal abnormalities of cerebral blood vessels, such as aneurysms. It might also reveal the displacement of blood vessels by a tumor.

Following cerebral angiography, ice should be applied to the sites of the injection to lessen edema and to help prevent bleeding. If the brachial artery or arteries have been used, frequent radial pulses should be taken. If the radial pulse is not palpable, the doctor must be notified.

Although complications following cerebral angiography are not common, respiratory distress due to edema or a hematoma near the trachea may occur. Because of this possibility, a tracheostomy set should be available. The patient should also be observed carefully for muscle weakness, which would appear if trauma to the brain had occurred during the test.

ELECTROENCEPHALOGRAM

The electroencephalogram is a record of the electrical impulses of the brain. Varying numbers of electrodes are placed on the patient's scalp and the brain waves are recorded by a machine operated by a technician. Although the test may take as long as 2 hours, depending upon how extensive an examination is required, the procedure is not painful and there are no complications associated with the examination nor are any restrictions placed on the patient's activity following the test. However, the electrode paste that was used should be removed from the patient's hair as soon as possible following the procedure.

ALTERATIONS IN LEVELS OF CONSCIOUSNESS

ANXIETY

Frequently anger is no more than a cloak to hide anxiety. Anxiety is different from fear in that a person who is afraid can usually identify the cause of the fear, but he may not be able to identify clearly the cause of his anxiety. Because of this, the person frequently feels helpless and overwhelmed. Anxiety interferes with a person's ability to assess his situation and to make adjustments to it. The anxious patient actually may not hear instructions or may forget to follow them. You should not scold such a person and indeed such an action on your part may make matters worse. You should instead listen to whatever the anxious patient has to say and accept what he says. If you can, you should comply with his requests; if not, the mere fact that you have taken time to listen implies that you respect and want to help him.

You should not avoid the anxious patient. Even if he has not summoned you, you might stop in to see him from time to time.

Such attention gives him the feeling that you will be available if he badly needs you. You may not approve of his behavior (and indeed it may be very bad), but you must approve and accept the person himself in order to be of any help.

Mild tranquilizers may be prescribed. Sometimes professional psychological counseling can be helpful. Of course, counseling should be done only by a professionally trained person. Comments such as, "You have nothing to worry about, everything will be alright," do not help the anxious person. Although such statements are intended to reassure the anxious person, instead, they indicate to him that his situation is not understood and that he is quite alone in dealing with his situation.

APPREHENSION

Patients who are in pain, bleeding, or short of breath are usually apprehensive. This apprehension may be sufficiently severe to cause confusion. Clearly, the remedy of apprehension that is a result of these obvious problems is the correction of the cause.

One of the most common causes of apprehension and restlessness in a confused patient is a full urinary bladder. However, it is often not easy to identify the cause of the apprehension. Your quiet presence and efforts to minimize environmental stimuli are greatly appreciated by the patient. If you can not stay with him, at least you can return frequently. You should always make sure that the patient can summon help and should provide some means to keep him from falling out of bed.

CONVULSIONS

Convulsions are the result of extreme irritability of the nervous system. There are many different causes of this irritability such as brain damage, epilepsy, and central nervous system infections. Young children who have very high fevers often have convulsions.

Regardless of the cause of the convulsions, the most important aspect of the patient's care is to prevent injury during the convulsion. You can do little except be a careful observer. Your accurate observations may be of great value to the doctor in identifying the cause of the convulsion.

You should time the convulsion with a watch. An important observation to note is the fact that sometimes, although not always, the convulsion starts in a particular part of the body, such as an arm or leg. You should also report whether the convulsion is accompanied by loss of consciousness or by incontinence, or is followed by a deep, prolonged sleep.

You should subject a patient who is likely to have convulsions to as little stimuli as possible. For example, noise and bright lights are stimuli to be eliminated.

You must not attempt to restrain the movements of a patient who is convulsing. Because it is impossible to stop a convulsion in this way, both you and the patient are likely to be injured if you attempt to hold him down. Ideally, you should place some soft pad, such as a folded wash cloth, in the patient's mouth between his back teeth to prevent him from biting his tongue. If the convulsion is already in progress, it may be impossible to insert the mouth gag safely. You should never put your fingers in the patient's mouth, as you can be seriously bitten. Human bites are often more serious and more likely to become infected than the bites of other animals.

DELIRIUM

Delirium is a state of confusion caused by interference with metabolic processes in the brain. It is usually temporary and reversible when the cause has been removed. The delirious patient may be disoriented as to time, place, or person, or possibly to all three. He may have illusions or hallucinations. An illusion is an inaccurate interpretation of some stimuli in the environment. Hallucinations are inaccurate subjective sensory experiences that occur without stimulation from the environment.

The delirious patient will have defective memory and judgment and he will be very restless. Your chief concerns are to ensure safety and to keep sensory stimuli to a minimum. You should keep any instructions brief and simple. Remind him frequently where he is, what day it is, and who you are. If the patient is an elderly person do not call him "Pop" (he is not your father) or use his first name. Dignity is important to everyone. The patient's situation has deprived him of much of his dignity and you can help him by showing respect.

For his safety, the patient may have to be restrained if someone cannot be present with him most of the time. If it is necessary to use physical restraint, such as tying him in bed, confusion will no doubt increase.

COMA

There are many causes of coma. Coma is evidenced by varying degrees of loss of consciousness. Patients are completely dependent upon their helpers. You may assist the doctor in estimating the depth of the coma. For example, the coma is not as deep if corneal reflexes are present as it is if they are absent. A patient in a very deep coma will not respond to the pain of supraorbital (digital pressure applied over the eye) pressure.

You should never say anything in the presence of an unconscious patient (whether in coma or under general anesthesia) that would be inappropriate to say to a conscious person. The sense of hearing is the last to leave and the first to return. It is not uncommon for a patient who has recovered from what appeared to be a very deep and prolonged coma to report accurately much of what had been said in his presence during the coma.

SPECIFIC DISEASES OF THE NERVOUS SYSTEM

MENINGITIS

Meningitis is an inflammation of the coverings of the brain and cord, the meninges. It can be caused by a variety of bacteria, viruses, and fungi. The most common organisms are the pneumococcus, streptococcus, staphylococcus, and meningococcus. Children are more prone to meningitis than are adults. The onset is usually sudden. The patient has a severe headache, high fever, stiffness and pain in the neck, nausea, vomiting, and visual disturbances. He is frequently comatose and may have convulsions. The treatment is the specific antibiotic to which the organism is sensitive, and supportive, symptomatic care directed toward maintaining the patient's resistance and normal defenses against disease.

ENCEPHALITIS

Encephalitis is an infectious disease of the central nervous system with pathological changes in both the gray and white matter of the brain and cord. It can be caused by the same organisms that cause meningitis, but also it may be caused by some chemicals such as lead, arsenic, or carbon monoxide. The care and symptoms are similar to those of meningitis; however, the patient may have a much higher fever. Fevers as high as 107° F. are not uncommon in encephalitis. The patient is also more likely to have respiratory difficulties than is the patient with meningitis. These patients usually recover but because of complications the prognosis is more guarded with encephalitis than with meningitis.

POLIOMYELITIS

Poliomyelitis or infantile paralysis is a viral disease that may cause flaccid paralysis of skeletal muscles. The virus attacks the anterior horns of the spinal cord. A very severe type involves the motor cortex and the brainstem.

The incidence of poliomyelitis has declined greatly since the Salk and Sabin vaccines have been available. Physiotherapy and supportive measures to prevent the complications of immobility arc important aspects of the treatment of poliomyelitis.

PARKINSON'S DISEASE

Parkinson's Disease is a slowly progressive disease of the central nervous system characterized by stiffness, fine tremors of resting muscles, a masklike expression, and a peculiar shuffling gait. The pathology is in the basal ganglia.

Patients may suffer their disease for 20 years or more. Because of their disability, they are susceptible to respiratory disease that may prove fatal.

Aldopa is a relatively new drug helping many patients. Several types of surgical procedures may also be helpful, particularly for younger patients. Surgery attempts to destroy a small area of the basal ganglia. When successful, these procedures give symptomatic relief and very likely there will not be a recurrence of symptoms even though the damaged areas in the central nervous system will not regenerate.

CEREBRAL VASCULAR ACCIDENTS

A cerebral vascular accident, or CVA, is caused by vascular damage in the brain—usually in the internal capsule—that is either a thrombosis of the vessels or a hemorrhage. The results of the damage are similar in both. The patient experiences weakness or paralysis on the side of the body opposite to the damage, depending upon the extent of the damage.

If a stroke is due to a cerebral hemorrhage, it usually occurs during a time of exertion or excitement. A cerebral thrombosis stroke usually occurs when a person is at rest or asleep.

In addition to the problem of having a hemiplegia (paralysis of one side of the body), patients frequently have speech problems. They may not be able to talk at all. This is called aphasia. If the motor speech area is severely damaged, they might not even be able to write. Sometimes they say something entirely different from what they mean and yet are aware of the fact that they are saying the wrong thing. Fortunately, through physiotherapy and speech therapy together with kind, respectful, supportive care, many patients can look forward to partial and sometimes complete recovery. Unfortunately, however, the very circumstances that led to a first stroke (hypertension or arteriosclerosis) may still be present. It is not uncommon for patients to suffer repeated strokes.

MULTIPLE SCLEROSIS

Multiple sclerosis is a disease that usually causes gradual paralysis and disturbances in speech, vision, and mentation (some deterioration of judgment and ability to concentrate). Emotional problems related to dependency and inability to move about are quite common in patients whose disease is advanced.

There are patchy areas of destruction of the myelin sheath throughout the central nervous system. Scar tissue forms in those places where the myelin has been destroyed. In the early stages of the disease, the patient has periods of remissions and exacerbations. In later stages, he may be severely handicapped.

No specific diagnostic tests exist for multiple sclerosis and often the diagnosis is very difficult to establish early in the disease process. Although much research concerning the cause and treatment of multiple sclerosis is in progress, at the present time treatment is only symptomatic and supportive. Patients should get plenty of rest, avoid situations that are emotionally upsetting, and have a nourishing diet.

MYASTHENIA GRAVIS

Myasthenia gravis is a relatively uncommon disease characterized by pronounced muscular weakness. The onset is gradual; however, the course of the disease varies greatly from one patient to another. Some patients live with the disease for many years, whereas in others the disease is rapidly fatal. Difficulty in swallowing or coughing may lead to aspiration of mucous secretions or food. If the muscles of respiration are involved, the prognosis is likely to be poor.

Drugs such as neostigmine give symptomatic relief. Relief is so dramatic that the drug can be used for diagnostic purposes. You should recall that acetylcholine is released at nerve endings to aid in the transmission of the nerve impulse to muscle tissue. Cholinesterase is then released to destroy the acetylcholine. Patients with myasthenia gravis have normal amounts of cholinesterase but a deficiency of acetylcholine; thus, nerve impulses to cause muscle contraction do not easily reach the muscles. Neostigmine inactivates cholinesterase and therefore causes the deficiency of acetylcholine to be balanced by a decrease in cholinesterase.

Recently, other drugs have been found to be helpful in the treatment of myasthenia gravis. Mestinone and mytelase help some patients to regain almost normal muscular activity.

MYOTONIA

There are several types of myotonia, most of which are hereditary. All are characterized by progressive muscle wasting and weakness. Some types also are complicated by endocrine disturbances and emotional problems. At the present time, no specific treatment for these conditions exists. Fortunately myotonia is a relatively rare disease.

CORD INJURIES

Complete transection of the spinal cord results in loss of sensation and paralysis of the skeletal muscles below the level of the transection. Following the cord transection there will be spasms of the paralyzed muscles. The patient and his family might interpret these spasms as indicating return of muscle function. Not only is it cruel to allow the patient to believe this, but it will also retard his acceptance of his disability and therefore delay successful rehabilitation.

Unfortunately, patients frequently have recurring urinary complications such as bladder and kidney infections. Treatment of these complications will be discussed in Chapter 18.

In spite of the severe disability resulting from cord injuries, many patients can be rehabilitated to the extent that they not only are able to accomplish the ordinary tasks of daily living but can be gainfully employed and enjoy a normal social life.

HYDROCEPHALUS

Hydrocephalus is usually evident in early infancy. Patients suffer an abnormal accumulation of cerebrospinal fluid, and frequently some obstruction in the circulation pathway of the fluid. The fluid remains in the ventricles of the brain pushing the soft brain tissue outward and causing cranial sutures and fontanels to bulge. The infant's head is greatly enlarged and very heavy.

The excess fluid can be surgically drained from a ventricle of the brain into a large vein such as the superior vena cava by means of inserting a shunt (a small plastic tube) between these structures. If the hydrocephalus is the result of an obstruction in the circulation pathway of the fluid, it may be possible to surgically remove the obstruction.

TABLE 12–1

Composition and Characteristics of Cerebrospinal Fluid in Health and in Disease

Normal Range	Conditions in Which Variations from Normal May Occur
Appearance: Clear and colorless	Hazy: with WBC count of 300–600 cells/mm.^{3} Turbid: with WBC count over 600 cells/mm.^{3} Red: fresh bleeding into the subarachnoid space, traumatic tap Xanthchromic (yellow- or amber-tinged): blood present for more than 4 hours, protein content greater than 100 mg./100 ml., bile pigment present
Volume: approximately 130 ml.	Increased: hydrocephalus, degenerative processes in which neural tissue is decreased Decreased: temporarily following lumbar or cisternal puncture
Pressure: 60–150 mm. H_2O	Increased: intracranial tumors, hemorrhage, hydrocephalus, meningitis, uremia Decreased: head injury, subdural hematoma, spinal tumors
Glucose: 50–80 mg./100 ml.	Increased: conditions that increase blood sugar Decreased: meningitis, poliomyelitis
Urea nitrogen: 6–23 mg./100 ml.	Increased: nephritis, uremia
Nonprotein nitrogen: 20–30 mg./100 ml.	Increased: nephritis, uremia
Uric acid: 0.6–0.7 mg./100 ml.	Increased: meningitis, nephritis, uremia
Protein: 20–40 mg./100 ml.	Increased: meningitis, central nervous system syphilis, cerebral hemorrhage, cerebral thrombosis, paralysis agitans, dementia praecox, encephalitis, poliomyelitis
Calcium: 2.3–2.8 mEq./liter	Increased: meningitis
Chloride: 120–130 mEq./liter	Increased: uremia Decreased: meningitis, central nervous system syphilis, brain tumors, encephalitis, poliomyelitis
Bicarbonate: 24–29 mEq./liter	Increased: compensating central nervous system acidosis
Sodium: 142–150 mEq./liter	Abnormal values not commonly observed
Potassium: 2.3–3.2 mEq./liter	Abnormal values not commonly observed
Magnesium: 2.5–3.0 mEq./liter	Abnormal values not commonly observed

From Shirley R. Burke, *The Composition and Function of Body Fluids* (St. Louis, Mo.: The C.V. Mosby Co., 1972).

SUMMARY QUESTIONS

1. What equipment is needed to assist the doctor with a neurological examination?
2. What equipment is needed to do a lumbar puncture?
3. How is a Queckenstedt test done?

4. List some common complications that may follow a pneumoencephalogram.
5. For what purposes might a cerebral angiogram be done?
6. Discuss observations needed and patient care during a convulsion.
7. Differentiate between meningitis and encephalitis.
8. List some of the symptoms of Parkinson's Disease.
9. What conditions might lead to a cerebral vascular accident?
10. What pathological changes take place in multiple sclerosis?

SUGGESTED READINGS

Astor, Gerald. "From Sitting On Top of the World to Sitting in a Wheelchair," *Today's Health*, March 1973.

Booth, B. J. "Physical Therapy Treatment for the Patient with Hemiplegia," *The Professional Medical Assistant*, March/April 1973.

Braney, M. L. "The Child with Hydrocephalus," *American Journal of Nursing*, May 1973.

Cassidy, F. A. "Adult Hydrocephalus," *American Journal of Nursing*, March 1973.

Crary, W. G., and Crary, G. C. "Depression," *American Journal of Nursing*, March 1973.

Field, W. E., and Ruelke, W. "Hallucinations and How to Deal with Them," *American Journal of Nursing*, April 1973.

Johnson, D. F. *Total Patient Care, Foundations and Practice*. 3rd ed. St. Louis, Mo.: The C. V. Mosby Co., 1972.

Leidig, R. M. "Narcolepsy," *American Journal of Nursing*, March 1973.

Renshaw, D. C. "Psychiatric First Aid in an Emergency," *American Journal of Nursing*, March 1972.

Stackhouse, J. "Myasthenia Gravis," *American Journal of Nursing*, September 1973.

13 Sense Organs

VOCABULARY

Dendrite
Eustachian tube
Meatus
Olfactory
Papillae

OVERVIEW

I TASTE
- **A.** Nerve receptors
- **B.** Nerve pathway
- **C.** Cortical localization

II SMELL
- **A.** Nerve receptors
- **B.** Nerve pathway
- **C.** Cortical localization

III HEARING AND EQUILIBRIUM
- **A.** External ear
- **B.** Auditory meatus
- **C.** Tympanic membrane
- **D.** Middle ear
- **E.** Inner ear
  - **1.** Bony labyrinth
  - **2.** Membranous labyrinth
    - **a.** Utricle and saccule
    - **b.** Semicircular canals
    - **c.** Cochlea
    - **d.** Organs of Corti

IV VISION
- **A.** Lacrimal apparatus
- **B.** Sclera and cornea
- **C.** Canal of Shlemm
- **D.** Iris
- **E.** Choroid
- **F.** Ciliary body
- **G.** Lens
- **H.** Retina and optic nerve
- **I.** Processes necessary for vision
  - **1.** Refraction
  - **2.** Regulation of pupil size
  - **3.** Convergence

V CUTANEOUS SENSATIONS

The special senses are sight, hearing, equilibrium, and smell. Cutaneous sensations such as touch, heat, cold, and pain and visceral sensations such as hunger, nausea, and thirst are considered general senses. For the interpretation of a sensation, there must be a functioning receptor, a nerve pathway, and brain cortex. Most conventional forms of anesthesia are based on interrupting the functions of any of these structures. For example, some local anesthesics block the pathway, while others merely inhibit the nerve receptor.

TASTE

For the sensation of taste, a substance must be in solution in order to contact the receptors, the taste buds located on the sides of the papillae of the tongue. A person who practices poor oral hygiene will probably not be able to taste his food because the papillae are covered over with a coating (sordes). The pathway is either the facial or the glossopharngeal nerves. The cortical centers for interpretation of the sensation of taste are located in the temporal lobes of the brain.

SMELL

The sensation of taste is closely related to that of smell because their cortical centers are close to each other. The stimulus for the sensation of smell must be a gaseous substance. The receptors are the olfactory cells in the upper part of the nasal mucosa, and the pathway is the olfactory nerve.

HEARING AND EQUILIBRIUM

There are many structures involved in the sensation of hearing. The external ear is relatively unimportant, although it does help to funnel the sound waves into the more internal structures concerned with hearing. The external auditory, or acoustic, meatus is lined with skin and is directed inward, forward, and downward. The tympanic membrane is located at the end of the canal. The function of this membrane is to vibrate in response to the sound waves and to transmit the vibrations into the middle ear.

In the middle ear there are three small bones (malleus, incus, and stapes) that move in response to the vibrations of the tympanic membrane. Also in the middle ear is the opening of the eustachian tube, which connects the middle ear with the nasopharynx. This tube helps equalize the pressure in the middle ear with that of the atmosphere. Because the mucous membrane of the throat is continuous with that of the middle ear, ear infections often follow respiratory infections. The mastoid sinuses drain into the middle ear. (You should recall that these sinuses are spaces in the temporal bones.)

From the middle ear the sound waves enter the inner ear through the oval window. The inner ear, called the labyrinth, is concerned with the sensations of both hearing and equilibrium. The bony labyrinth is hollowed out of the petrous portion of the temporal bone. It contains a small amount of fluid called perilymph. Parts of the bony labyrinth are the cochlea, vestibule, and three semicircular canals.

The membranous labyrinth lines the bony labyrinth and is about the same shape. Parts of the membranous labyrinth include two small sacs (the utricle and saccule) located in the vestibule, three semicircular canals, and the membranous cochlea. All these membranous structures are filled with a watery fluid called endolymph.

The vibration of the stapes against the oval window causes waves in the endolymph. These waves stimulate the organs of Corti, which are located on the basilar membrane within the membranous cochlea. The organs of Corti are the dendrites of the cochlear branch of the auditory nerve. The nerve impulse resulting from the stimulation of the organs of Corti then travels to the temporal lobe of the brain, which interprets the sound. You should note that as the stapes pushes the oval window inward, there must be a corresponding outward bulge, because fluid cannot be compressed. The outward bulge is accomplished by the round window that is located just below the oval window (see Figure 13–1).

The structures in the inner ear concerned with equilibrium are the semicircular canals that lie in the three different planes, the utricle, and possibly the saccule. Within the utricle are tiny hair cells that have otoliths (small stones) attached to their free surface. These hair cells bend backward when you begin to move forward. Their action gives you a sensation of falling off balance in a backward direction. As a result you will bend forward to correct the sensation. A runner leans forward as he begins the race also to correct the feeling of being off balance.

There are also hair cells in the semicircular canals. These hair cells are stimulated by the movement of the endolymph in the canal. The result, a nerve impulse that travels over the vestibular branch of the cochlear nerve, is interpreted as motion and assists us in maintaining our balance. Figure 13–1 shows some of the major structures concerned with hearing and equilibrium.

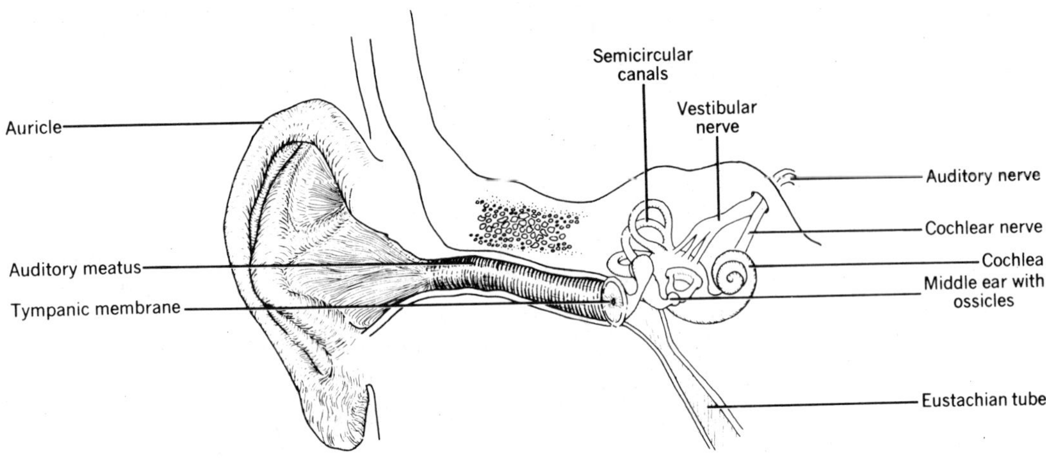

FIGURE 13-1
Some of the major structures of the outer, middle, and inner ear.

VISION

The receptors for vision are located in the retina of the eye, its pathway is over the optic nerves, and the center for cortical localization is located in the occipital lobe of the brain.

The eyeball occupies only the anterior 1/5 of the orbital cavity. Also found in the orbital cavity are the extrinsic muscles for movement of the eye, the lacrimal glands, and a large amount of adipose tissue. The lacrimal gland, which lies at the upper outer angle of the eye, secretes tears constantly to keep the cornea of the eye moist. These tears run down into the nose through the lacrimal duct. The lacrimal duct is located at the medial aspect (inner canthus) of the eye.

The sclera, which is dense fibrous membrane, is the outermost covering of the eyeball. The cornea is the transparent anterior part of this outer coat. The cornea contains no blood vessels and so must be nourished by the fluids that bathe its surface.

At the junction between the cornea and sclera is the canal of Schlemm. This canal drains the fluid (aqueous humor) that fills the space between the cornea and the lens.

Just posterior to the canal of Schlemm is the iris, which is the colored part of the eye. The iris functions to regulate the amount of light that enters the eye. In dim light the pupil dilates (increases in diameter) so that maximum light can enter, whereas in bright light the pupil constricts (decreases in diameter) and limits the amount of light entering the eye.

Beneath the sclera is a vascular coat, the choroid. Anteriorly, the choroid is thickened to form the ciliary body, much of which is made up of the ciliary muscle. This muscle is attached by ligaments to the lens. The action of the ciliary muscles changes the shape of the lens. The lens becomes thicker for near vision and thinner for distant vision. The diameter of the pupil constricts for near vision and dilates for distant vision.

The innermost layer of the eyeball is the retina, which contains nerve tissue. It is in the retina that the rods and cones are located. The cones function for daylight and color vision, and the rods for dim light vision. The retina is held firmly to the choroid by a clear jelly-like substance called the vitreous humor. Posteriorly, the retina is continuous with the optic nerve. The optic disc corresponds to the entrance of the optic nerve. Because there are no rods or cones in this area, it is called the blind spot. You should study Figure 13–2, which shows the major anatomic features of the eye.

All the processes necessary for vision must be in perfect coordination. Refraction is the bending of light rays. The speed of the light rays varies inversely with the density of the medium through which it passes. The refractive media are the anterior surface of the cornea, the aqueous humor, the lens, and vitreous humor. The normal eye has refractive media of such strength that an object 20 feet away will form a clear image on the retina.

Accommodation is the process by which the thickness, and therefore the strength, of the lens is changed for viewing objects close by or far away. If objects close at hand are to be viewed, the ciliary muscle contracts, pulling the choroid forward and lessening the tension on the suspensory ligaments of the lens. This con-

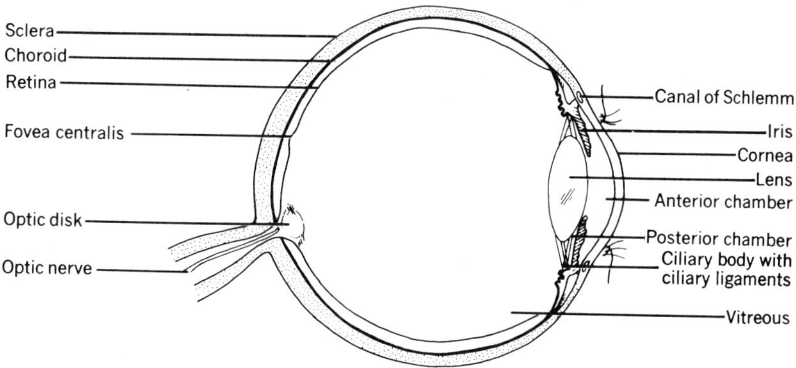

FIGURE 13-2
A transverse section of the eyeball.

traction causes the lens to bulge, or to become more convex. For distance vision, the ciliary muscle relaxes and the lens flattens.

Another process necessary for good vision is the regulation of the size of the pupil, which is the function of the muscles of the iris of the eye. To focus on near objects, the pupil must constrict. Constriction of the pupil also protects the retina from bright lights.

Convergence is the process by which you see with both eyes but you do not see double because the two images fall on corresponding parts of each retina. The nearer the object to be viewed, the more the eyes converge. Hold a pencil for one of your friends to view and slowly move the pencil toward his eyes and then some distance away. You will be able to observe the process of convergence.

CUTANEOUS SENSES

Touch, pressure, cold, heat, and pain are cutaneous senses. Receptors for these senses are widely distributed in the skin. The number of receptors for each of these sensations varies greatly; for this reason not all parts of the body are equally as sensitive. For example, the tip of your tongue and your finger tips are quite sensitive, but the back of your neck is much less sensitive.

The receptors for touch and pressure are located in nerve endings around the hair follicles and in the papillary layer of the skin. The pathway for the nerve impulse is found in the cranial nerves going directly to the brain or in spinal nerves leading to the spinal cord and then to the brain. Regardless of the pathway, if the sensation is to be interpreted, it must reach the general sensory area of the brain located just posterior to the central fissure.

Receptors for the sensation of cold are near the surface of the skin, whereas those for heat lie deep in the skin. The afferent pathways for these nerve impulses as well as their cortical localization are similar to those for touch and pressure.

Chapter 2 discussed visceral pain. Pain receptors in the skin are the most numerous of all the cutaneous receptors. These receptors are important because the sensation of pain warns you of danger so that you can protect yourself from extensive injury. The afferent pathways and cortical localization for cutaneous pain are similar to those for the other cutaneous sensations. You may find it helpful to review the discussion in Chapter 11 concerning the tracts within the spinal cord for a more thorough understanding of how these sensations travel through the cord to the brain.

SUMMARY QUESTIONS

1. Where is the lacrimal duct located and what is its function?
2. Where are the semicircular canals and with what are they concerned?
3. Where is the canal of Schlemm and what is its function?
4. List the refractive media of the eye.
5. What structures are located in the membranous labyrinth?
6. What is the function of the organs of Corti?
7. Where is the eustachian tube and what is its function?
8. Where are the receptors for the sensation of taste located?
9. What is the nerve pathway for the sensation of smell?
10. Into what structure do secretions from the mastoid sinuses drain?
11. List in sequence the structures through which a sound wave passes until the sound is finally interpreted by the cerebral cortex.
12. Discuss cutaneous sensations.
13. Where are the receptors for vision located?
14. What is the function of the iris?
15. Where is the vitreous humor and what is its function?
16. What processes are involved in near vision?

SUGGESTED READINGS

Mann, Ida, and Pirie, Antoinette. *The Science of Seeing.* New York, N. Y.: Penguin Books, 1950.

Southall, James P. C. *Introduction to Physiological Optics.* New York, N. Y.: Dover Publications, 1961.

Vail, Derrick. *The Truth About Your Eyes.* New York, N. Y.: Farrar, Straus, & Cudahy, 1959.

Van Dergeijk, W. A., *et al. Waves and the Ear.* New York, N. Y.: Doubleday & Co., Inc., 1960.

Von Duddenbrook, Wolfgang. *The Senses.* Ann Arbor, Mich.: University of Michigan Press, 1958.

14 Disorders of the Special Senses

VOCABULARY

Allergen
Analgesic
Antihistamine
Aqueous humor
Arteriosclerosis
Canal of Schlemm
Conjunctiva
Optic disc
Retina
Sebaceous
Vitreous humor
Zygomatic bone

OVERVIEW

I DIAGNOSTIC TESTS
- **A.** Hearing tests
- **B.** Eye examinations

II HEARING DEFECTS
- **A.** Conduction deafness
- **B.** Nerve deafness
- **C.** Swiss cheese deafness

III IMPAIRED VISION AND DISEASES OF THE EYES
- **A.** Hyperopia
- **B.** Myopia
- **C.** Astigmatism
- **D.** Presbyopia
- **E.** Cataracts
- **F.** Glaucoma
- **G.** Detached retina
- **H.** Conjunctivitis
- **I.** Hordeolum
- **J.** Chalazion

DIAGNOSTIC TESTS

HEARING TESTS

Simple tests such as whispering words to the patient in such a way that he is unable to see the examiner's lips and having the patient repeat the whispered words are helpful in screening large numbers of people for hearing defects. Tuning forks also help to determine whether a hearing defect is conductive (an obstruction in the sound wave in the outer or middle ear), or neural (a defect in the inner ear). The doctor or his assistant places the vibrating tuning fork against the patient's forehead; the patient tells in which ear the sound is louder. The sound will be louder in the weak ear if there is some degree of conduction deafness, and in the better ear if the impairment is the result of nerve damage.

One can perform more accurate hearing tests by using the audiometer, an instrument that produces pure tones of controlled loudness and pitch. The patient uses earphones and sits in a soundproof room. He signals when he hears a tone and when he no longer hears it.

EYE EXAMINATIONS

A physician's assistant can perform some eye examinations. Others are done by an ophthalmologist, a medical doctor specializing in the diagnosis and treatment of diseases of the eye. An optometrist is highly skilled and qualified to diagnose and to treat errors of refraction; he is not a medical doctor.

The Snellen chart consists of a series of letters or symbols of different sizes. People with unimpaired vision can read the largest letters at a distance of 200 feet. The smaller letters can be seen at 100, 50, and 20 feet. The patient usually sits 20 feet from the chart and must read with one eye the smallest line visible to him. If he can read the letters that people with normal vision can read at 20 feet, his vision is described as 20/20. If he can read at a distance of 20 feet letters no smaller than those of a person with normal vision can see at 30 feet, his vision is described as 20/30. Vision that is reduced to 20/200 is usually considered legally blind.

To determine whether the refracting media of the eye bend light rays to focus normally on the retina, the ophthalmologist utilizes various lenses. This procedure can determine for patients who have less than 20/20 vision which lenses will offer the most effective correction of the refractive error. Frequently, before the refraction is done, the physician or his assistant instills into the patient's eye eyedrops that temporarily paralyze the muscles of accommodation. To administer eye drops, with your index finger you hold the upper lid firmly against the frontal bone, and with your thumb you hold the lower lid firmly against the zygomatic. Then you instill the drops in the lower conjunctival sac and instruct the patient to keep his eyes closed for a few minutes. Following the examination, the patient should wear dark glasses until his accommodation returns to normal. You should never use these drops for patients who have glaucoma because they will dilate the pupils and thereby make pressure on the canal of Schlemm (see below under "Glaucoma").

The ophthalmologist can measure intraocular pressure with a tonometer (see Figure 14–1). Before the test, he instills into the eyes some local anesthetic such as proparacaine. This test is a valuable diagnostic aid in determining glaucoma, which is evidenced by increased intraocular pressure. Other than increased pressure, there may be no other signs of the disease. Because glaucoma is one of the most common causes of blindness, all people who are middle-aged or older should have their intraocular pressure measured annually.

With an ophthalmoscope the physician can examine the interior of the eye. He usually performs this examination in a dark room. If the optic disc is cupped (bulges outward), increased intraocular pressure is present. If the disc is chocked (appears

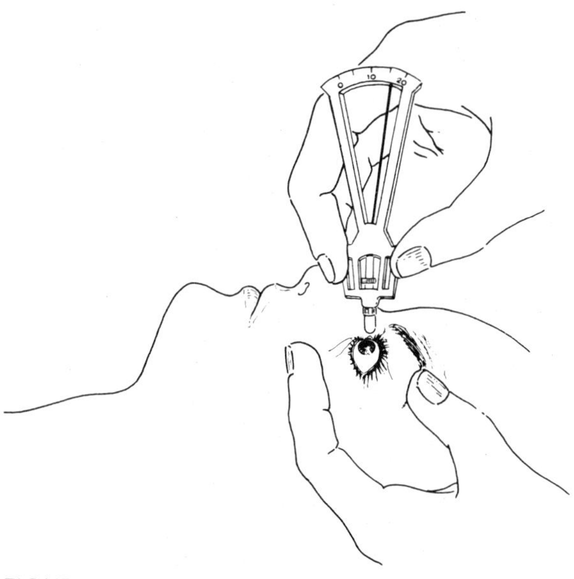

FIGURE 14-1
Measuring intraocular pressure with a Schiøtz-type tonometer.

squeezed), increased intracranial pressure may be present. The doctor can also examine the condition of the blood vessels in the eye, a check that can be helpful in the early diagnosis of such diseases as arteriosclerosis.

HEARING DEFECTS

CONDUCTION DEAFNESS

Conduction deafness is caused by injury or disease of the outer or middle ear. A ruptured ear drum, otitis media (infection of the middle ear), or otosclerosis can cause conduction deafness. Otosclerosis is a condition in which the ossicles of the middle ear do not move freely and therefore cannot transmit the sound waves into the oval window. Large amounts of wax in the ear canal can also impair hearing. This wax can be removed by the doctor. Attempts on the part of the patient to clean the wax out of his ear will most likely result in pushing the wax further into the ear and in possible damage to the tympanic membrane. Regardless of the cause of conduction deafness, it usually can be corrected by repairing the defect or by a hearing aid.

NERVE DEAFNESS

Nerve deafness involves pathology in the labrynth or in the auditory nerve. It can not be helped by a hearing aid. Patients can be taught to lip read. Some words are much easier to make out by the lip reader than others. If you are not understood by someone who must read lips, you should rephrase the statement.

SWISS CHEESE DEAFNESS

Swiss cheese deafness is fairly common in elderly people. In this type of deafness a person hears some words and messages but not others. He is able to hear low-pitched sounds much better than high pitched sounds. He may also have problems hearing certain consonants. This is not a psychological disorder. How cruel it is to imply that such an elderly person hears only what he wants to hear when he wants to hear it. You should never shout at such a patient as he is, or course, sensitive about his hearing problem. If you shout, you will very likely elevate the pitch of your voice and make your words more difficult to hear. You should repeat your message in a lower pitch. If this does not help, you might keep your pitch low and rephrase the message to eliminate the consonants that are causing problems.

IMPAIRED VISION AND DISEASES OF THE EYES

HYPEROPIA

Hyperopia is farsightedness. In this condition the eyeball is too short and thus the light rays focus at a theoretic point behind the retina. Lenses prescribed by an optometrist can correct this refractive error.

MYOPIA

Myopia (nearsightedness) usually results from an elongation of the eyeball. The light rays focus at a point in the vitreous humor in front of the retina. Proper lenses can also correct myopia.

ASTIGMATISM

Astigmatism results from an irregularity in the shape of the cornea, or, sometimes, of the lens. Vision is blurred and the patient will need lenses to correct the refractive error.

PRESBYOPIA

Presbyopia is caused by the gradual loss of elasticity of the lens and a weakening of the ciliary muscles as we grow older. The result is a decreased ability to accommodate for near vision. A person must hold his book far away in order to see clearly. He will need bifocals. The lower part of these glasses is used for near vision and the top for distant vision. People who have never worn glasses frequently want "reading glasses" instead of bifocals, as they believe that bifocals will be a difficult adjustment for them. However, reading glasses will blur their distant vision, and they will find that they are constantly taking their reading glasses off and putting them back on. Thus, the short time of adjustment to bifocals is well worth their effort.

CATARACTS

Cataracts result from an opacity of the lens. This condition can be treated surgically by removing the lens. Following surgery the patient may have to wear very thick lenses; however, some people are able to wear contact lenses. Frequently, people have difficulty judging distances at first and must be particularly cautious in walking up and down stairs. A word of caution about this may prevent falls that not uncommonly result in fractures, particularly in the elderly.

Phaco-emulsification is a relatively new procedure for the treatment of cataracts. In this procedure, which takes only a few minutes, the lens is removed by suction. Following the procedure it is not necessary for the patient to remain in bed and few limitations are imposed on his activities.

GLAUCOMA

Glaucoma results from a disturbance of the normal balance between the production and the drainage of aqueous humor. The increased intraocular pressure, if it is not relieved, may result in blindness. Although it can occur at any age, it is most common in people over 40. Early diagnosis and treatment are most important in preventing loss of vision.

In acute glaucoma the patient has attacks of severe pain in and around his eyes, sees halos, particularly around lights, and has blurred vision. These attacks usually occur suddenly. The patient should be given miotics (drugs that constrict the pupil) to relieve the pressure on the canal of Schlemm. Other drugs, such as carbonic anhydrase inhibitors, are given to slow the production of the aqueous humor. Patients need analgesics and complete rest. They usually must have surgery to prevent further attacks.

Chronic glaucoma occurs more frequently than the acute type. Patients often have no symptoms until their disease process is fairly well advanced; then the symptoms are similar to those of acute glaucoma. They need miotics and carbonic anhydrase inhibitors. Sometimes they may also require surgery.

All patients with glaucoma should avoid coffee, tea, and other caffeine products and must also limit their total fluid intake. It is important that they avoid lifting heavy objects that can increase intraoccular pressure and limit activities that cause eye strain and fatigue.

DETACHED RETINA

In a detached retina the nerve layer of the retina becomes separated from the pigmented layer and deprives the nerve layer of blood supply. Vision is thus lost in the affected area. The symptoms depend upon the size of the affected area, but generally the patient reports that he sees flashes of light or a sensation of spots or moving particles before his eyes. The detachment may be treated surgically, or, if it is a relatively small area, it may be corrected by photocoagulation. Photocoagulation is a procedure in which a beam of light directed toward the area of detachment causes the retina at that point to adhere to the choroid.

CONJUNCTIVITIS

Conjunctivitis (pink eye) is an inflammation of the conjunctiva due to allergy or to microorganisms. If the etiology is allergy, the patient may be treated with antihistamines or perhaps can be desensitized to the particular allergen. If it is due to bacteria, it may be treated with the appropriate antibiotics.

HORDEOLUM

A hordeolum is a stye, or an infection at the edge of the eyelid in a lash follicle. Hot, wet compresses help to localize the infection. Within a few days the lesion will drain and will require no further treatment. Occasionally, it is necessary to incise and to drain the lesion. If a person is subject to frequent recurrences of this type of infection, he should wash his face three or four times a day with an antibacterial soap. Because this preventative measure will very likely cause dry skin, he should know that the use of cold cream to combat the unpleasant dryness will also counteract the affect of his efforts to minimize the population of bacteria on his skin.

CHALAZION

Small sebaceous glands are located within the upper eyelids. An infection of these glands, called a chalazion, can be quite painful and will require surgical treatment.

SUMMARY QUESTIONS

1. Differentiate between hyperopia and myopia.
2. What causes presbyopia?
3. What treatment is used for patients with cataracts?
4. What is one diagnostic test used to detect glaucoma?
5. What causes astigmatism?
6. How is a hordeolum treated?
7. What are symptoms of a detached retina?

8. Explain how hearing tests are done.
9. What is the difference between conduction deafness and nerve deafness?
10. How can you best communicate with a patient who has Swiss Cheese deafness?

SUGGESTED READINGS

Raiford, L. M. *Contact Lens Management*. Boston, Mass.: Little, Brown, & Co., 1962.

Smith, J. F., and Nachayel, D. P. "Retinal Detachment," *American Journal of Nursing*, Sept. 1973.

Weinstack, F. J. "Tonometry Screening," *American Journal of Nursing*, April 1973.

15 The Gastrointestinal System

VOCABULARY

Deaminization
Disaccharide
Endocrine
Enzyme
Exocrine
Flora
Hormone
Hydrolysis
Mitochondria
Monosaccharide
Polysaccharide
Sordes

OVERVIEW

I GENERAL FUNCTIONS OF THE DIGESTIVE SYSTEM

II FOODSTUFFS

- **A.** Carbohydrates
- **B.** Fats
- **C.** Proteins
- **D.** Vitamins
- **E.** Minerals
- **F.** Water

III DIGESTIVE PROCESSES

- **A.** Mechanical
- **B.** Chemical

IV ALIMENTARY CANAL, ACCESSORY ORGANS AND THEIR FUNCTIONS

- **A.** Mouth, tongue, teeth, and salivary glands
- **B.** Pharynx
- **C.** Esophagus
- **D.** Stomach
- **E.** Duodenum
- **F.** Jejunum
- **G.** Ileum
- **H.** Cecum and appendix
- **I.** Colon
- **J.** Liver
- **K.** Gall bladder
- **L.** Pancreas

The general function of the digestive system is to break down foods. Foods are any substance taken into the body for the purpose of yielding energy, building tissues, or regulating body processes. These substances must be rendered into simple diffusible forms, which can be used by the body cells. About two thirds of all foods are composed of water, which is most vital for all cellular activity.

FOODSTUFFS

Carbohydrates are the most abundant and economical of the foods. They are economical not only from the point of view of the price we pay for them, but also from the point of view of the ease in which the body can utilize them and the energy that can be produced. Fats are a concentrated form of fuel that are not as easily utilized by the body but do provide an important reserve food supply under circumstances in which a person is unable to consume adequate amounts of food. Proteins are mostly used for building new tissues such as enzymes, hormones, and, in the case of children, growing tissues.

In most diets, most of the protein is probably converted to glucose and used for fuel rather than building new tissues. Minerals are vital to metabolic processes and many function as coenzymes or enzyme activators. Vitamins, which are essential for normal growth and development, also function as coenzymes. Vitamins will be discussed in more detail at the end of this chapter.

The digestive processes are both mechanical and chemical. In mechanical digestion, the food is ground to smaller and smaller particles, liquefied, and moved along at the proper speed. In chemical digestion, enzymes assist in breaking the foodstuffs down into their simplest diffusible forms. With enzyme action carbohydrates are rendered into glucose, galactose, and fructose, which are monosaccharides, the simplest form of carbohydrate. Proteins are broken down to amino acids, and fats to fatty acids and glycerol. Finally, these simple substances can be absorbed. Absorption depends upon the surface area, the length of time the foodstuffs are exposed to the surface area, and the rich capillary blood network of the mucosa of the small intestines.

The digestive system is composed of the alimentary canal, which is about 30 feet long, and accessory organs such as the teeth, salivary glands, pancreas, and liver. For a visual presentation of these structures, you can refer to the transparencies in Chapter 1. The canal has four layers: a mucous membrane lining; a submucous coat that is very vascular; a smooth muscle layer that has both circular and longitudinal muscle layers; and, in the case of the stomach, an additional oblique layer of smooth muscle fibers. The outer coat of the alimentary tract above the diaphragm is fibrous tissue, and below the diaphragm is serous membrane, the peritoneum.

You should be familiar with some special parts of the peritoneum. The greater omentum is attached to the greater curvature of the stomach and hangs like an apron over the intestines. Varying amounts of fat are deposited in the greater omentum. It provides protection and insulation to the structures that it covers. The lesser omentum extends from the lesser curvature of the stomach to the liver. The mesentery is a fan-shaped membrane that attaches the loops of small intestine together and to the posterior abdominal wall. There is also a peritoneal cavity, which actually is only a potential cavity, because normally the two layers of the peritoneum (the visceral and the parietal) are separated by only a small amount of lubricating serous fluid. The functions of the peritoneum are to prevent friction between the organs, to hold the organs in place, and, in the case of the greater omentum, to store fat and to help insulate the organs.

ALIMENTARY CANAL AND ACCESSORY ORGANS

MOUTH

The mouth is chiefly concerned with grinding the food into smaller pieces and lubricating it so it will have a smooth passage into the lower parts of the tract. The salivary glands of the mouth pour salivary juices into the mouth for lubrication of the food. These juices contain salivary amylase, which is capable of breaking down carbohydrates into simpler sugars. However, as a rule, we do not keep our food in our mouths long enough for this to happen.

The parotid gland is located anterior and inferior to the ear. Its duct opens into the upper jaw where the dentist puts the cotton roll to absorb the saliva from the parotid gland. There are also submaxillary, located near the inner surface of the mandible, and sublingual salivary glands, located beneath the tongue. All of these glands secrete saliva.

The tongue contains taste buds. These taste buds are located on the sides of the small papillae of the tongue. For this reason, a patient whose tongue is coated with sordes will be unable to taste his food and is quite likely to have anorexia. The tongue assists in chewing and swallowing as well as in speech.

We are born with buds for two sets of teeth. The milk or deciduous set has 20 teeth. These start to erupt by the age of about 6 months, and all 20 are usually in by the age of 2 years. The permanent set has 32 teeth, which begin to appear at about 5 to 7 years and are usually complete with the wisdom teeth at about 25 years of age.

Bacteria multiply rapidly in the moist environment of the mouth, particularly if particles of food are lodged between the teeth. Although brushing the teeth after eating is very important, a toothbrush scarcely reaches between the teeth. For this reason you should use dental floss to prevent decay of the teeth and, more important, to prevent gum infections. The importance of good dental care for children can hardly be overemphasized, not only because it can help to form lifelong habits of good dental hygiene, but also because premature loss of the deciduous teeth can lead to deformity and poor occlusion of the permanent teeth.

PHARYNX AND ESOPHAGUS

After leaving the mouth, the food goes to the pharynx, the esophagus, and then to the stomach. For all practical purposes, no chemical changes take place in the pharynx or esophagus. It takes about 4 to 8 seconds for the food to go from the mouth to the stomach.

STOMACH

The stomach, which is a hollow muscular structure, is found in the left upper quadrant of the abdomen. It contains gastric fluid and mucus at all times. The cardiac sphincter is located between the stomach and the esophagus, the central portion or major part of the stomach is called the body. The fundus of the stomach is the rounded, upper portion above the esophageal opening. At the distal portion is the pyloric valve, which remains closed until the stomach has completed its work on the food.

The mucous membrane lining of the stomach has many folds, or rugae, in it to allow for expansion of the stomach. This membrane secretes a great deal of mucus and some gastric juices to aid in digestion. The gastric juice contains hydrochloric acid, which provides a low pH unfavorable to the growth of bacteria and also functions to activate pepsinogen to become pepsin. Pepsin begins the chemical breakdown of proteins. The gastric juice also contains gastric lipase and rennin. The gastric lipase can start the digestion of emulsified fats. However, as a rule, we do not eat many emulsified fats. The rennin is probably only of importance in infants. It works on the protein in milk to make it more solid and to delay the passage of the milk from the stomach.

The stomach absorbs very little food because (with the exception of water, inorganic salts, and alcohol) the food is not yet in its diffusible form. In addition, the mucus coating the lining of the stomach is so thick that little food comes into contact with the stomach wall. The mucus is important in preventing the hydrochloric acid from eroding the stomach wall. Peptic ulcers may be due to an increased quantity of hydrochloric acid or to a lack of protective mucus in the stomach.

Normally an average meal will stay in the stomach about 3 to 4 hours. During this time it is churned and mixed with the gastric juices and becomes quite liquid. The acid liquid is called chyme. The churning is accomplished by the contractions of the smooth muscles of the stomach walls. The innermost layer of smooth muscle has oblique fibers, the middle layer is made of circular fibers, and the outer muscle layer has longitudinal fibers. The circular and longitudinal fibers are mostly for peristalsis, and the obliques do most of the churning.

SMALL INTESTINES AND ACCESSORY ORGANS

After the pyloric valve opens, the food leaves the stomach and enters the small intestines, a narrow tube about 23 feet long. It has three divisions: the duodenum, the jejunum, and the ileum. The mucosa of the small intestine is very vascular and is arranged in circular folds. On these folds are small protuberances, called villi, which give the lining a velvety appearance.

In each villus is a blood capillary loop for the absorption of monosaccharides and amino acids; and a lymph capillary, or lacteal, for the absorption of fatty acids and glycerol. These folds and villi give the small intestine a tremendous surface area, which is important in the absorption of foodstuffs. This inner surface is about 600 times as great as the serosal or outer surface.

Accessory organs of the small intestine include the liver, the gall bladder, and the pancreas. The liver, the largest and busiest gland of our bodies, is often referred to as the chemical capital of the body. It is covered with a tough fibrous covering (Glisson's capsule) for protection. The metabolic functions of the liver are so numerous and intricate that no attempt will be made to go into complete detail concerning liver functions. However, a few that are important for you to know are listed below. The liver helps to regulate the blood sugar level by converting fructose and galactose to glucose. Although fructose and galactose are simple sugars in order to be utilized they must be changed to glucose. It changes glucose to glycogen, which it stores and releases as glucose when the blood sugar level decreases. The liver also regulates amino acid concentrations. It deaminizes amino acids that are not needed for protein synthesis and makes glucose out of these proteins. This process is called gluconeogenesis. The liver forms bile, which it sends to the gall bladder for storage. It helps in the destruction of old red blood cells, saves the iron for reuse, and eliminates the remainder of the red cells in the bile. The liver also forms blood proteins such as albumin, globulin, fibrinogen, and prothrombin. A final and very important function of the liver is to detoxify harmful substances such as alcohol that may be ingested.

The liver has a dual blood supply. The hepatic artery is nutrient to the liver, and the portal vein brings blood rich in the products of digestion so that the liver can perform its functions on these foodstuffs. Both the portal vein and the hepatic artery lead to the liver sinusoids, where their blood comes into contact with the liver cells. From the sinusoids the blood goes into the central vein, then to the hepatic vein, and finally into the inferior vena cava.

In Figure 15–1 you can see how the hepatic artery blood and that from the portal vein mixes in the liver sinusoids, and also the relationship of the bile capillaries to the liver cells and the blood sinusoids. The bile ducts unite to form the hepatic bile duct, which carries the bile to the cystic duct and then to the gall bladder.

The gall bladder is a small pear-shaped sac that can hold about 50 ml. of bile. It stores and concentrates the bile. When food enters the duodenum, a hormone (cholecystokinin) is released from the duodenal mucosa and is sent to the gall bladder to cause it to evacuate the bile. The common bile duct receives the bile from the cystic duct and empties the bile into the duodenum through the Ampulla of Vater. Bile has a detergent action that will emulsify dietary fats so that they can be hydrolyzed by lipases. The results

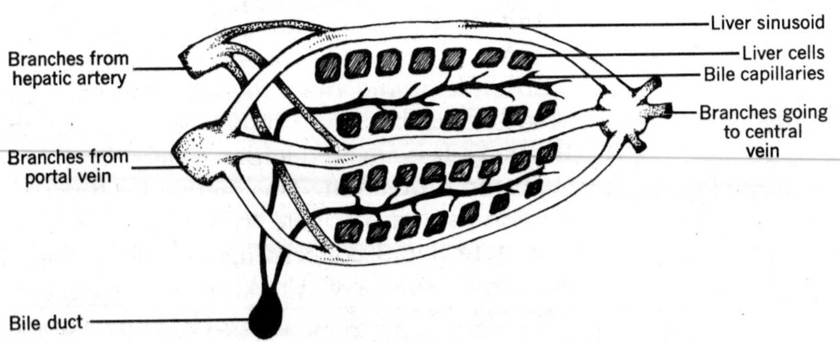

FIGURE 15-1
A highly diagrammatic drawing showing the relationship of the bile capillaries, liver cells, and liver sinusoids.

of this hydrolysis are fatty acids and glycerol; these substances then can be absorbed into the lacteals of the intestinal villi.

The pancreas is located behind the stomach at about the level of the first and second lumbar vertebrae. The head of the pancreas is found in the curve of the duodenum, and the tail of the pancreas extends laterally to the spleen. The pancreas is both an endocrine gland and an exocrine gland. The exocrine portion produces sodium bicarbonate to alkalinize the acid chyme, and secretes digestive enzymes: trypsin to help digest proteins, amylase for carbohydrates, and lipase for fat digestion. All these substances enter the duodenum through a duct that opens at the Ampulla of Vater. The endocrine portion produces insulin to lower the blood sugar level by causing increased utilization of carbohydrates, stimulates the storage of glycogen and the formation of fat from glucose, and decreases gluconeogenesis. Insulin is degraded in the liver by glutathione transhydrogenase and to a somewhat lesser degree by the kidney. Insulin is antagonized by epinephrine, by the glucocorticoids, and by thyroxine. Another part of the endocrine portion of the pancreas produces glucagon, which helps to convert liver glycogen to glucose. You can receive insulin and glucagon by injection. Insulin is given to lower the blood sugar level, and glucagon is given to elevate the blood sugar level in the event of an insulin reaction.

The mechanical actions in the small intestines are called peristalsis. The chemical actions are numerous. The bile emulsifies the fats, and the presence of bile in the large intestine enables bacteria there to produce vitamin K. In the absence of bile, a patient's stools will be clay-colored and he may have bleeding tendencies because of the deficiency of vitamin K.

The sodium bicarbonate from the pancreas produces an alkaline medium, so that the otherwise acid chyme will not cause erosion of the mucosa of the small intestines. This mucosa is not protected by as much mucus as is found in the stomach.

Pancreatic trypsin breaks proteins down to amino acids. Pancreatic amylase reduces polysaccharides to disaccharides, and pancreatic lipase reduces emulsified fats to fatty acids and glycerol.

The intestinal glands also produce enzymes. Erepsin can convert proteins to amino acids. Carbohydrases are produced by the intestinal mucosa to complete the digestion of carbohydrates, reducing them to monosaccharides (glucose, fructose, and galactose).

The greatest amount (about 80 percent) of the absorption of food takes place in the small intestines, because it is here that the foodstuffs are in their diffusible form, the surface area is so great, and peristalsis is slow enough to allow time for absorption.

LARGE INTESTINE

The large intestine, which is about 5 feet long, begins at the ileocecal valve and extends to the anus. The cecum is the first part of the large intestine. The appendix, which is about 3 inches long, is a narrow tube closed at one end and attached to the cecum. The relatively little blood supply to the appendix makes it vulnerable to infection.

The colon is one continuous tube subdivided into four parts: the ascending colon, transverse colon, descending colon, and sigmoid colon. Most of its action is that of peristalsis and the absorption of water and electrolytes. Severe diarrhea cannot only result in serious dehydration but also in electrolyte imbalance. In addition to peristalsis and absorption of fluids, a great deal of bacterial activity causes the putrefaction of proteins and the production of vitamin K. Very little chemical action takes place in the large intestine.

Failure to evacuate the bowel promptly may result in constipation because too much water is absorbed and the feces becomes hard. Cathartic salts are relatively nonabsorbable and tend to pull water into the bowel and soften the feces. Mineral oil acts as a lubricant and therefore is laxative. Some other types of laxatives work because they act as nonabsorbable bulk and stimulate peristalsis.

The sigmoid colon leads to the rectum, which is about 5 inches long and empties into the anal canal. The rectal sphincter protects the external orifice.

VITAMINS

Vitamins, organic compounds synthesized by plants and animals, are necessary for normal metabolism and also function as coenzymes. Some are fat soluble and some are water soluble. Vitamins A, D, E, and K are fat soluble. Mineral oil interferes with the action of all fat-soluble vitamins. Proper fat metabolism is essential in the utilization of the fat soluble vitamins. Vitamin A is found in milk, butter, egg yolk, liver, and cod liver oil. It prevents night blindness and helps to restore epithelium and therefore increases our resistance to infection. Vitamin D is the antirachitic vitamin, promoting the absorption of calcium from the gastrointestinal tract. The best sources of vitamin D are fish liver oil, egg yolk, butter, and milk. Although sunlight causes the synthesis of some vitamin D by the skin, overexposure to sunlight is not without its hazards. Much evidence now supports the theory that prolonged overexposure to sunlight not only causes early aging but also probably predisposes one to skin cancer. Vitamin K, found in green leafy vegetables, is also synthesized by intestinal bacteria. It is needed for normal blood clotting and proper functioning of the cellular mitochondria. Circumstances that might lead to a deficiency of vitamin K are the absence of bile, the suppression of bacterial flora with antibiotics, and ulcerative colitis with poor absorption of the vitamin. The newborn is K-deficient and unless its mother has received an injection of vitamin K during labor, the newborn should be given an injection of the vitamin. Vitamin E has long been controversial; however, it is now believed that this vitamin helps to retard the aging process. Sources of vitamin E are wheat, corn, and leafy vegetables.

Vitamins B and C are water soluble. Vitamin C, or ascorbic acid, is important in protein and iron metabolism. It helps in the formation and maintenance of supporting tissues (bone, cartilage, and the intercellular substances of capillaries). A deficiency of vitamin C will cause scurvy, which is characterized by bleeding tendencies and poor wound healing. Sources of vitamin C are chiefly citrus fruits and tomatoes. If you do not eat breakfast, you are probably C-deficient. The degree of the deficiency may not be sufficient to cause scurvy, but it may be sufficient to reduce your ability to utilize dietary iron. Many iron deficiency anemias may not be related to insufficient iron intake, but instead to the inability to utilize the iron.

There are about a dozen members of the vitamin B complex, all of which are fairly stable when cooked. They are found in grain cereals, pork, milk, eggs, fruit, and vegetables. Deficiencies are characterized by weight loss, weakness, sores on the lips, and (in the case of a deficiency of vitamin B-1) neuritis.

SUMMARY QUESTIONS

1. What foodstuffs function as enzyme activators?
2. What is the purpose of mechanical digestion?
3. What is the purpose of chemical digestion?
4. What enzymes work on carbohydrates?
5. What enzymes work on fats?
6. What enzymes work on proteins?
7. Name the simplest forms of proteins, carbohydrates, and fats.
8. Beginning with the mouth, list in sequence the structures of the alimentary canal.
9. What is the function of the greater omentum and where is it located?
10. What type of membrane lines the alimentary canal?
11. What functions does the hydrochloric acid in the stomach accomplish?

12. What chemical digestive processes take place in the stomach?
13. Why is there so little absorption of foodstuffs in the stomach?
14. Discuss at least five functions of the liver.
15. What are two functions of bile?
16. What are the secretions from the exocrine portion of the pancreas and what does each of these secretions accomplish?
17. What are the secretions from the endocrine portion of the pancreas and what are the functions of these hormones?
18. Absorption of the products of digestion depends upon what factors?
19. List the functions accomplished by the large bowel.
20. What vitamins are fat soluble?
21. What vitamins are water soluble?
22. What might result from a deficiency of vitamin C?
23. What might result from a deficiency of vitamin D?
24. What might result from a deficiency of vitamin A?
25. What might cause a deficiency of vitamin K and what are the symptoms of a deficiency of vitamin K?

SUGGESTED READINGS

Hall, R. L. "Food Additives," *Nutrition Today,* July/August 1973.

Jacob, S. W., and Francone, C. A. *Structure and Function in Man.* Philadelphia, Pa.: 3rd W. B. Saunders Co., 1974.

Lowenberg, Miriam E. *Food and Man.* New York, N.Y.: John Wiley & Sons, Inc., 1968.

Macleod, C. C. "Dietary Intake of Older People," *Nutrition,* Spring 1970.

Paulley, J. W. "The Psyche Has a Finger—Sometimes a Fist—in Every Pie," *Nutrition,* Summer 1970.

16 Diseases of the Gastrointestinal System

VOCABULARY

Anorexia
Antimetabolites
Ascites
Colostomy
Diuretic
Epistaxis
Hematemesis
Ileostomy
Jaundice
Leukocytosis
Parasites
Sclera

OVERVIEW

I COMMON DIAGNOSTIC PROCEDURES
- **A.** Gastric analysis
- **B.** Stool examinations
- **C.** X-ray examinations and fluoroscopy
  - **1.** Upper gastrointestinal series
  - **2.** Barium enema
  - **3.** Cholecystogram
- **D.** Endoscopic examinations
  - **1.** Esophagoscopic examination
  - **2.** Gastroscopic examination
  - **3.** Sigmoidoscopic examination
  - **4.** Proctoscopic examination

II NAUSEA AND VOMITING
III DIARRHEA AND CONSTIPATION
IV PEPTIC ULCERS
V CANCER OF THE BOWEL
VI APPENDICITIS
VII INTESTINAL OBSTRUCTION
VIII PILONIDAL SINUS
IX CIRRHOSIS OF THE LIVER
X HEPATITIS

COMMON DIAGNOSTIC PROCEDURES

A gastric analysis is frequently ordered to determine the amount of free hydrochloric acid in the gastric secretions. The specimen for this analysis is usually obtained by passing a nasogastric tube through the nose and into the stomach, so that the gastric secretions can be aspirated through this tube by a syringe. Of course, a specimen can be obtained from vomitus. If the patient has vomited, it is unnecessary for him to be subjected to the somewhat unpleasant procedure of having the tube passed into his stomach.

The absence of free hydrochloric acid in the stomach may indicate that the patient has pernicious anemia or cancer of the stomach. If the gastric contents contain food eaten the previous day, it is likely that the patient has a pyloric obstruction.

The patient should have no food or fluids for at least 8 hours before the test is done. For this reason, it is usually more convenient for the test to be done early in the morning. The number of specimens and the interval at which they are withdrawn may vary in different hospitals and with different doctors. Each specimen should be labeled with the patient's name and the time the specimen was obtained. Sometimes it is necessary to give histamine, 0.5 mg. by subcutaneous injection to stimulate the flow of gastric juice. However, the drug may cause a sudden fall in blood pressure, pallor, rapid, weak pulse, and loss of consciousness. If histamine is to be used, epinephrine should be available for immediate use in the event of histamine-induced shock.

Stool specimens may be needed to determine the presence of various pathogens, such as the typhoid bacillus, or of intestinal parasites. Stools that are to be examined for parasites must be warm and fresh, so that the motion of the parasites can be seen through a microscope.

Often the stool is examined for the presence of occult blood (blood not visible to the naked eye). To prepare for this examination, the patient should not eat red meat for 24 hours before the specimen is taken.

If some type of malabsorption syndrome is suspected, the stools may be examined for fat content. If this type of examination is done, it may be necessary to collect the stools for 24 hours, or sometimes for several days in succession.

X-ray examinations and fluoroscopy are quite helpful in diagnosing many types of gastrointestinal diseases. When lesions of the upper part of the gastrointestinal tract are suspected, an upper gastrointestinal series may be ordered. Prior to a morning examination, the patient must fast after midnight and probably will not be allowed to have lunch until some time in the afternoon. He is required to swallow barium while the doctor observes its passage to the stomach through a fluoroscope. The speed with which the barium passes through the tract and the appearance of the organs are noted. In addition to a fluoroscopic examination, X-rays will be taken for a permanent record of the appearance of the organs. Normally, the barium leaves the stomach within 6 hours. Additional X-rays are taken 6 hours after the barium swallow, and during this time the patient is not allowed anything to eat. This is an

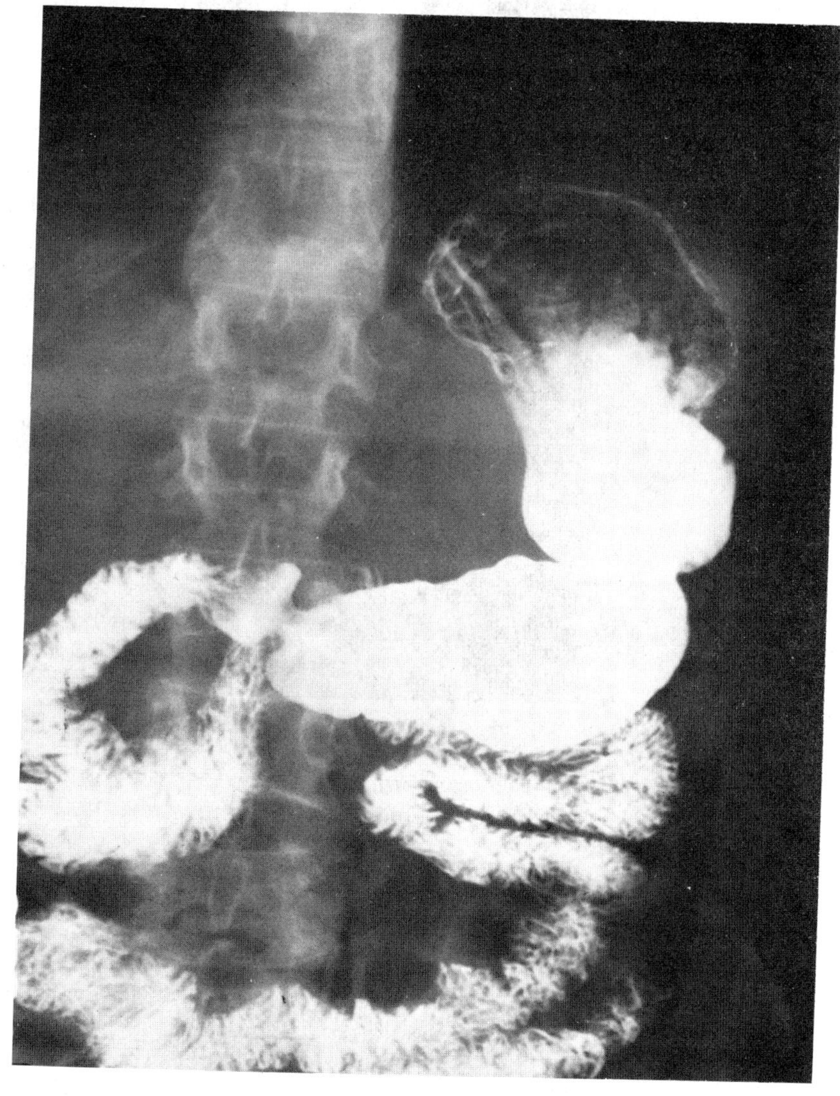

FIGURE 16-1
An upper gastrointestinal tract X-ray. (Courtesy of Mercy Hospital Radiology Department, Pittsburgh.)

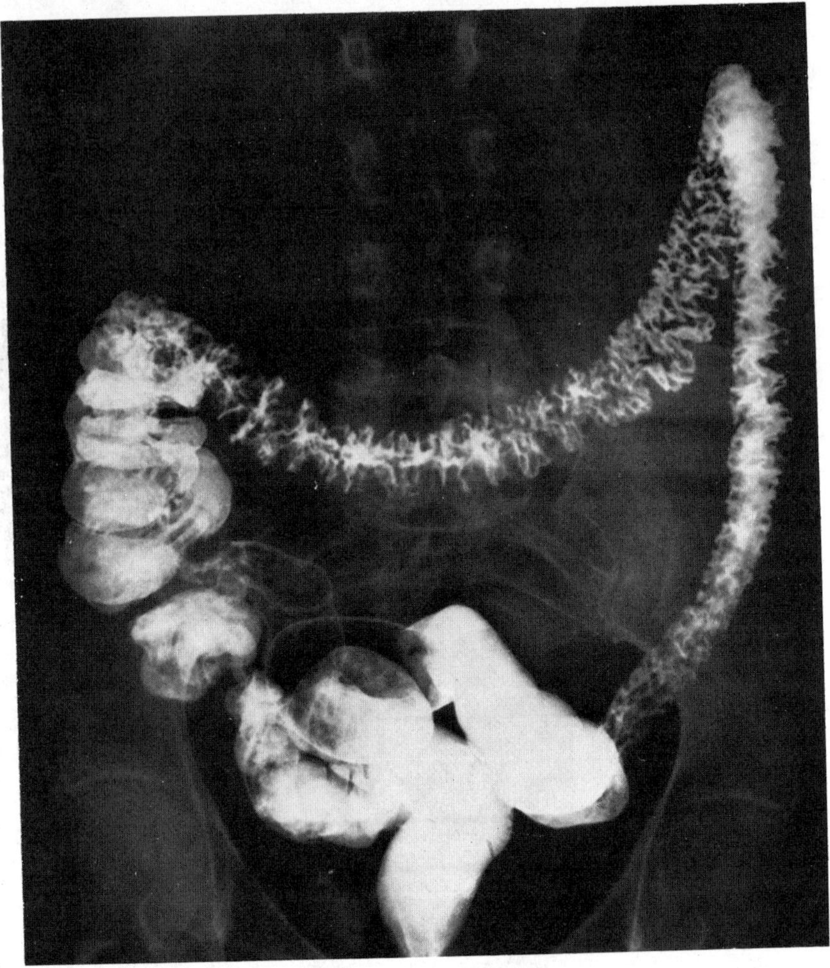

FIGURE 16-2
A lower gastrointestinal tract X-ray. (Courtesy of Mercy Hospital Radiology Department, Pittsburgh, Pa.)

unpleasant procedure, not only because it takes so long, but also because the barium has a chalky taste and the patient must assume various positions while the series of films is being taken.

To prepare a patient for X-ray examinations of the large bowel, he should receive only a clear liquid diet for 24 hours prior to the examination. If this is not practical, the evening meal prior to the examination should consist of toast and clear liquids. He must be given laxatives and enemas prior to the examination. Frequently, two or three enemas are necessary to cleanse completely the bowel of feces. In the X-ray department, the patient will receive a barium enema and must retain this barium while the films are being taken.

Whenever barium is given, it is important that a cathartic be given to the patient following the examination. Any retained barium can become a hard mass and cause an impaction in the rectum, or possible intestinal obstruction. After barium has been used, it is necessary to observe whether the barium has been passed and whether the patient is having regular bowel movements.

A cholecystogram is a series of X-rays used to study the gall bladder. In preparation for a morning examination, the patient is given several tablets (depending upon his body weight) to swallow the evening before the examination. Before breakfast the next morning the first set of X-rays is taken. Then the patient is given a fatty meal and a second set of X-rays is taken. The normal gall bladder will have evacuated much of its bile and will appear smaller in response to the fatty meal. Gallstones (cholelithiasis) can also be visualized by the cholecystogram.

Endoscopic examinations for gastrointestinal studies are used to visualize directly the internal organs through a hollow instrument passed through either the mouth or the rectum. An esophagoscopy examination is used to examine the esophagus. Through a gastroscope the internal surface of the stomach can be examined. By means of a sigmoidoscopy, the sigmoid, rectum, and anus can be examined. A proctoscopy is done to examine the rectum and anus. All of these procedures are uncomfortable and sometimes quite painful. In addition to using endoscopic examinations to examine these structures visually, the procedure may also be used to remove a specimen of tissue from the organ for microscopic examination. Rectal polyps (small tumors in the rectum) may be removed while a proctoscopy is being done.

DISORDERS OF THE GASTROINTESTINAL SYSTEM

NAUSEA AND VOMITING

Nausea and vomiting are common symptoms of almost any abnormality of the gastrointestinal system. It is important that the emesis be observed for blood (hematemesis) as this may be indicative of a bleeding ulcer. The time of the vomiting with respect to meals and the quantity of the emesis should also be noted. If the nausea and vomiting continue over a prolonged period, they can lead to weakness, weight loss, anorexia, and metabolic alkalosis. If the vomiting is caused by gastritis, however, metabolic acidosis may occur. Although the gastric juices are low in base sodium, the

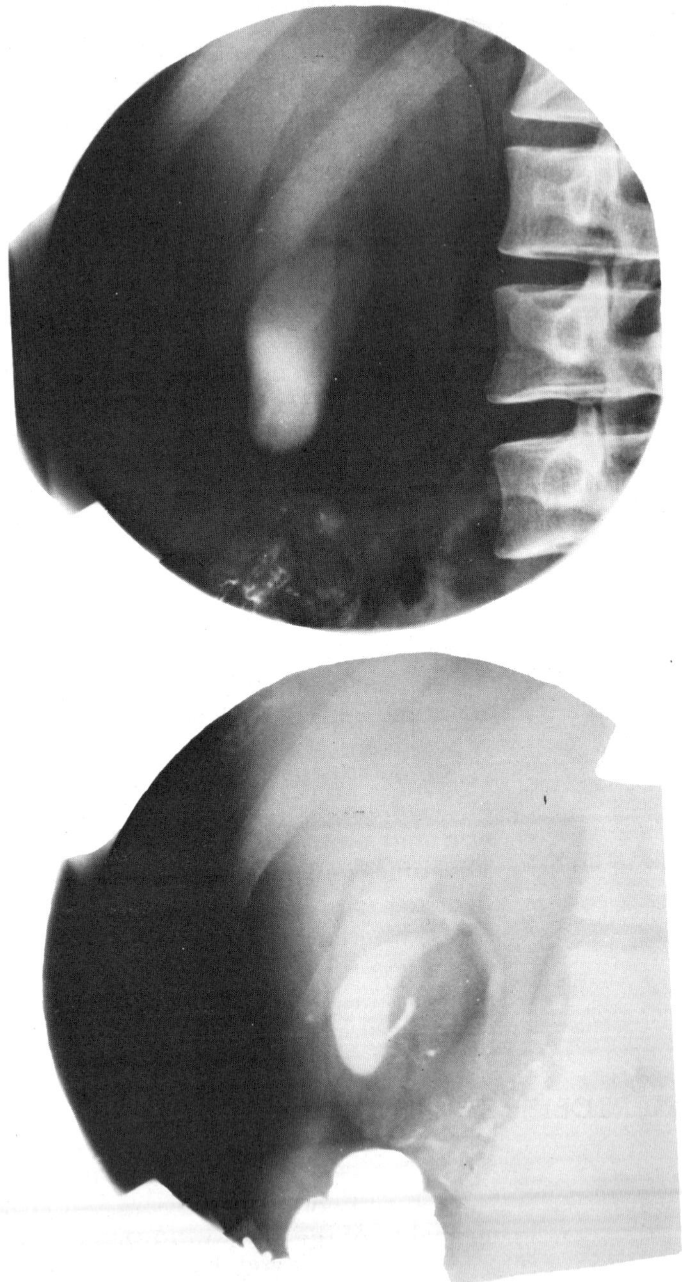

FIGURE 16-3
Normal gall bladder series. The film at the top shows a full gall bladder. The lower film was taken following a fatty meal. Much of the bile has been evacuated from the gall bladder. (Courtesy of James W. Lecky, M.D., Professor of Radiology, University of Pittsburgh.)

gastric mucus is high in sodium. With gastritis, the gastric mucosa produces more and more mucus in an attempt to soothe the irritated mucous membrane, this depletes the blood level of sodium and metabolic acidosis may result.

DIARRHEA AND CONSTIPATION

Diarrhea and constipation are often related to emotional stress; however, they are also found in many other conditions. Alternating diarrhea and constipation may indicate tumors of the bowel. Severe diarrhea, regardless of the cause, can lead to metabolic acidosis because much base is lost in the stools. With severe diarrhea, particularly in young infants, dehydration can be a serious problem. Patients with ulcerative colitis have severe diarrhea that may contain blood and much mucus. These patients have marked weight loss, fever, and severe electrolyte imbalance. Cancer of the colon is much more frequent in people with ulcerative colitis than it is among the general population. Patients should be given a bland diet and small frequent feedings. They will probably need supplemental vitamins, because their diet must be deficient in raw fruits and vegetables.

Constipation can result from poor dietary or bowel habits. Simple constipation is best treated by increasing the amount of roughage in the diet and taking liberal amounts of fluids. Every effort should be made to establish a regular time for defecation. Constipation can also be a sign of intestinal obstruction. In this event, the patient is usually greatly distended with gas and has abdominal pain.

PEPTIC ULCERS

Peptic ulcers are erosions in the mucous membrane lining of the stomach or duodenum. Patients complain of burning in the epigastric region. The pain, which usually occurs one to several hours after meals, is usually relieved by ingestion of protein foods such as milk or by taking some aluminum hydroxide preparation such as Gelusil or Maalox. Cholinergic blocking agents that retard peristalsis and hydrochloric acid secretion may also be helpful. Patients need a bland diet and should avoid stressful situations which seem to aggravate their condition.

Ulcers of the stomach are much more likely to become malignant than ulcers of the duodenum. For this reason, if conservative medical treatment does not result in a marked improvement within a relatively short period of time, surgical consultation should be sought. The surgical treatment for peptic ulcers involves the removal of the ulcer, or, in the case of an extensive gastric ulcer, a partial gastrectomy.

Complications of peptic ulcers include hemorrhage, which will be evidenced by hematemesis, or the passing of tarry stools. The ulcer may perforate all the way through the wall of the organ, in which case the patient will have peritonitis. This condition constitutes a real surgical emergency. Aside from the severe pain that occurs with the perforation, the patient's abdomen will have a boardlike rigidity.

CANCER OF THE BOWEL

Cancer of the bowel almost never has symptoms early enough for the patient to be able to avoid very major surgery. If the condition is diagnosed early enough, the surgeon might be able to resect the cancerous portion of the bowel and to anastamose (connect) the normal segments. Usually, however, the patient must have a colostomy or ileostomy. With this type of surgery, the patient will no longer have bowel movements through the rectum but rather through a stoma on the abdominal wall. If an ileostomy has been performed, the feces will be quite liquid and the patient will have to wear a bag over the stoma to collect the liquid feces. A patient having a colostomy, particularly if it is on the descending colon, has a much better chance of having bowel movements that are of almost normal consistency and therefore are easier to control. Good skin care around the stoma is very important, because feces are irritating. Once the skin is abraded, it is difficult to heal and also it is difficult to fix a bag to the raw area. Liberal applications of tincture of benzoin to the raw area may help to protect it from further irritation and help in the healing process. Like other cancers, malignancies of the gastrointestinal tract may also be treated with radiation and antimetabolites.

APPENDICITIS

Appendicitis is one of the most common surgical emergencies. Because the appendix is a blind pouch and has a relatively poor blood supply, it is particularly prone to infection. The attack is usually accompanied by severe abdominal pain, usually fairly generalized throughout the abdomen. Alternatively, it may be localized around the umbilicus. The patient will have a moderate fever and leukocytosis; he may also experience nausea and vomiting. The treatment is an appendectomy, which a surgeon usually can do by merely separating the muscle fibers without cutting the abdominal muscles. If the muscles are not cut the patient will be able to resume normal activities in a relatively short period of time. It is common practice to remove the appendix when a patient has any type of abdominal surgery even though there is no pathology of the appendix, because the appendix serves no useful purpose and is susceptible to infection.

INTESTINAL OBSTRUCTION

Intestinal obstruction can result from hernias, tumors, severe constipation, or from interference with intestinal innervation inhibiting peristalsis. This interference is called paralytic ileus and is a fairly common complication of any abdominal surgery. Proximal to the obstruction, the bowel is distended; whereas distal to the obstruction, the bowel is empty. Peristalsis is very forceful proximal to the obstruction and causes severe, intermittent cramps. You can hear these peristaltic waves with a stethoscope. Distal to the obstruction there will be no bowel sounds. The patient will experience severe vomiting, which after a time is foul smelling. If the obstruction is very low, however, he may not vomit. The treatment is the surgical removal of whatever is causing the obstruction.

Paralytic ileus, however, may be relieved by intestinal intubation: Either a Cantor tube or a Miller-Abbot tube is introduced through the nose into the stomach and on down into the small intestines. The tube is then attached to suction so that the contents of the distended bowel are removed. The tube is left in place usually for several days until peristalsis returns. During this period the patient takes nothing by mouth and is given intravenous fluids. During this time the patient should receive frequent and thorough mouth care. His mouth will be dry and the ducts of the salivary glands may become obstructed, in which case the glands may become infected.

PILONIDAL SINUS

A pilonidal cyst is usually no problem until it becomes infected and a draining sinus develops. In such a case the sinus can be very painful. The sinus tract comes from the cyst at the base of the spine. The cyst contains hair and pus. People whose occupations require long hours of sitting are predisposed to pilonidal sinuses. The lesion should be treated several times a day with warm water compresses. With this treatment the lesion will probably heal in a matter of a few days to a week. Unfortunately, the sinus frequently recurs.

The surgical treatment of a pilonidal sinus is usually quite extensive, considering the fact that the visible lesion is so small. The tract is laid open with a deep and wide V-shaped incision and packed with gauze to keep it open so it will heal from the bottom up. It will take weeks for the incision to heal.

CIRRHOSIS OF THE LIVER

Cirrhosis of the liver is a hardening of the liver and a blocking of the liver sinusoids. The liver is usually quite enlarged. The symptoms may include ascites, abdominal pains, epistaxis, hematemesis, dilated veins, scanty body hair, jaundice, and an increased bleeding tendency. The skin is dry and inelastic, the pulse is rapid, and breathing is shallow. Diet is very important in the treatment of cirrhosis. It should be high in protein, carbohydrates, and vitamins and low in fats. Alcohol must be avoided. Diuretics may help. Sedatives, particularly opiates and barbiturates, should be avoided because they are likely to cause coma. Because of his bleeding tendency, the patient will probably need supplemental vitamin K. The ascites can be temporarily relieved by a paracentesis. A paracentesis is a relatively simple procedure in which a trochar (a sharp hollow tube) is inserted into the peritoneal cavity and the fluid is allowed to drain out. Sometimes as much as a gallon of fluid is withdrawn. This procedure will greatly relieve the patient's discomfort.

Cirrhosis is a serious disease and the prognosis is particularly grave when jaundice and ascites occur. A serious and often fatal complication of cirrhosis is that of esophageal varicies. If these varicosed veins of the esophagus rupture, the patient may hemorrhage to death before the bleeding can be controlled.

HEPATITIS

There are at least two kinds of hepatitis, homologus serum hepatitis and viral hepatitis. In the serum hepatitis, the virus is present predominantly in the bloodstream; in the infectious type, it is in both the bloodstream and in the gastrointestinal tract. Serum hepatitis has an incubation period of 60 to 160 days, but infectious hepatitis has an incubation period of only about 20 to 40 days.

In both types, patients experience malaise, yellow sclerae, anorexia, headache, chills, fever, nausea and vomiting, dark urine, and light-colored stools. Jaundice may also be present. Most patients recover; however, they may go into coma that usually results in death. The treatment is primarily directed toward improving the patient's resistance. Bed rest is important. Fluids should be encouraged. Diet should be high in proteins, vitamins, and carbohydrates.

CHOLECYSTITIS AND CHOLELITHIASIS

Cholecystitis is an inflammation of the gall bladder, and cholelithiasis is stone formation in the gall bladder. Both of these conditions should be treated with a low fat diet. In the case of cholelithiasis, the patient will probably have to have a cholecystectomy. Gallstone colic has an abrupt onset and may last for several hours. The patient vomits and has severe pain that usually radiates to the back or to the right shoulder. Morphine and atropine may help. During an attack, the patient may not even be able to tolerate a physical examination until he has had some relief of his pain.

SUMMARY QUESTIONS

1. For what purpose is a gastric analysis done?
2. What precautions must be taken if the patient is to receive histamine as a part of the gastric analysis procedure?
3. Give three reasons why it may be necessary to examine feces.
4. What preparation is necessary for a patient who is to have X-ray studies of the upper part of his gastrointestinal tract?
5. What preparation is necessary for a patient who is going to have a barium enema?
6. What is a cholecystogram?
7. Name four types of endoscopic examinations and indicate the purpose of each.
8. What term is used to indicate blood in vomitus?
9. Alternating constipation and diarrhea may be a symptom of what serious illness?
10. What type of diet would you recommend for a patient with ulcerative colitis?
11. What type of peptic ulcer is most likely to be malignant?
12. Discuss the medical treatment of a patient with a peptic ulcer.
13. Name two complications of peptic ulcers and describe the symptoms of each.
14. What is a colostomy?
15. Why is appendicitis so common?
16. List some of the causes of intestinal obstruction.
17. What is paralytic ileus and how is it treated?

18. What conservative treatment is appropriate for a pilonidal sinus?
19. List several symptoms of cirrhosis of the liver.
20. What is ascites?
21. What diet would you recommend for a patient with hepatitis?
22. What is cholelithiasis and what is the usual treatment?

SUGGESTED READINGS

Girdwood, P.H. "Problems in Assessment of Vitamin Deficiencies," *The Proceedings of the Nutrition Society,* May 1971.

Lieber, C.S., and DeCarli. "Quantitative Relationship Between Amount of Dietary Fat and Severity of Alcoholic Fatty Liver," *American Journal of Clinical Nutrition,* April 1973.

Mitra, M.L. "Confusional States in Relationship to Vitamin Deficiency in the Elderly," *Journal of the American Geriatrics Society,* June 1971.

Rutledge, F. "Gastric Diet for Gastric or Duodenal Ulcers," *Journal of the Canadian Dietetic Association,* Summer 1973.

Sarles, H., *et al.* "Diet, Cholesterol, Gallstones and Composition of Bile," *American Journal of Digestive Diseases,* March 1970.

Sherlock, S. "Carbohydrate Changes in Liver Disease," *American Journal of Clinical Nutrition,* April 1973.

Willelt, R. "Real Protection from Food-Borne Hazards," *The Cornell Hotel and Restaurant Administration Quarterly,* August 1971.

17 The Urinary System

VOCABULARY

Acidosis
Adipose
Adrenal gland
Cation
Catheterize
Electrolyte
Fascia
Metabolism
Peritoneum
Pituitary
Plasma
Rugae
Uterus
Vagina
Vasomotor

OVERVIEW

I KIDNEYS
  A. Location
  B. Support
  C. Renal cortex
    1. Glomerulus
    2. Renal tubules
    3. Acidification of urine
  D. Renal medulla
  E. Renal pelvis
II URETERS
III URINARY BLADDER
IV URETHRA
V MICTURITION

The organs of the urinary system include the kidneys, ureters, urinary bladder, and urethra. Functions of this system include the elimination of some of the soluble waste products and the regulation of water and of electrolyte balance.

KIDNEYS

The kidneys are located on the posterior wall of the abdominal cavity between the level of T-12 (twelfth thoracic vertebra) and L-3 (third lumbar vertebra). They are bean shaped and their concave border is directed toward the midline. The ureters and blood vessels enter and leave from the hilum, which lies in the middle of the concave border. The kidneys are supported by renal fascia, the peritoneum, and adipose tissue. With severe weight loss, adipose support may be lacking and the kidneys may become displaced downward; this condition is called nephroptosis. Nephroptosis can result in stasis of urine and predisposition to renal calculi (kidney stones).

CORTEX

The outer part of the kidney is called the renal cortex. On cross section the cortex has a granular, reddish-brown appearance. This part of the kidney contains most of the structures of the microscopic nephron units. Most of the work of the kidney is done by the nephron units. You should know the various parts of the nephron unit and understand the functions of these structures. Figure 17–1 shows the major anatomic parts of a nephron unit and the related structures. Note that the loop of Henle extends into the medullary portion of the kidney.

Glomerulus

The afferent arteriole, which is derived from the renal artery, leads into a tuft of glomerular capillaries. The blood pressure in these capillaries is much higher than it is on the capillary level in most other areas of the body. For this reason, a great deal of fluid and crystalline particles is filtered out of the glomerular capillaries. The glomerular filtrate has essentially the same composition and pH as does the blood plasma; however, the filtrate normally does not contain proteins. From the glomerular capillaries blood flows into an efferent arteriole, then to the peritubular capillaries, and finally back to the renal vein.

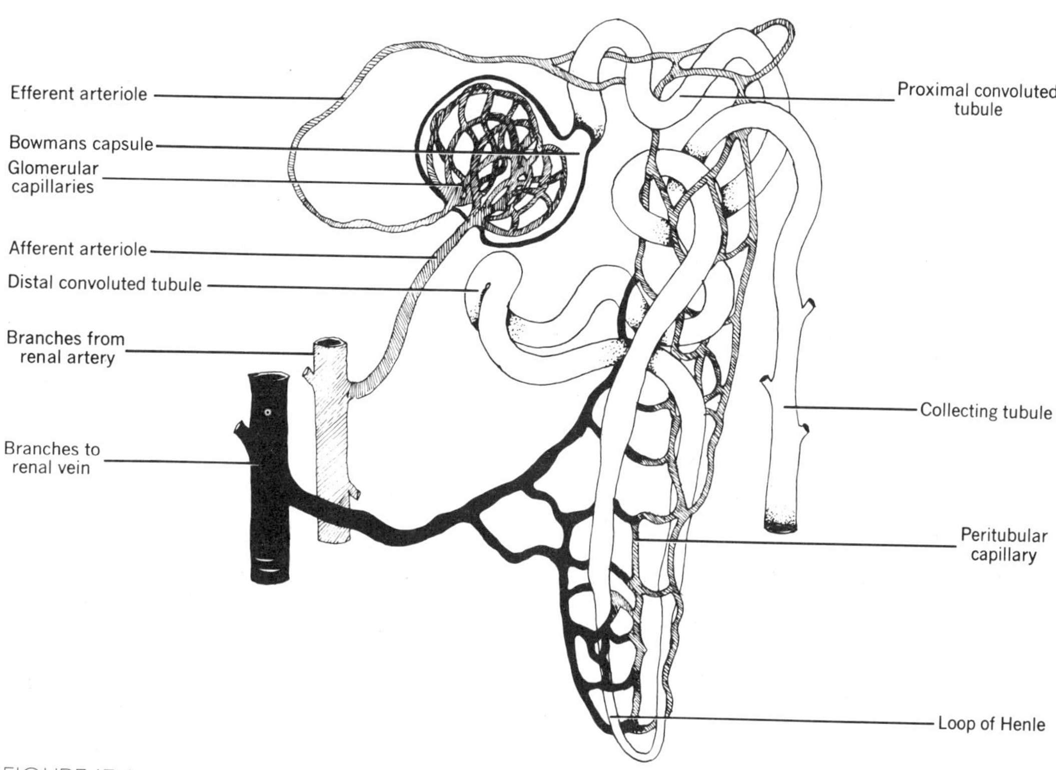

FIGURE 17-1
A nephron unit and related structures.

Renal Tubules

The fluid and crystalline particles that have been filtered out of the glomerular capillaries enter Bowman's capsule and then are carried to the renal tubules. It is in the renal tubules that substances such as glucose and sodium that are needed by the body are returned to the bloodstream via the peritubular capillaries. Some fluid and waste products, such as urea, are not returned to the bloodstream. These substances will be carried to the collecting tubules and ultimately excreted as urine. Urine has a much lower pH than that of the plasma from which it was derived. We shall now consider how the acidification process takes place.

Acidification of Urine

The acidification of urine is accomplished by the reabsorption of bicarbonate and the conversion of sodium monoamine hydrogen phosphate in the tubular urine to acid sodium dihydrogen phosphate. Both processes are possible because the kidney tubular cells, like any other cells in the body, produce carbon dioxide and water in the process of their metabolism. The carbon dioxide and water combine to form carbonic acid, which ionizes to form hydrogen and bicarbonate. Because hydrogen is a cation, it makes an even exchange with the sodium in the tubular urine. The sodium that enters the tubular cells combines with the bicarbonate and then returns to the peritubular capillaries. The hydrogen in the tubular urine increases the acidity of the urine, and the sodium that was saved helps to preserve the base pH of the plasma.

By a similar mechanism, base sodium monoamine hydrogen phosphate in the tubular urine is converted to acid phosphate by the exchange of a sodium from the tubular urine for a hydrogen from the tubular cell.

The kidney also can help in maintaining acid–base balance by converting neutral urea to ammonia. Conversion is most active when, in the presence of acidosis, the kidney is compensating by eliminating more and more acid in the urine. To prevent the urine from becoming too acid, it converts the urea to base ammonia and eliminates it in the form of ammonium.

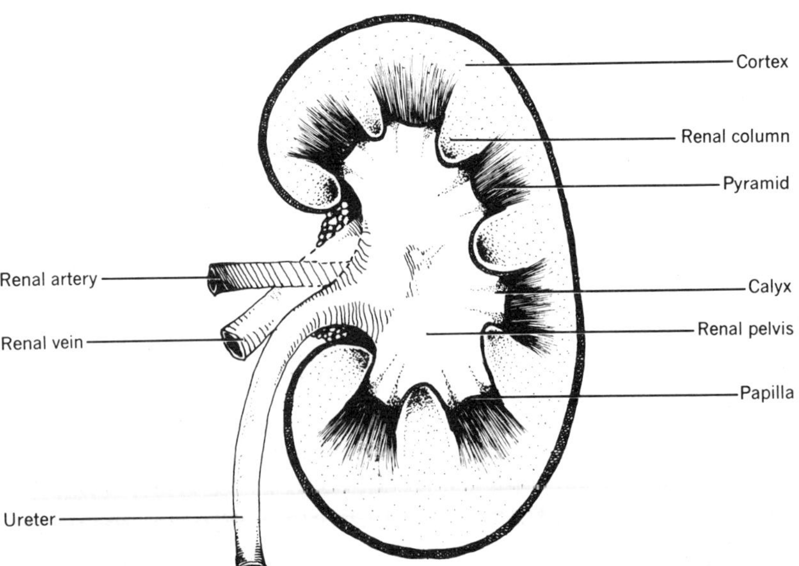

FIGURE 17-2

Grossly observable parts of the kidney as viewed when the kidney has been sectioned transversly. Note the blood vessels and ureter at the hilum.

RENAL MEDULLA

The collecting tubules are located in the medullary portion of the kidney and give the pyramids their striated appearance (see Figure 17–2). In the collecting tubules, depending upon the body's need to conserve fluid, some of the water can be returned to the bloodstream and the urine becomes more concentrated.

The return of water depends upon a hormone from the posterior pituitary. This hormone, ADH or vasopressin, is released in response to tissue dehydration. Vasopressin causes the collecting tubules to become more permeable to water. The water is returned to the bloodstream and ultimately is used to hydrate the tissues.

RENAL PELVIS

From the collecting tubules the urine then goes into the renal pelvis, which has a smooth, yellowish-white appearance. The pelvis merely serves as a basin for the collection of urine before the urine is passed on to the ureters.

URETERS

The ureters are narrow tubes leading from the kidney to the urinary bladder. The mucous membrane that lines the ureters is continuous with the mucous membrane lining of the urinary bladder and the urethra.

URINARY BLADDER

The urinary bladder is a hollow muscular organ located in the pelvis, posterior to the pubic bones. In the male it is anterior to the rectum, and in the female it is anterior to the uterus and vagina. A full urinary bladder may rise up into the abdominal cavity and usually can be palpated above the symphysis pubis.

The urinary bladder is lined with mucous membrane. The lining has rugae to allow for expansion. At the inferior and posterior part of the urinary bladder is an area that has no rugae. This area, the trigone, is a triangular structure where urine from the two ureters enters the bladder and urine leaves the urinary bladder to enter the urethra.

URETHRA

The female urethra is about 1.5 inches long and curves obliquely down and forward from the bladder. It has only an excretory function.

In the male the urethra is about 8 inches long and has an S shaped curve. One of these curves must be straightened out in order for a man to void or to be catheterized. The male urethra has both an excretory and reproductive function.

MICTURATION

Micturation is a reflex act of expelling urine. The stimulus is the stretching of the bladder wall when approximately 250 to 300 ml. of urine has accumulated. Although urinary output varies considerably, depending upon fluid intake, the average 24-hour output of urine is about 1.5 liters.

Urine is formed constantly and drains from the kidneys into the ureters and bladder drop by drop. Normally, less urine is formed at night than in the daytime. Several factors other than fluid intake influence urine formation. Because urine is derived from the blood plasma, the formation of urine depends upon an adequate blood supply to the kidney. Vasomoter nerve fibers in the walls of the blood vessels supplying the kidney control the circulation of blood to the kidney and therefore help regulate kidney function. Several hormones influence urine production. The mineral corticoids from the adrenal cortex are released in response to a decrease in vascular volume and cause the kidney tubular cells to return more water and sodium to the peritubular capillaries and thus to decrease urine production. The antidiuretic hormone from the posterior pituitary also causes a decrease in urinary output. For a more detailed discussion of hormones that influence urine production see Chapter 21.

SUMMARY QUESTIONS

1. Where are the kidneys located?
2. Name the parts of the nephron unit and discuss the function of each of these parts.
3. What structures convey urine from the kidney to the urinary bladder?
4. How are the kidneys supported?
5. List several factors that influence urine production.

6. Describe the structure of the urinary bladder.
7. Where are the collecting tubules located?
8. How does the appearance of the renal pelvis differ from that of the renal cortex and renal medulla?
9. Explain the processes involved in the acidification of urine.

SUGGESTED READINGS

Berger, A. J. *Elementary Human Anatomy.* New York, N.Y.: John Wiley & Sons.

Best, C. H., and Taylor, N. B. *The Human Body: Its Anatomy and Physiology.* New York, N.Y.: Holt, Rinehart & Winston, 1963.

Burke, S. R. *The Composition and Function of Body Fluids,* St. Louis, Mo.: C. V. Mosby Co., 1972.

18 Diseases of the Urinary System

VOCABULARY

Albuminuria
Biopsy
Catheterize
Cytology
Dysuria
Electrolyte
Estrogen
Hematuria
Hydrocortisone
Lithotomy
Meatus
Nephrectomy
Nephrostomy
Osteoporosis
Pyuria
Radiopaque
Ureter
Urethra

OVERVIEW

I DIAGNOSTIC TESTS
- **A.** Urinalysis
- **B.** Intravenous pyelogram
- **C.** Retrograde pyelogram
- **D.** Cystoscopy
- **E.** Phenolsulfonphthalein test
- **F.** Blood chemistry tests

II URINARY OBSTRUCTIONS
- **A.** Calculi
- **B.** Strictures

III TUMORS

IV INFECTIONS
- **A.** Cystitis
- **B.** Pyelonephritis
- **C.** Glomerulonephritis

V HEMODIALYSIS

VI KIDNEY TRANSPLANTS

DIAGNOSTIC TESTS

URINALYSIS

The examination of urine is a simple procedure that is always done as a part of a urological examination and of any routine physical examination. Depending upon the purpose of the urinalysis, the specimen to be examined may be a single voided specimen, a 24-hour urine specimen, a catheterized urine specimen, or a clean-caught specimen.

If urinary tract infection is suspected, the specimen should either be a clean-caught specimen or be obtained by catheterization. Twenty-four hour specimens are usually ordered when endocrine disorders are suspected. This type of urinalysis will be discussed in Chapter 22. The simple voided specimen is commonly used as a diagnostic aid for urinary diseases other than infections and as a part of a routine physical examination.

Catheterized Specimens

To obtain a catheterized specimen all equipment must be sterile. Necessary supplies include an antiseptic solution to cleanse the area around the urinary meatus, sponges, a container for the specimen, catheters, and sterile gloves. After the area around the urinary meatus is thoroughly cleansed, the catheter is inserted through the urinary meatus into the urinary bladder. About 60 ml. of urine is an adequate specimen. If the patient has a full urinary bladder, the urine obtained toward the end of the procedure should be used for the specimen.

Clean-Caught Specimens

For a clean-caught specimen, the area around the meatus is cleansed and the patient is instructed to void directly into a sterile container. Although this procedure is much easier than that for a catheterized specimen, contamination of the specimen is more likely. In spite of this disadvantage, some doctors prefer clean-caught specimens to catheterized specimens because, if faulty technique is used during the catheterization, bacteria may be introduced into the urinary tract. Clean-caught specimens and catheterized specimens are then examined for the presence of bacteria, and the sensitivity of the organism to a variety of antibiotics is tested.

Appearance

Urine is also examined for appearance. Normally, fresh voided urine is clear yellow. Cloudiness may indicate pus in the urine. If the urine is dark yellow, it probably has a high specific gravity.

Specific Gravity

The specific gravity of urine is measured with a urinometer. Enough urine to float the urinometer is placed in a cylinder. The reading is taken at the point where the scale touches the surface of the urine. The average normal range for specific gravity is 1.010 to 1.020. If the specimen is the first urine voided in the morning, the specific gravity is likely to be high normal or slightly above. When the kidneys are damaged, the ability to concentrate or to dilute urine may be impaired. Patients with diabetes mellitus may have a high urine specific gravity because of the glucose content in their urine, even though the urine appears quite dilute.

Composition of Urine

Lab sticks are used to examine the urine for the presence of albumin, blood, glucose, and acetone, and to determine the pH of the urine. The lab stick is dipped into the specimen and then compared with the color scale provided. Normally urine contains no albumin, blood, acetone, or glucose. The urine pH is normally acid; however, the pH varies considerably depending upon metabolic activity and other factors.

INTRAVENOUS PYELOGRAM

An intravenous pyelogram includes an X-ray examination of the kidney following the injection of a radiopaque dye. The X-ray films will reveal the outline of the kidneys.

The patient should be given a cathartic the evening before the examination, and an enema the morning of the examination so that the contents of the bowel do not interfere with the visualization of the kidneys on the X-ray film. The patient should have nothing by mouth for 12 hours prior to the examination. Fasting dehydrates the patient so that the dye used for the examination will be concentrated.

Ordinarily, an X-ray film is taken of the abdomen before the dye is injected. This film may show the presence of any radiopaque stones in the urinary tract, and it can be used as a control when it is compared with subsequent films.

TABLE 18–1

Composition of Urine in Health and Disease

Normal Range	Conditions in Which Variations from Normal May Occur
Volume in 24-hr. 1200–1500 ml. (varies greatly with fluid intake)	Increased: diabetes insipidus, absorption of large quantities of edema fluid, diabetes mellitus, certain types of chronic renal disease, tumors of brain and spinal cord, myxedma, acromegaly, tabes dorsalis Decreased: dehydration, diseases that interfere with circulation to kidney, acute renal failure, uremia, acute intestinal obstruction, portal cirrhosis, peritonitis, poisoning by agents that damage kidneys
pH 4.7–8.0	Increased: compensatory phase of alkalosis, vegetable diet Decreased: compensatory phase of acidosis, administration of ammonium chloride or calcium chloride, diet of prunes or cranberries
Specific gravity 1.010–1.020	Increased: dehydration, administration of vasopressin tannate, glycosuria, albuminuria Decreased: diabetes insipidus, chronic nephritis
Constituents (24-hour specimen)	
Urea 20–30 Gm.	Increased: tissue catabolism, febrile and wasting diseases, absorption of exudates as in suppurative processes Decreased: impaired liver function, myxedema, severe kidney diseases, compensatory phase of acidosis
Uric acid 0.60–0.75 Gm.	Increased: leukemia, polycythemia vera, liver disease, febrile diseases, eclampsia, absorption of exudates, X-ray therapy Decreased: before attack of gout, but increased during attack
Ammonia 0.5–15.0 Gm.	Increased: diabetic acidosis, pernicious vomiting of pregnancy, liver damage Decreased: alkalosis, administration of alkalies
Creatinine 0.30–0.45 Gm.	Increased: typhoid fever, typhus, anemia, tetanus, debilating diseases, renal insufficiency, leukemia, muscular atrophy
Calcium 30–150 mg.	Increased: osteitis fibrosis cystica Decreased: tetany
Phosphates 0.9–1.3 Gm.	Increased: osteitis fibrosa, alkalosis, administration of parathormone
Chlorides 110–250 mEq.	Increased: Addison's disease Decreased: starvation, excessive sweating, vomiting, pneumonia, heart failure, burns, kidney disease
Sodium 43–217 mEq.	Increased: compensatory phase of alkalosis Decreased: compensatory phase of acidosis
17-Ketosteroids Men 5–27 mg. Women 5–15 mg.	Increased: Cushing's syndrome, adrenal malignancy, administration of cortisone, administration of ACTH, ovarian tumors

		Decreased: hypopituitarism, pituitary tumors, Addison's disease, myxedema, hepatic disease, chronic debilitating diseases
Aldosterone	up to 15 μg.	Increased: adrenal malignancy, conditions associated with excessive sodium loss, cardiac failure, nephrosis, hepatic cirrhosis Decreased: Addison's disease, eclampsia
Pressor amines Norepinephrine 5–100 μg. Epinephrine 11.5 μg.		Increased: essential hypertension, phoechromocytoma, severe stress, insulin-induced hypoglycemia
Amylase 8,000–30,000 Wohlgemuth units		Increased: early in acute pancreatitis, perforated duodenal ulcer, stone in common bile duct, carcinoma of the pancreas or bile duct, salivary gland disease Decreased: some liver diseases and some renal diseases

From Shirley R. Burke, *The Composition and Function of Body Fluids* (St. Louis, Mo.: The C. V. Mosby Co., 1972).

TABLE 18–2

Abnormal Constituents of Urine

Constituent	**Conditions in Which Variations from Normal May Occur**
Bence-Jones protein	Multiple myeloma, bone metastases of carcinoma, osteogenic sarcoma, osteomalacia
Albumin	Transient albuminuria may occur during pregnancy or prolonged exposure to cold, or following strenuous exercise; albuminuria present in nephritis, nephrosis, nephrosclerosis, pyelonephritis, amyloidosis, renal calculi, bichloride of mercury poisoning, and sometimes with blood transfusion reactions
Acetone	Diabetes mellitus, eclampsia, starvation, febrile diseases in which carbohydrate intake is limited, pernicious vomiting of pregnancy
Glucose	Unusually high carbohydrate intake, diabetes mellitus
Bilirubin	Obstructive jaundice, hemolytic jaundice, hepatitis, cholangitis
Urobilin and urobilinogen	Hemolytic jaundice, pernicious anemia, hepatitis, eclampsia, portal cirrhosis, lobar pneumonia, malaria
Erythrocytes	Glomerulonephritis, pyelonephritis, tuberculosis of kidneys, tumors of kidney, ureter, and bladder, polycystic kidneys, calculi, hemorrhagic diseases, occasionally with anticoagulant therapy
Leukocytes	Increased in urethritis, prostatitis, cystitis, pyelitis, pyelonephritis (a few leukocytes are found in normal urine)
Casts	Glomerulonephritis, nephrosis, pyelonephritis, febrile diseases, eclampsia, amyloid disease, poisoning by heavy metals

From Shirley R. Burke, *The Composition and Function of Body Fluids* (St. Louis, Mo.: The C.V. Mosby Co., 1972).

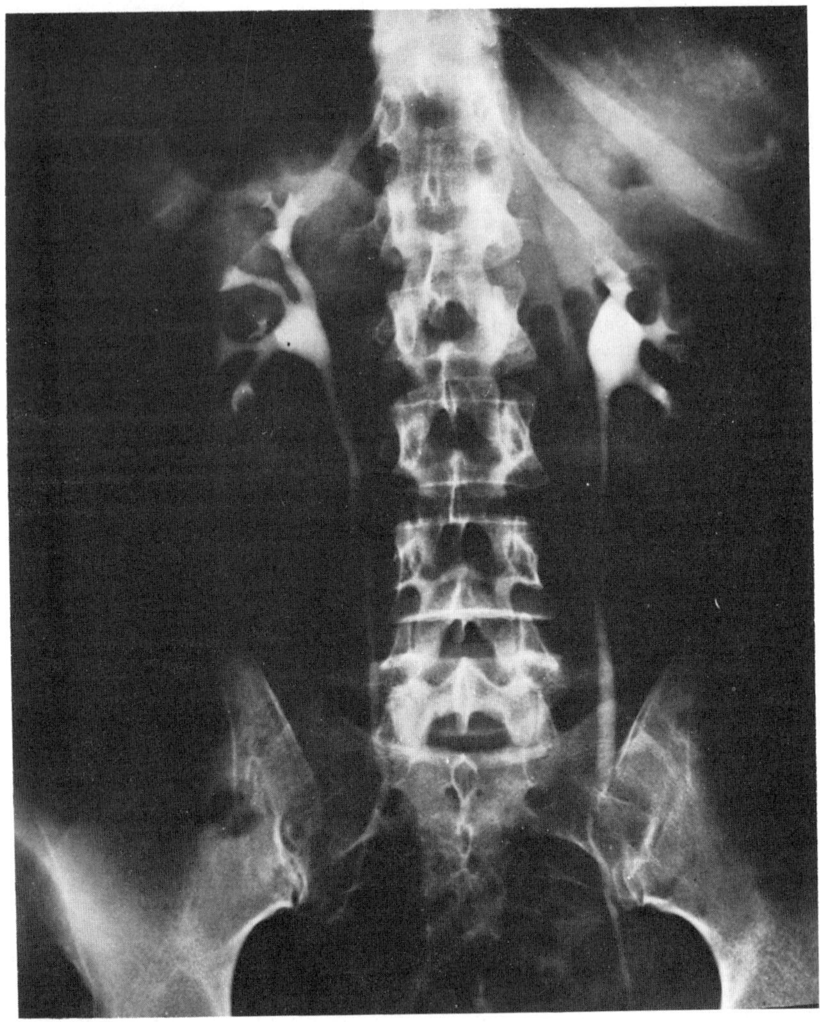

FIGURE 18-1
Normal intravenous pyelogram. Note the calyces, renal pelvis, and ureters are clearly visible. (Courtesy of James W. Lecky, M.D., Professor of Radiology, University of Pittsburgh.)

Because dyes used for this examination contain iodine, to which the patient may be allergic, a skin test with the dye should be done before it is injected intravenously. After the dye is injected, the patient frequently feels warm and is flushed. These symptoms should pass away in a few moments. Antihistamines or hydrocortisone should be readily available in the event of an allergic response. Postinjection films are taken at 5-, 10-, 15-, and 30-minute intervals.

Following the examination, the patient should be encouraged to take fluids liberally to flush any remaining dye from the urinary tract and to overcome the dehydration.

RETROGRADE PYELOGRAM

The retrograde pyelogram is similar to the intravenous pyelogram. The patient requires the same preparation; however, a sedative is usually ordered prior to the test. In this examination the dye is injected directly into the pelvis of the kidney through ureteral catheters. In order to insert ureteral catheters, the patient will have to be cystoscoped.

CYSTOSCOPY

A cystoscopic examination is done to visualize directly the bladder, to take specimens for biopsy, to obtain urine specimens from each kidney, or to insert ureteral catheters for a retrograde pyelogram. The patient should be encouraged to take liberal amounts of fluids prior to the examination so that adequate urine will be in the ureters for specimens. The examination is usually done under local anesthesia; however, if general anesthesia is used, intravenous fluids will be ordered. Usually, food is withheld prior to the examination because the discomfort of the procedure may cause nausea. A sedative or narcotic is often given just before the examination. The patient must be informed of the nature of the examination and should sign a permission form.

The patient is placed on the examination table in the lithotomy position (see Figure 18–2). The external genitalia are cleansed with antiseptic solution, and a local anesthetic is instilled into the urethra and bladder. The aseptic technique must be maintained throughout the procedure.

The cystoscope is lubricated and inserted into the patient's urethra. Urine in the bladder is removed, and sterile warm fluid is used to irrigate the bladder and to remove any material that might interfere with visualization. When the view is clear, the bladder is distended with fluid.

If kidney urine specimens are needed, small ureteral catheters are inserted into the pelvis of each kidney. Sterile test tubes labeled "Urine from the right kidney" and "Urine from the left kidney" are attached to the respective catheters.

Following the examination, the patient should be encouraged to take fluids to dilute the urine and to lessen the irritation of the lining of the urinary tract. Voiding will be painful for about a day following the examination, and some hematuria (blood in the

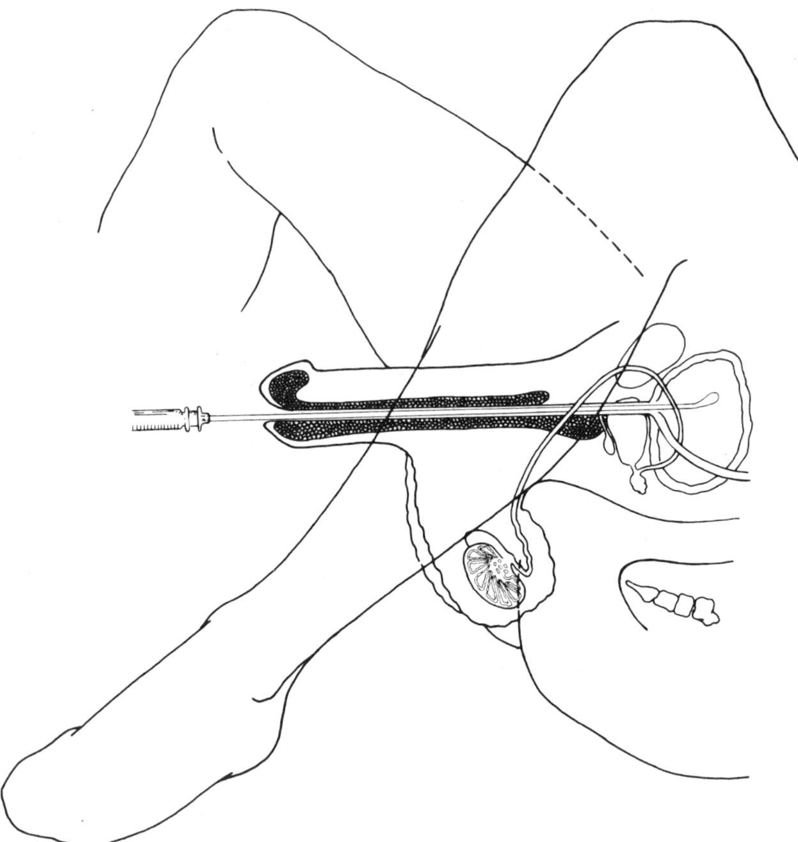

FIGURE 18-2
A cystoscopic examination.

urine) is to be expected. Because this procedure is frequently done on an out-patient basis, the patient should be thoroughly instructed concerning what is to be expected. He should be told to return to the doctor if he experiences excessive pain, fever, gross hematuria, or inability to urinate.

PHENOLSULFONPHTHALEIN TEST

The phenolsulfonphthalein (P.S.P.) test determines how well the kidney can eliminate a dye that has been injected intravenously. Prior to the injection of the dye, the patient should be given two or three glasses of water. The patient must drink at least a pint of fluid in order that he will be able to provide the urine specimens required. The first voiding is discarded, and the dye is then injected intravenously. After the time of the injection is recorded, complete voided urine specimens are collected at intervals of 15 minutes, 30 minutes, 1 hour, and 2 hours. The specimens are labeled with the exact time of the voiding. Normally 80 percent of the dye will be eliminated in the 2-hour period.

BLOOD CHEMISTRY

When kidney disease is suspected, blood is examined for the amount of blood urea nitrogen (BUN) and the nonprotein nitrogen (NPN) content. Normally, the BUN is 12 to 25 mg. per 100 ml. of blood and the normal NPN is 15 to 35 mg. per 100 ml. of blood. These values may be greatly increased in certain kidney diseases.

DISEASES OF THE URINARY SYSTEM

URINARY OBSTRUCTIONS

Obstructions of the urinary tract can occur any place in the tract. Obstructions may be caused by strictures, tumors, stones, spasms of the ureters, cysts, or a kink in the ureter. Regardless of the cause of the obstruction, if it is not corrected, it can result in kidney damage.

When urine cannot bypass the obstruction, it backs up and will cause distention of the structures above the blockage. For example, if a stone is lodged in a ureter, distention of the ureter—called a hydroureter, occurs. If the stone cannot be passed or is not removed, it can lead to hydronephrosis. In hydronephrosis, the pelvis of the kidney becomes distended; the pressure of the urine in the pelvis of the kidney compresses the medullary and cortical portions of the kidney as well as the blood vessels. This condition can result in permanent kidney damage.

If the obstruction is low in the urinary tract—for example, a urethral stricture—there may be bilateral hydroureters and hydronephrosis. Figure 18–3 shows examples of the results of obstruction in a ureter.

In addition to obstructions causing hydronephrosis and hydroureters, the stasis of urine can lead to infections of the urinary tract. The microorganisms causing the infection may enter the urethra from the outside or may be bloodborne. The most common organism that causes infection by entering the urethra is Escherichia coli. Bloodborne infections are usually caused by streptococci, staphylococci, or pneumococci.

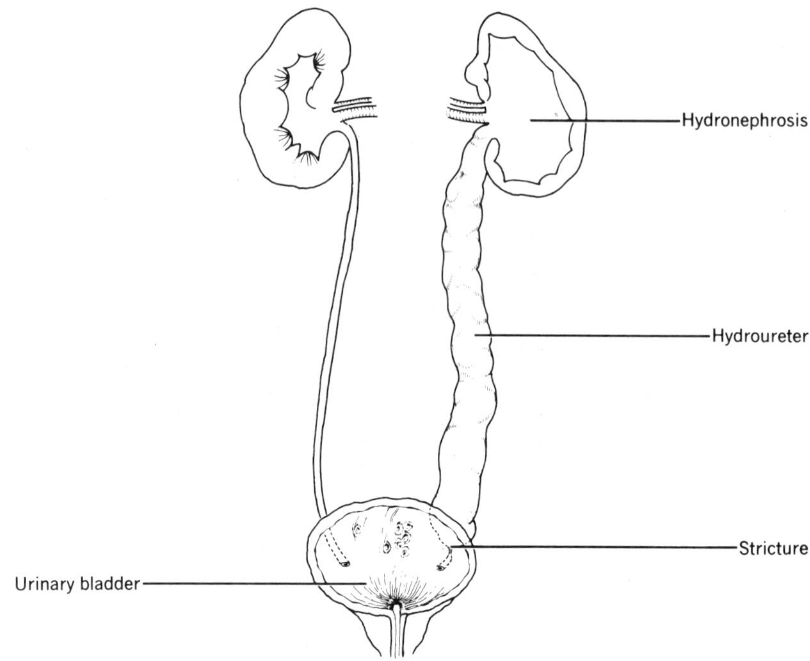

FIGURE 18-3
Hydroureter and hydronephrosis resulting from a stricture of the ureter.

The aim of the treatment for urinary tract obstructions is first to establish free flow of urine and then to remove the obstruction. These may be accomplished by cystoscopy and by the insertion of a ureteral catheter to relieve the pressure within the kidney. The ureteral catheter usually has to be left in place for several days. The drainage tubing attached to the ureteral catheter and the collection bottle must be sterile.

If the obstruction is so complete that a ureteral catheter cannot be passed, it may be necessary to insert a tube into the kidney pelvis through a skin incision. This procedure is called a nephrostomy.

After the acute phase has subsided, surgery will be done to remove the obstruction.

CALCULI (STONES)

Stone formation in the urinary tract is caused by any condition that leads to the precipitation of salts in the urine. Normally, these salts remain in solution; however, if the urine tends to be alkaline, stone formation may occur. Most fruits and vegetables favor alkalinity of the urine, whereas meats and cereals favor acidity. Cranberry juice is helpful in making the urine more acid. Patients who have a tendency toward stone formation may have to eliminate citrus fruits and carbonated beverages from their diet, because these favor greater alkalinity of the urine. They should have a fluid intake of 2500 to 3000 ml. a day. Mandelamine is a drug used to increase the acidity of the urine.

Other conditions that predispose patients to stone formation are prolonged bed rest or inactivity, parathyroid disease, osteoporosis, and low estrogen levels that occur in post-menopausal women.

Patients frequently have hematuria, pyuria, retention of urine, and, if the stone is lodged in the ureter or passing through it, severe renal colic. Pain can be so severe that the patient cannot even be examined until it is relieved. Narcotics are given to relieve the pain, and some antispasmodic such as atropine is used to relieve the smooth muscle spasms of the ureter.

Sometimes the stone will pass spontaneously and no treatment other than the relief of pain is needed. A stone in the bladder may be removed by cystoscopy. In the case of renal calculi, a nephrotomy (an incision into the kidney) or even a nephrectomy (removal of the kidney) may be necessary.

STRICTURES

Strictures are bands of fibrous tissue that reduce the circumference of the urethra or ureter. They may be caused by infections or by trauma.

Symptoms of urinary tract strictures are a slow stream of urine, burning, frequency, retention of urine, and difficulty in voiding. The condition may be treated by dilating the tract with instruments and catheters. The procedure is quite painful, and it may be necessary for the doctor to repeat the dilatation several times. He inserts the largest size of catheter or instrument that will fit and then gradually increases the size until the tract can be dilated to near its normal circumference. The patient should be told that following this treatment some hematuria may occur. If the bleeding persists, he should return to the doctor. Voiding will be painful for several days. Warm baths help to relieve the discomfort.

If the stricture cannot be relieved by dilatation, the patient must undergo surgery. Surgical procedures involve cutting the bands of scar tissue and in some cases excising the constricted area and inserting a graft.

TUMORS

Tumors of the kidney or urinary bladder are usually first evidenced by painless hematuria. Occasionally, cytological examination of the urine will reveal tumor cells. Most kidney tumors are malignant and metastasize early. As for tumors of any organ, the treatment is surgery combined with radiation and/or antineoplastic drugs.

INFECTIONS

Cystitis

Cystitis is an inflammation of the urinary bladder. It is somewhat more common in women than in men because the female urethra is shorter and organisms can more easily enter the bladder. The symptoms of cystitis include frequency, burning, urgency (a feeling of needing to void although the bladder is not full), dysuria, and hematuria. If the infection is severe enough, the patient may have chills and fever.

In order to treat cystitis, the causative organisms must be identified and appropriate antibiotics used. Warm baths will help to relieve the discomfort. Unfortunately, cystitis frequently recurs and may become chronic. Although the symptoms of chronic cystitis are similar to those of acute cystitis, they are not usually as severe.

Pyelonephritis

Pyeloneprhitis is an infection of the kidney. The disease can be either acute or chronic. The acute disease usually is a complication of an infection elsewhere in the body. Chronic pyelonephritis develops if treatment of acute pyelonephritis is not successful.

The patient with pyelonephritis has fever, nausea and vomiting, flank pain, and pyuria. Bed rest is necessary and the patient should be encouraged to take fluids to keep the urine dilute. Drugs to help acidify the urine may be ordered. Once the causative organism has been identified and its sensitivity to antibiotics has been established, the appropriate antibiotic should be ordered.

Glomerulonephritis

Glomerulonephritis is a kidney infection characterized by inflammation of the glomeruli. It can be either acute or chronic. Although the reason is not clearly understood, acute glomerulonephritis frequently follows upper respiratory tract infections. Many patients with chronic glomerulonephritis have no history of having had the acute type.

Patients with both acute and chronic glomerulonephritis have generalized edema, fever, decreased urine output, headaches, hypertension, and visual disturbances. The urine contains albumin and blood. Frequently, patients are anemic. If the disease is severe, they become lethargic or even comatose.

The treatment for glomerulonephritis is mainly symptomatic. During the acute stages, the patient should be on bed rest, and be given a diet low in sodium and high in carbohydrates. Some doctors limit protein intake during the acute phase; however, some encourage proteins to replace the protein that is lost in the urine. Fluids are encouraged; however, if the edema is marked, fluids may be limited to balance urine output. Antibiotics may be given to prevent a superimposed infection. Patients, particularly those with chronic glomerulonephritis, must avoid exposure to infections.

HEMODIALYSIS

In the case of renal diseases that are far advanced and include a marked decrease in renal function, repeated dialysis may be necessary. Dialysis is a procedure whereby the electrolyte concentration of the blood can be maintained at near normal levels in the presence of greatly reduced renal function. Patients must be hospitalized for fairly long periods of time until they are able to accept the fact that for the rest of their lives they will be dependent upon dialysis equipment (frequently called artificial kidneys) and they learn how to use the equipment. Once this is accomplished they can go home with their artificial kidney and perform the treatments themselves. Patients must usually do the dialysis two or three times a week. Although they can perform their own treatments, they must have medical supervision for the rest of their lives. They are able to work and live relatively normal lives; however, they must follow a very strict diet. Usually their diets are limited to 1 gram of protein per kilogram of body weight a day. Sodium is usually restricted, particularly if the patient has high blood pressure. Potassium intake is also frequently restricted. Most fresh fruits are high in potassium; however, canned fruits served without the juice can be included in the diet. Potatoes and most cereals are high in potassium, as are cola drinks and chocolate.

Many types of machines are used for hemodialysis, but the underlying principle is the same in all of them. A cannula (hollow tube) is placed in a peripheral artery of the patient. The cannula is then connected to the machine, which is filled with fluid. The composition of the fluid is similar to the electrolyte composition of normal plasma; it also contains glucose. As the blood flows through the dialyzer, electrolytes that are in higher concentration in the blood than in the fluid diffuse from the bloodstream into the fluid. The glucose in the dialysate (the fluid in the machine) prevents the diffusion of glucose out of the patient's bloodstream. The blood is then returned to a peripheral vein of the patient. The time required to do this procedure varies depending upon the type of equipment used and the extent of the renal damage.

In addition to maintaining patients who suffer from chronic renal disease, hemodialysis may be life saving for people who have taken overdoses of diffusible drugs such as barbiturates or some other sedatives. In these cases, dialysis need only be done until the bloodstream is free of the drug.

KIDNEY TRANSPLANT

Some patients with greatly impaired renal function are fortunate enough to have a functioning kidney transplanted into their bodies. Ideally, the kidney is donated by a near blood relative because tissues of blood relatives are less likely to be rejected. During and after a transplant, the patient must be given drugs or radiation to suppress his immune system in order to help prevent the rejection of the transplanted organ. The patient is also very susceptible to infection. For this reason, great care must be taken to prevent the patient from being exposed to sources of infection such as relatives or hospital personnel who have respiratory infections.

SUMMARY QUESTIONS

1. What equipment is needed to obtain a catheterized urine specimen?
2. List some abnormal constituents of urine.
3. How is the specific gravity of urine measured?
4. How would a urine specimen for microscopic examination be obtained?
5. How does an intravenous pyelogram differ from a retrograde pyelogram?
6. What blood chemistry studies are done most commonly for patients with renal disease?
7. What is hydronephrosis and what might cause it?
8. What foods favor alkaline urine?
9. What are the symptoms of renal calculi?

10. What might cause strictures of the urinary tract and how are these strictures treated?
11. What are the symptoms of cystitis?
12. What is pyelonephritis and what might cause it?
13. Describe the diet for a patient with glomerulonephritis.
14. Briefly explain the principle of hemodialysis and tell how it is done.

SUGGESTED READINGS

Crim, M. H., and Calloway, D. H. "A Method for Nutritional Support of Patients with Severe Renal Failure," *Nutrition and Metabolism,* Vol. 12, No. 2, 1970.

Khan, A. J., and Pryles, C. V. "Urinary Tract Infections in Children," *American Journal of Nursing,* August 1973.

Papper, S. "The Effects of Age in Reducing Renal Function," *Geriatrics,* May 1973.

Trussell, M. E. "An Attempt at a Practical Diet in Renal Failure," *Nutrition,* Spring 1971.

Tuttle, E. P. "Hemodialysis, The Right to Live: The Price of Survival," *The Professional Medical Assistant,* November/December 1971.

19 Reproductive Systems

VOCABULARY

Conception
Ejaculation
Endoscope
Epithelium
Gynecology
Hemophilia
Insemination
Ligation
Mucous membrane
Muscular dystrophy
Nucleus
Obstetrics
Pigmentation
Pituitary
Serous membrane
Symphysis pubis
Umbilicus
Vasectomy

OVERVIEW

I FEMALE REPRODUCTIVE ORGANS
- **A.** Internal organs
  - **1.** Ovaries
  - **2.** Uterine tubes
  - **3.** Uterus
  - **4.** Vagina
- **B.** External organs
  - **1.** Mons pubis
  - **2.** Labia
  - **3.** Clitoris
  - **4.** Vestibule
- **C.** Mammary glands

II MALE REPRODUCTIVE ORGANS
- **A.** Testes
- **B.** Excretory ducts
  - **1.** Epididymis
  - **2.** Vas deferens
  - **3.** Ejaculatory ducts
  - **4.** Urethra
- **C.** Accessory organs
  - **1.** Scrotum
  - **2.** Spermatic cord
  - **3.** Seminal vesicles
  - **4.** Prostatic gland
  - **5.** Cowper's glands
  - **6.** Penis

III STAGES OF REPRODUCTIVE LIFE

IV PHYSIOLOGY OF REPRODUCTION
- **A.** Fertilization
- **B.** Parturition

V FERTILITY

VI CONTRACEPTION

VII GENETICS

Both the male and the female have organs for the formation of the sex cells and a route through which these cells can unite. We shall study the similarities and differences in the male and female reproductive systems and learn the locations, structures, and functions of their organs.

FEMALE REPRODUCTIVE SYSTEM

INTERNAL ORGANS

Ovaries

The ovaries are almond-shaped organs about 1.5 inches long. They are located lateral to the uterus near the sides of the pelvis. The ovaries are attached to the posterior surface of the broad ligaments that extend from the sides of the uterus across the pelvis to the pelvic wall and floor. You should especially note the relationship of the ovaries to the ovarian (Fallopian) tubes (see Figure 19–1, which shows the ovaries at the distal ends of the tubes and slightly inferior to the tubes).

The outer surface of the ovaries is germinal epithelium. Beneath the germinal epithelium is connective tissue, in which the ovarian follicles are formed. Each follicle contains one germ cell or oocyte. During the child-bearing years of a woman, usually only one follicle matures each month (except during pregnancy when none does). The mature follicle can be seen bulging from the surface of the ovary and will rupture and release a mature oocyte. This process is called ovulation and usually occurs 14 days before the menstrual period.

The mature oocyte has 23 chromosomes; 22 of these are regular chromosomes and one is an X chromosome or sex cell. The chromosomes contain genes that will help to determine the hereditary characteristics of the child in the event that the egg is fertilized. The immature oocyte contains 46 chromosomes, as do other cells of our body; however, during the maturation of the oocyte, the chromosomes undergo a reduction division resulting in only half the original number of chromosomes.

The follicles not only contain the oocytes but also produce a hormone, estrogen. Estrogen is essential for the normal development of the female reproductive organs and the development of the secondary sex characteristics of the female (distribution of body hair, development of the breasts, and characteristic female body build).

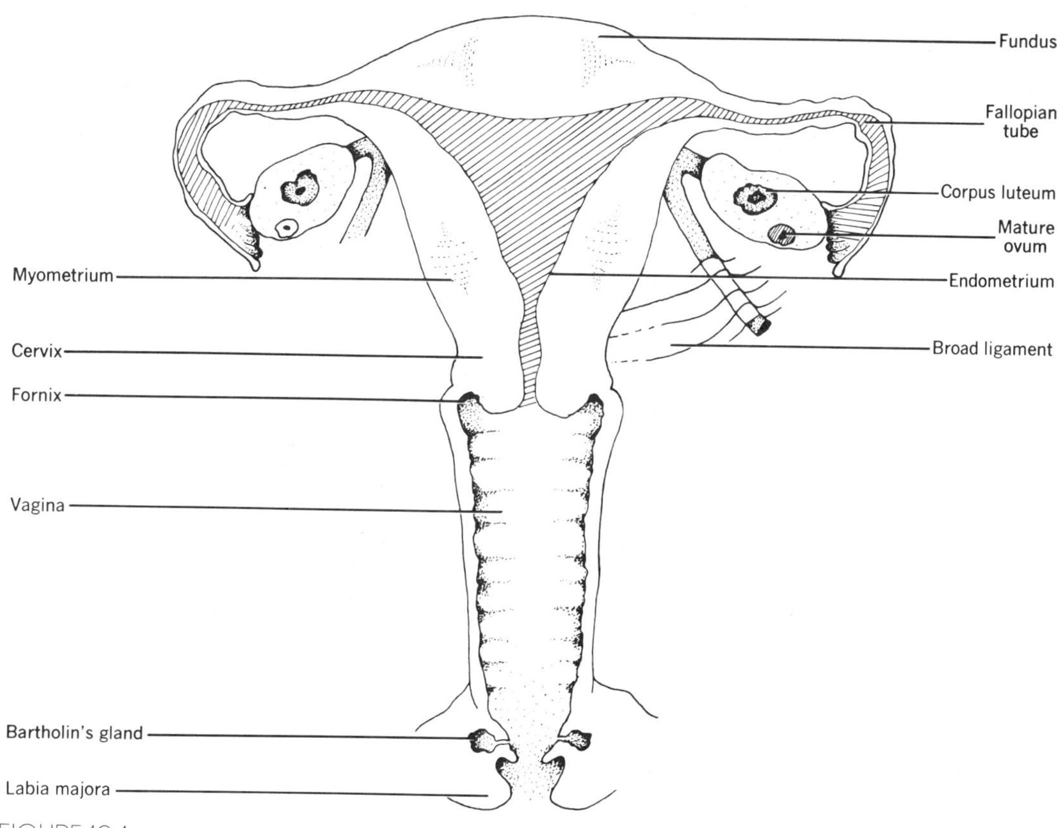

FIGURE 19-1
A posterior view of the female reproductive organs.

Corpus luteum
At the site where the follicle has ruptured, the corpus luteum (yellow body) develops. It produces progesterone and estrogen. The progesterone is essential for the preparation of the lining of the uterus to receive the fertilized egg. Under the influence of progesterone the lining of the uterus increases in thickness. If the egg is not fertilized, the corpus luteum remains for about 2 weeks and then is replaced with a white scar called the corpus albicans. If pregnancy occurs, the corpus luteum continues to develop and ovulation and menstruation cease. Late in pregnancy the progesterone from the corpus luteum promotes the development of the secretory tissue in the breasts.

Relationship of Ovarian Hormones to Pituitary Gonadotropic Hormones. The follicle-stimulating hormone (FSH) from the anterior pituitary as well as the luteinizing hormone (LH) is necessary for the production of estrogen. Of these two pituitary hormones, FSH dominates in causing the production of estrogen. Increasing levels of estrogen decrease the pituitary production of FSH and LH. The luteinizing hormone and perhaps the luteotropic hormone (LTH) cause the production of progesterone and estrogen following ovulation.

Large amounts of estrogen and small amounts of progesterone secreted by the corpus luteum decrease the output of FSH and LH from the pituitary but probably increase the output of LTH. As the corpus luteum begins to regress, a decreased production of estrogen and progesterone allows the anterior pituitary once again to begin production of FSH and LH, which will stimulate the development of a new follicle. Thus, a new cycle begins.

Uterine Tubes

The uterine or Fallopian tubes are located in the upper folds of the broad ligaments and are attached to the upper part of the uterus (see Figure 19–1). Each tube, which is about 4½ inches long, opens into the pelvic cavity near the ovary. The tubes have a serous covering, under which is found smooth muscle and a mucous membrane lining. The end of the tubes have a fringe or finger-like processes called fimbriae.

The function of the uterine tubes is to convey the oocyte toward the uterus. Fertilization usually occurs in the outer one third of the tube.

Uterus

The uterus is located in the pelvic cavity between the urinary bladder and rectum. It is a hollow, muscular, pear-shaped organ. During the child-bearing years of the woman's life it is about 3 inches long, 2 inches wide, and 1 inch thick. During pregnancy, obviously it greatly increases in size. The fundus of the uterus is the upper convex portion just above the entrance of the tubes. The body is the central portion and the cervix is the lower necklike portion (see Figure 19–2).

The external surface of the fundus and body of the uterus is serous membrane. The myometrium (muscle) layer is composed of an interlacing of longitudinal, circular, and spiral fibers. The myometrium is capable of the very powerful contractions necessary for the normal birth of an infant. The uterus is lined with mucous membrane (endometrium). The thickness of the endometrium varies during the menstrual cycle; it is thickest just before the menstrual period.

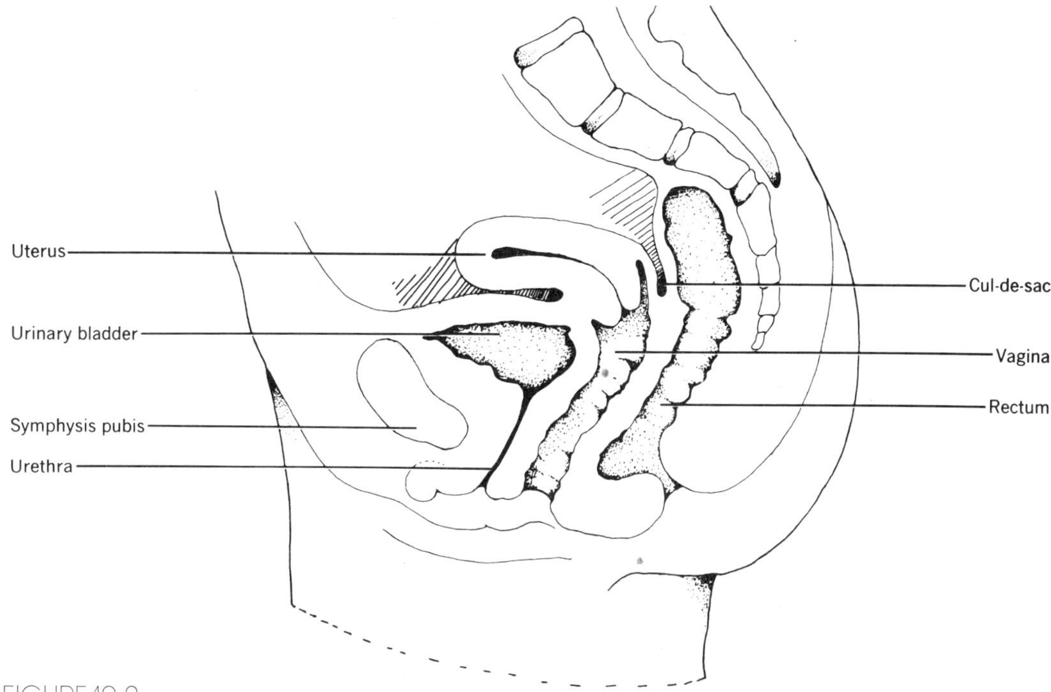

FIGURE 19-2
A lateral view of the female reproductive organs showing their relationship to other pelvic organs.

Ligaments of the Uterus. The ligaments of the uterus give it some support; however, its main support is provided by the pelvic muscles.

The broad ligaments that extend laterally from the uterus to the pelvis and pelvic floor enclose the tubes, round ligaments, blood vessels, and nerves. Two round ligaments pass from the lateral angles of the uterus below the entrance of the tubes and extend toward the sides of the pelvis, out the inguinal canals, and then into the labia majora and mons pubis. One anterior ligament extends from the uterus to the urinary bladder; a posterior ligament extends from the uterus to the rectum. The uterosacral ligaments extend from the posterior part of the cervix on either side to the sacrum.

Normally, these ligaments hold the uterus in a position in which the fundus tilts forward over the urinary bladder and the cervix points downward and back (see Figure 19–2). A full urinary bladder tilts the fundus of the uterus backward. For this reason, the bladder should be emptied prior to the gynecological examination.

The function of the uterus is to retain the fertilized egg during its growth and development. It sustains the growing fetus and during the birth process produces powerful contractions to expel the mature infant.

Vagina

The vagina is located behind the urinary bladder and urethra. It is anterior to the rectum. This musculomembranous tube extends down and forward from the uterus to the vulva (external genitalia). The lining of the vagina is mucous membrane arranged in many folds, or rugae. Mucus produced by this lining provides lubrication and has a low pH unfavorable to the growth of some bacteria. The hymen is a fold of connective tissue that may partially close the external orifice of the vagina.

The fornix is a recessed area around the cervix (see Figure 19–1). You should note that the posterior fornix is more recessed than the anterior fornix. It is through this posterior fornix that a small incision can be made to insert an endoscope to view the pelvic cavity. The cul-de-sac of Douglas (as seen in Figure 19–2) is the lowest point in the pelvic cavity.

The functions of the vagina are to serve as a passageway for menstrual flow, to receive the erect penis during sexual intercourse, and to serve as the birth canal.

EXTERNAL ORGANS

Mons Pubis

The mons pubis or mons veneris is a rounded eminence, or protuberance, anterior to the symphysis pubis. It is a fat pad covered over with skin and hair.

Labia

The labia majora are large longitudinal folds of skin and fatty tissue extending from the mons to the anus. The labia minora are smaller cutaneous folds between the labia majora. These folds meet anteriorly to form the prepuce.

Clitoris

The clitoris is a small body of erectile tissue analogous to the male penis. The clitoris becomes markedly distended during sexual activity.

Vestibule

The vestibule is the space between the labia minora. The urethral and vaginal openings are located in the vestibule, as are the Bartholin's glands, located on either side of the vaginal opening. These glands secrete lubricating fluid.

Perineum

Strictly speaking, the perineum is the entire external surface of the pelvic floor from the pubis to the coccygeal region. In obstetrical practice, however, the area between the vagina and the anus is called the perineum. Figure 19–3 shows the external structures of the female reproductive system.

Mammary Glands

The mammary glands, or breasts, are located anterior to the pectoralis major between the second and sixth ribs. These compound

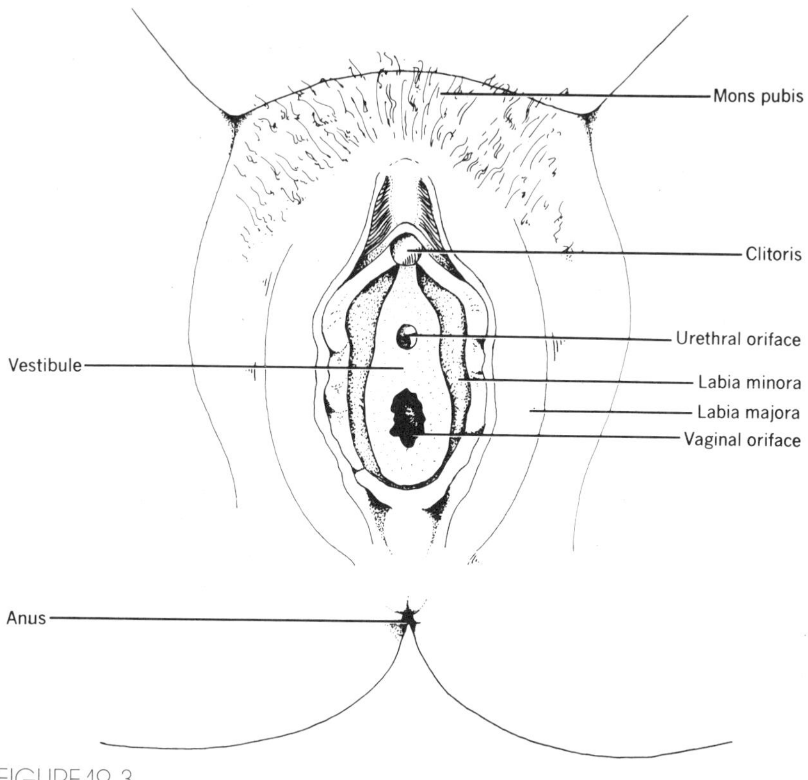

FIGURE 19-3
External female genitalia.

glands are lateral to the sternum and extend over to the axilla. The nipples, located just above the fifth rib, are smooth muscle tissue covered over with pigmented skin.

Only slight changes in the mammary tissue take place from infancy until the approach of puberty. After the onset of the menses, there are changes in the developing mammary glands with each period. In the premenstrual phase, vascular engorgement and an increase in the glands take place. During the postmenstrual phase the glands regress and remain in an inactive stage until the next premenstrual phase.

After the second month of pregnancy a visible enlargement of the breasts and increased pigmentation of the nipples take place. For the first three days after the birth of the infant the breasts produce colostrum, a small amount of thin yellowish fluid. The secretion of true milk begins on the third or fourth day and continues through the nursing period.

MALE REPRODUCTIVE SYSTEM

In our study of the male reproductive system we shall follow the same general pattern as with the study of the female reproductive system. First we shall consider the place of the sex cell formation and then the route to the outside of the body and the accessory organs along this route.

TESTES

The male testes correspond to the female ovaries. Located in the scrotum, these organs are oval-shaped structures enclosed in a fibrous capsule. Inside the testes are lobules that contain the seminiferous tubules with germinal cells to produce the spermatozoa. These tubules open into the epididymis.

In fetal life the testes are formed in the abdomen, near the kidneys. As the fetus grows, the testes move downward through the inguinal canal and usually enter the scrotum before birth. The descent of the testes into the scrotum is important to the proper functioning of these glands. Spermatogenesis, the formation of the male sex cells, can occur only at temperatures lower than that within the abdominal cavity. For this reason, if the testes do not descend into the scrotum where the temperature is relatively low, sterility will result. Failure of the testes to descend from the abdomen is called cryptorchidism.

Just like the mature oocyte, the mature sperm has 23 chromosomes. The immature spermatocyte has 46 chromosomes, one of which is an X chromosome and one a Y chromosome. When the reduction division forming the mature sperm takes place, the Y, or male, chromosome passes to one of the sperm and the X chromosome passes to the other. If an oocoyte is fertilized by a sperm with an X chromosome, the combination leads to the formation of a female. On the other hand, combination of a sperm containing a Y chromosome with an oocyte creates an XY pattern and causes the development of a male child.

In addition to the production of sperm, the testes produce a male hormone, testosterone. This hormone is essential for the development of the male secondary sex characteristics such as growth of hair on the face and body, an increase in skeletal muscle mass, and the growth of the larynx that causes the deeper pitch of the male voice. Unfortunately, the precise action of the testosterone on spermatogenesis is not known; however, in the absence of testosterone sperm will not develop.

EXCRETORY DUCTS

Epididymis

The epididymis is an elongated, triangular-shaped structure located at the upper and posterior part of each testis. It receives the sperm from the tubules, and the maturation of the male sex cells is completed here (see Figure 19–4).

Vas Deferens

The vas deferens are muscular tubes about 18 inches long. They lead one from each epididymis up through the inguinal canal into the pelvic cavity, cross to the inferior surface of the urinary bladder, and unite with the ducts of the seminal vesicles to form the ejaculatory ducts.

Ejaculatory Ducts

The ejaculatory ducts are two short tubes that descend through the prostatic gland and empty into the prostatic portion of the urethra (see Figure 19–4).

Urethra

The male urethra has both an excretory and reproductive function. The urethra in the male is about 7 or 8 inches long and leads from the bladder to the external opening. It has three portions: the prostatic, the membranous, and the cavernous urethra.

ACCESSORY STRUCTURES

Scrotum

The scrotum is a pouch of thin, dark skin continuous with the skin of the groin and perineum. When it is cold, the smooth muscles within the walls of the scrotum contract and bring the testes closer to the warmth of the body. When it is hot, the muscles relax so that the sperm being formed and stored can be kept at an optimum temperature.

Spermatic Cords

The spermatic cords extend from the testes through the inguinal canal and terminate at the internal inguinal ring. These cords contain the vas deferens, blood vessels, and nerves.

Seminal Vesicles

The seminal vesicles are two membranous pouches directly behind the urinary bladder. They produce a thick alkaline secretion that aids in the motility of the sperm. The ducts of the seminal vesicles unite with the vas deferens to form the ejaculatory ducts (see Figure 19–4).

Prostatic Gland

The prostatic gland lies directly below the urinary bladder and surrounds the prostatic portion of the urethra. It also adds an alkaline secretion to the sperm to aid in their motility and to neutralize the acidity of the urethra. The prostate produces prostaglandins (see page 308) to facilitate ejaculation.

Cowper's Glands (Bulbourethral Glands)

Cowper's glands are two small glands just below the prostate on either side of the membranous urethra. Their ducts, which open into the cavernous urethra, secrete a small amount of alkaline fluid into the urethra just before the sperm reach this point in the pathway.

Penis

The penis, which is suspended from the front and sides of the pubic arch, is composed of three cylindrical masses of cavernous (erectile) tissue. The spaces in this cavernous tissue become congested with blood during sexual activity and cause an erection.

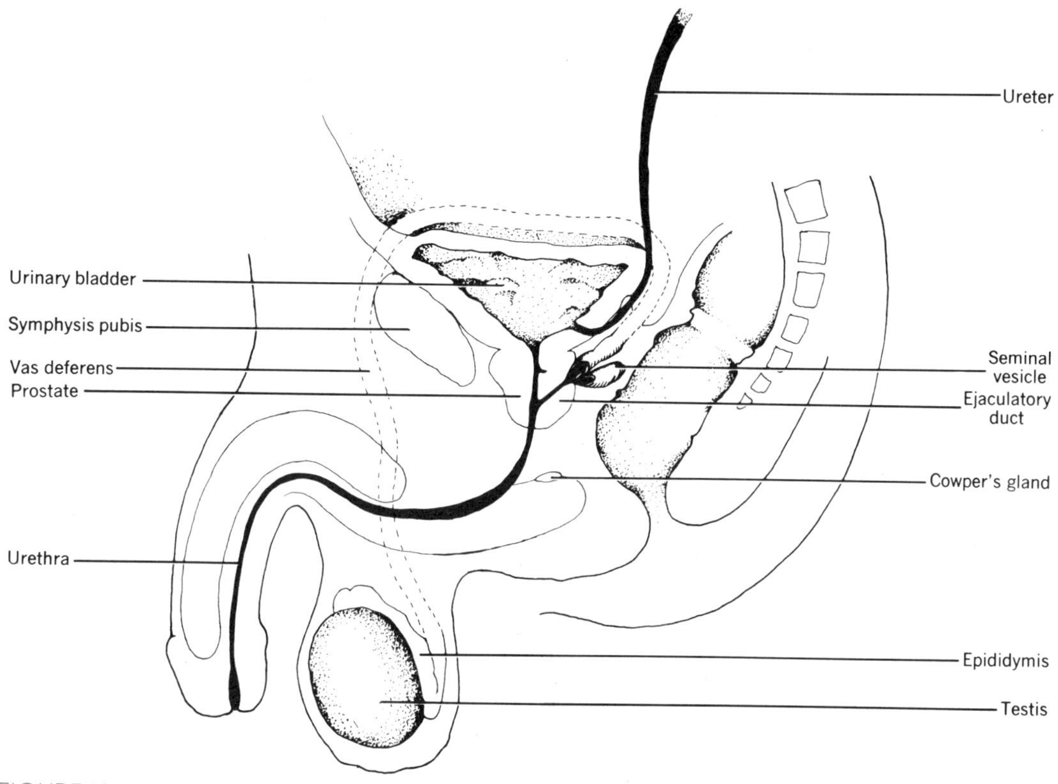

FIGURE 19-4
A lateral view of the male reproductive organs and the relationship of these structures to others in the male pelvis.

The glans penis, a cone-shaped enlargement at the end of the penis, is covered over by a fold of skin called the prepuce. Circumcision is the surgical removal of the prepuce.

Semen

Semen is a thick grayish-white fluid composed of the secretions of the various glands and about 30 to 200 million sperm per ml. During intercourse approximately 4 to 6 ml. of semen are discharged. Only one sperm can fertilize an oocyte; the remainder disintegrate.

STAGES OF REPRODUCTIVE LIFE

PUBERTY

In the United States males reach puberty about the age of 14 to 16. At this time they can ejaculate functional sperm. The larynx increases in size and creates changes in the voice. The external genitals grow rapidly and facial and body hair appear.

The onset of female puberty is marked by the first menstrual period, which usually occurs about the age of 12. The breasts begin to develop, the pelvis widens, and fat pads develop on the hips. Hair grows on the mons pubis, on the labia majora, and in the axilla.

Menstruation

Menstruation is a periodic discharge of bloody fluid and some endometrial tissue from the uterine cavity. Ordinarily, the menstrual period occurs every 25 to 32 days from the onset of puberty to menopause, except during pregnancy and lactation. It is not uncommon, however, for young women, particularly when they are experiencing unusual changes in their lives, such as making the transition between high school and college life, to go for several months without having a menstrual period.

In describing the menstrual cycle, we shall start from the first day of the menstrual period. The first phase, called the period of degeneration (menstruation), usually lasts 4 or 5 days. Endometrial epithelium is cast off and some capillary bleeding from the endometrium takes place. Frequently, there is also a tendency for other mucous membranes such as the gums to bleed or to become slightly endematous (swollen).

The postmenstrual period lasts for 2 or 3 days. During this time new epithelium is being formed. Following this period is an interval that lasts about 14 days. During this interval the ovarian follicle is growing rapidly and ovulation occurs. Regardless of the length of the entire menstrual cycle, ovulation occurs 13 to 15 days before the next period of degeneration.

The premenstrual period is 5 to 6 days preceding menstruation. During this time the endometrium thickens and congests with blood. It is not uncommon for the young woman to experience feelings of emotional stress during this premenstrual period.

Menopause

Menopause—the end of menstruation, ovulation, and the childbearing period of life—occurs about the age of 45 to 50. However, it is not uncommon for women who are beginning menopause to continue to ovulate although menstruation has not occurred for many months. In such cases pregnancy may occur.

Because of a decrease in the amount of estrogen atrophy of the breasts, uterus, ovaries, and vaginal mucosa takes place. Hot flashes, alternating vasoconstriction and dilatation, are common complaints of women who are undergoing menopause. Sometimes emotional upsets, depression, and a decreased tolerance of stress occur. Because without estrogen there tends to be a desalting of the bones, fractures may occur with relatively little trauma.

It may be possible to avoid all of these post menopausal problems by the administration of estrogen. If a woman is on estrogen therapy, she should be under the careful observation of a gynecologist because the proper estrogen dosage varies somewhat.

PHYSIOLOGY OF REPRODUCTION

Spermatogenesis begins about the age of 12 in the seminiferous tubules under the influence of gonadotropic hormones from the pituitary and continues throughout adult male life. Viable sperm may be stored in the epididymis and vas deferens for about 6 weeks. They can survive at body temperature for 24 to 72 hours. The sperm is about 3/100 mm. in diameter and is shaped like a tadpole.

Primary oocytes are already formed at the birth of the female and have begun the first reduction division. This division is finished at puberty. The mature oocyte is about 1/10 mm. in diameter.

FERTILIZATION AND PREGNANCY

Penetration of the sperm into the oocyte and the union of the nuclei of the two cells are possible because the sperm releases an enzyme, hyaluronidase, that allows the outer layer of the oocyte to be more permeable. Fertilization usually takes place in the outer one third of the uterine tube a few hours after ovulation and insemination.

The zygote is a fertilized oocyte with 46 chromosomes and all the potentials of the new individual: size, hair color, sex, and so forth. As stated earlier, the female will have received an X chromosome from both the mother and the father. The male will have received an X chromosome from the mother and a Y chromosome from the father.

The zygote divides and subdivides rapidly and soon appears as a mulberry-shaped mass of fluid-filled balls of cells. This zygote is moved along the uterine tube toward the uterus. It takes the zygote about 3 or 4 days to reach the uterus, where it will be implanted. Implantation occurs about a week after fertilization.

At the beginning of the third week of pregnancy the zygote develops into an embryo and by the ninth week into a fetus. The fetus is surrounded by a thin transparent sac (the amnion) filled with fluid. Around the amnion chorionic villi develop and penetrate the endometrium.

The fetus is nourished by the placenta. The placenta not only serves as a nutritive and excretory organ for the fetus but after the third month secretes estrogen, progesterone, lactogen, and relaxin. Relaxin increases the mobility of the sacroiliac joint and the symphysis pubis.

The umbilical cord connects the fetus to the placenta. It contains two arteries going from the fetus to the placenta and one vein from the placenta to the fetus. Figure 19–5 shows various stages in the development of the embryo and fetus as well as the relationship of an advanced fetus to the placenta and membranes.

PARTURITION

Parturition, or the birth of the child, is accomplished by periodic uterine contractions. These contractions (labor pains) can be palpated and should be timed. When the contractions begin to occur at regular intervals of a few minutes apart, the birth of the child can usually be expected to occur within a short period of time.

The first stage of labor occurs with the dilatation of the cervix. Normally, at this time the amnionic sac ruptures and the fluid escapes through the vagina. The second stage of labor is the descent and delivery of the infant. The umbilical cord is tied and a few minutes later the placenta is delivered (the third stage of labor).

Following the birth of the child the fundus of the uterus can be palpated about the level of the mother's umbilicus. It takes about 6 to 8 weeks for the uterus to return to near its original size. Breast feeding helps in this involution.

FERTILITY

Fertility depends upon normal functioning of the male and female reproductive organs. In addition to this, several factors contribute to fertility at the time of insemination.

Following ovulation, the oocyte remains viable for about 12 to 24 hours. It is during this period that female fertility is greatest. The sperm retain their fertilizing power for about 36 hours. The sperm must be able to produce sufficient hyaluronidase to allow for penetration of the oocyte. The sperm count of the semen should be about 70 million per ml. Sperm counts of 30 million may result in infertility. The sperm must be sufficiently mobile to travel to the uterine tube, and the semen must have a high enough pH to allow the sperm to survive. Clearly, any obstruction in the pathway (either of the male or female structures) will result in infertility.

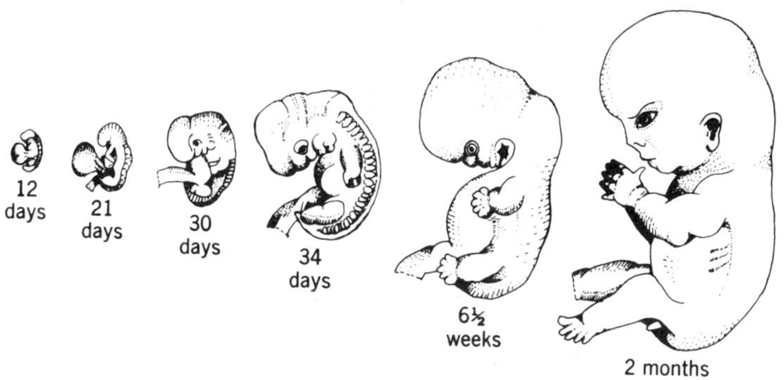

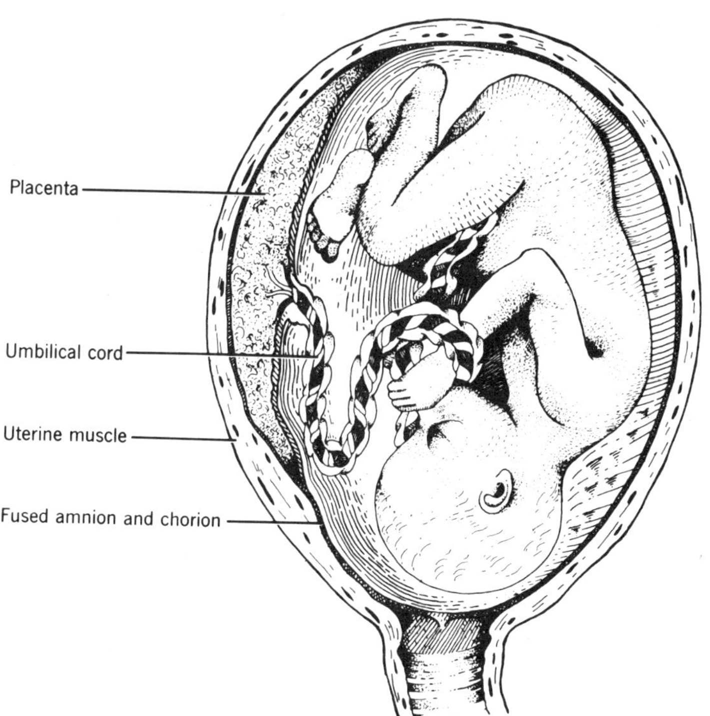

FIGURE 19-5

Various stages of development of the embryo and a section of the uterus showing the relationship of an advanced fetus to the placenta and fetal membranes. (Modified from E. E. Chaffie and E. M. Greisheimer, *Basic Physiology and Anatomy,* Philadelphia, Lippincott.)

CONTRACEPTION

There are many methods of preventing conception and the effectiveness of these methods varies considerably. Ideally, the method used should be esthetically acceptable to both partners, simple to employ, inexpensive, and generally speaking should not result in permanent infertility. Ligation (tying) of the uterine tubes and vasectomy (surgical resection of a segment of each vas) will result in permanent infertility. Vasectomy valves are available; in cases where these are implanted, the device does not cause permenent infertility.

Information concerning methods of contraception and various contraceptive devices, as well as instruction for their use, can be obtained from the local Planned Parenthood Association.

A national sample of 6752 married or formerly married women of reproductive age who participated in a 1970 National Fertility Study reported on the methods of contraception they used and the effectiveness of these methods.

Only 4 to 5 percent of the women who used oral contraceptives or intrauterine devices failed to prevent unwanted pregnancy over a year's time. Twenty percent of those who used foams or rhythm failed to prevent pregnancy. Of those who used diaphragms 17 percent failed. Ten percent of the condom users experienced failures.

In India women who have not had "educated" instructions concerning the prevention of unwanted pregnancies nonetheless have for generations practiced a method of contraception that has a remarkably rational basis. They hollowed out a half of a lime and inserted it in their vagina in such a way as to cover their cervix.

GENETICS

Your phenotype (how you appear) depends upon several factors. The genetic makeup that is determined by the genes you have received from your parents is obviously of great importance. However, certain conditions can cause permanent changes in the hereditary material.

Examples of causes of mutations or permanent changes in hereditary material are: exposure of a parent to radiation, viral infections (such as measles in the mother during the early stages of pregnancy), and probably many drugs taken by mothers-to-be during pregnancy. Much evidence shows that the age of the parents may affect the infant. Generally speaking, the older the parents, the more likelihood of bearing an infant with hereditary defects.

There are several fundamental patterns by which the inherited characteristics of an individual are determined. The genes on the chromosomes are responsible for certain characteristics. One gene can affect more than one trait, or many genes can affect a single trait, as is the case in muscular dystrophy.

A person can be homozygos, in which case he has received from each parent the same gene for a trait; or he may be heterozygos, if he has received from each parent a different type of gene for a particular trait. In this instance, the phenotype will be that of the dominant gene.

Simple dominance is the expression of one contrasting gene over another for the same trait. For example, the gene for tall is dominant over the gene for short stature. If a person who is homozygos tall mates with a person who is homozygos short, the offspring will be tall; however, the genetic makeup for height is heterozygos. If two individuals mate who are heterozygos for the genes determining height (they are both tall—however, each has a recessive gene for short stature), three quarters of their children will be tall and one quarter will be short. This is not to say that, if they have four children, three will be tall and one will be short: With each pregnancy the chances of receiving the genes for the short stature are the same, one in four.

With codominance, the genes have equal expression. Blood types for type A blood and type B blood are codominant. Both A and B types are dominant over type O. If a person who is homozygos for type A mates with a person who is homozygos for type B, the offspring will have type AB blood. If, however, these parents are phenotype for A and B types but their genotypes are different (they have a recessive gene for one of the other blood types), the offspring will not necessarily have type AB blood.

Many hereditary characteristics are X-linked. The X chromosome is larger than the Y chromosome. In a male who has both an X and a Y chromosome, the genes on the upper part of the X that he received from his mother have no pair for them on the shorter Y chromosome and therefore will be expressed. Because the female with two X chromosomes will have another gene to pair with, the trait will not be expressed. However, she carries the trait and can transmit it to her offspring. It is by this mechanism that hemophilia and many other hereditary diseases are transmitted. The female carries the trait and transmits it to some of the male offspring. The trait will not be expressed in a female unless she has received a defective X chromosome from both her mother and her father.

The criteria for determining characteristics that are X-linked are that they are expressed more frequently in males. The traits are transmitted from an affected male through his daughters to half of the daughters' sons. X-linked traits are not transmitted directly from father to son.

Whether hereditary material is transmitted by simple dominance or codominance or is X-linked, the trait is not always expressed at birth. Each gene has its characteristic time for action. Muscular dystrophy does not appear for 10 or 15 years. Hereditary baldness (X-linked) does not appear for 25 to 30 years. Unfortunately, Huntington's chorea, a degenerative disease of the cerebral cortex, does not appear for 30 to 40 years, well after the affected individual has had opportunity to pass the trait on to an offspring.

In the past, genetic counseling for the most part could do little more than tell a couple the probability of passing on to their children an undesirable trait. Today it is possible to sample the amnionic fluid early in pregnancy and to determine whether the trait will be present in the child or not. Depending upon the seriousness of the defect, the parents can decide whether or not to have the pregnancy terminated.

SUMMARY QUESTIONS

1. Where are the oocytes formed?
2. Where are the sperm formed?
3. List the hormones produced by the ovary and discuss the functions of these hormones.
4. List the hormones produced by the testes and discuss the functions of these hormones.
5. Describe the structure of the uterus.
6. Where does the fertilization of the oocyte usually take place?
7. List some factors that may cause infertility.
8. Describe the first, second, and third stages of labor.
9. What are the functions of the placenta?
10. Which female hormone dominates during the first half of the menstrual cycle?
11. What causes this hormone to be produced?
12. Name and describe the ligaments of the uterus.
13. Describe the normal position of the uterus.
14. What is the function of the smooth muscle of the scrotum?
15. Where is the epididymis located and what is the function of this structure?
16. What is the function of the vas deferens?
17. Where is the prostate located?
18. What structures are contained within the spermatic cord?
19. What causes the erection of the penis?
20. List the phases of the menstrual cycle.
21. List in sequence all of the structures through which the sperm must travel from the place of sperm formation until fertilization occurs.

22. List some possible causes that may result in permanent damage to the hereditary material.
23. Explain homozygos and heterozygos.
24. In terms of genetics, what is the difference between simple dominance and codominance?
25. How are X-linked hereditary characteristics transmitted?

SUGGESTED READINGS

Berger, A. J. *Elementary Human Anatomy.* New York, N.Y.: John Wiley & Sons, 1964.

Best, C. H., and Taylor, N. B. *The Human Body: Its Anatomy and Physiology.* New York, N.Y.: Holt, Rinehart & Winston, 1963.

Grim, L. M., and Kowalski, K. "Changing Patterns of Obstetric Care," *American Journal of Nursing,* October 1973.

Manisoff, M. T. "Intrauterine Devices," *American Journal of Nursing,* July 1973.

Robinson, A. M. "Infertility and What Can Be Done About It," *RN Magazine,* April 1973.

Robinson, A. M. "Birth Control Through Sterilization," *RN Magazine,* April 1973.

Sasmor, J. L., *et al:* "The Childbirth Team During Labor," *American Journal of Nursing,* January 1973.

Wilbur, C., and Aug, R. "Sex Education," *American Journal of Nursing,* January 1973.

20 Diseases of the Reproductive Systems

VOCABULARY

Analgesic
Anticoagulant
Biopsy
Caesarean section
Cystoscopy
Cytology
Diplococcus
Endocrine
Endometrium
Estrogen
Hematuria
Hirsutism
Metastasis
Motility
Ovulation
Peritonitis
Pitocin
Placenta
Postpartum
Proteinuria
Septicemia
Serology
Serosanguinous
Spirochete

OVERVIEW

I DIAGNOSTIC PROCEDURES
- **A.** Tests to determine fertility
  - **1.** Semen examinations
  - **2.** Determination of time of ovulation
  - **3.** Rubin Test
- **B.** Gynecological examinations
  - **1.** Cervical biopsy
  - **2.** Dilatation and curettage
- **C.** Venereal disease examinations
- **D.** Examination of the unborn child

II DISORDERS OF THE FEMALE REPRODUCTIVE SYSTEM
- **A.** Disorders of menstruation
- **B.** Abortions
- **C.** Ectopic pregnancies
- **D.** Infections
- **E.** Tumors
- **F.** Uterine displacements
- **G.** Diseases of the breasts

III COMPLICATIONS OF PREGNANCY
- **A.** Toxemia
- **B.** Infections
- **C.** Problems associated with labor and delivery
  - **1.** Prolonged labor
  - **2.** Placental abruption
  - **3.** Placenta previa
  - **4.** Postpartum hemorrhage

IV DISEASES OF THE MALE REPRODUCTIVE SYSTEM
- **A.** Benign prostatic hypertrophy
- **B.** Malignancies
- **C.** Infections

V VENEREAL DISEASES

DIAGNOSTIC PROCEDURES

TESTS TO DETERMINE FERTILITY

Semen Examinations

Several procedures help in diagnosing the causes of infertility. One of the most simple is a microscopic examination of fresh semen. In addition to doing a sperm count, the doctor notes the motility of the sperm and any abnormality in their shape. If the sperm count is low, it is well that the couple have intercourse only during the period of time when ovulation is most likely to occur.

Determination of Time of Ovulation

A woman who has an irregular menstrual period may determine her time of ovulation by taking daily measurements of her basal body temperature. She should take the temperature early in the morning before any activity. A slight drop in the body temperature will occur near the time of ovulation, and a slight rise in the temperature—of about 0.3° to 0.5°F.—after ovulation.

At the time of ovulation the glucose content of the vaginal secretions is higher than it is at other times during the menstrual cycle. As another test, a woman can insert a strip of Testape (the same tape that is commonly used by diabetics to determine the glucose content of their urine) into her vagina and touched to the cervix. She should keep a record of the results of the test so that she can determine when her glucose level is at its peak.

Rubin Test

The Rubin test is used to determine whether the ovarian tubes are patent (open) or closed. The doctor usually performs the test right after the menstrual period ceases. Prior to the examination, the doctor may prescribe an injection of atropine to decrease the tubal spasm during the examination. He introduces a sterile cannula into the uterus and then forces gas (carbon dioxide) through the uterus and ovarian tubes and into the pelvic cavity. The doctor, or his assistant, listens with a stethoscope for a swish that indicates that the gas has escaped into the pelvic cavity. It is necessary to watch the pressure that is needed to force the gas through the tubes. If it reaches 200 mm. Hg, the tubes are probably occluded. Pressure is usually not increased above this level because of the danger of rupturing a tube. If a tube is patent, the patient will experience referred pain in her shoulder on the side of the tube.

Following the Rubin test, the patient assumes a knee-chest position for a short period of time so that the gas will rise in her pelvis. By following this procedure, she will be more comfortable sooner than if she stands up immediately.

Although the test is considered a diagnostic procedure, in some instances the gas may help to blow out an obstruction of the tube and result in fertility.

GYNECOLOGICAL EXAMINATIONS

To prepare a patient for a gynecological examination, the patient should be instructed not to douche prior to the examination, to empty her bladder, and to remove her clothing. The assistant should give her a loose gown or a sheet to wrap around her until she is positioned on the examining table. During a gynecological examination done by a male physician, a female assistant should be present in the room.

The most common position for a gynecological examination is the lithotomy position (see Figure 20–1). Although the vagina is not sterile, the equipment used for this examination must be sterilized each time it is used.

The doctor inspects and palpates the abdomen and breasts. He also inspects the external genitalia for signs of irritation or abnormal discharge from the vagina. He then does a digital examination of the vagina by inserting one or two fingers of his gloved hand into the vagina. With his other hand, he can palpate the lower abdominal wall. Between his two hands, he can palpate the position, size, and contour of the uterus and other pelvic structures.

The doctor performs a visual examination of the vaginal walls and cervix by inserting a lubricated bivalve speculum into the vagina (see Figure 20–2). With the speculum in place, he can take a

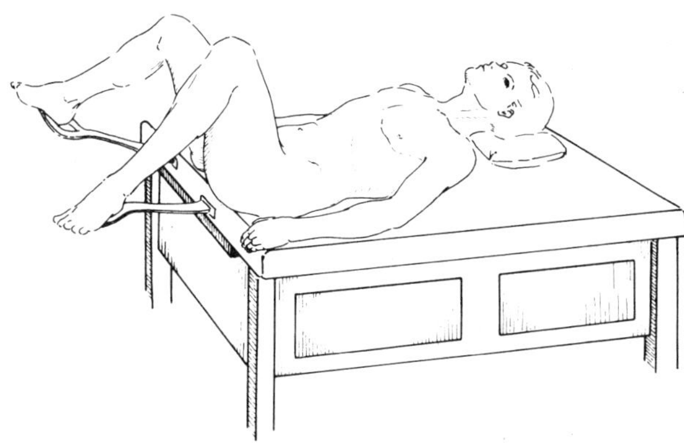

FIGURE 20-1
Lithotomy position.

cervical smear to obtain a specimen for cytological examination. This examination (Papanicolaou Test) is done to detect any abnormal cells that may indicate malignancy. If the result of the Papanicolaou test is positive or questionable, the patient should have a cervical biopsy or a dilatation and curettage done.

Cervical Biopsy

A cervical biopsy can be performed in the doctor's office or a clinic. Although the patient may experience some discomfort, the procedure is not painful as the cervix does not have pain receptors. The specimen, properly labeled, is sent to the laboratory in a bottle of 10 percent formalin. If the patient has had vaginal packing inserted, she should be instructed not to remove it for 24 hours. Some vaginal discharge and slight bleeding may occur for a few days following the procedure. If there is excessive bleeding, the patient should return to the doctor. Because of the danger of infection, the patient should not use tampons, douche or have sexual intercourse for about a week after the biopsy.

Dilatation and Curettage (D and C)

Preparation of a patient who is to have a dilatation and curettage is similar to that of any patient who is to have a general anesthetic. The procedure itself involves dilating of the cervix and the removal of endometrial tissue. A specimen of the tissue will be sent to a pathologist for examination. Following the surgery the patient usually experiences no discomfort. She will have some serosanguinous discharge for a short period of time.

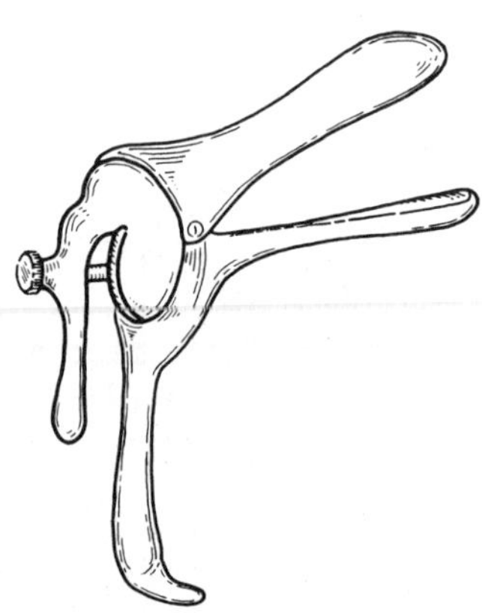

FIGURE 20-2
A bivalve speculum used for a gynecological examination.

Venereal Disease Examination

Examinations used to detect venereal diseases include microscopic examinations and serological tests. The gonococcus that causes gonorrhea is a gram negative diplococcus. Gonorrhea can also be detected by placing a drop of a person's serum on a slide coated with gonorrhea antigen. If the person has gonorrhea, agglutination (clumping) will occur within 2 minutes. In a small percentage of cases, agglutination may occur in the absence of gonorrhea; however, this simple test is of great value in screening large numbers of people. If the test is positive, microscopic examination can be done to confirm the diagnosis.

Serology tests, such as the Wasserman or Kahn tests, are used to diagnose syphilis. Sometimes these tests will result in a false positive in people who have collagen disease or have had a recent inoculation. If a doctor suspects central nervous system syphilis, he may have to examine the cerebrospinal fluid.

EXAMINATION OF THE UNBORN CHILD

Amniocentesis

Amniocentesis can be done as early as the twelfth week of pregnancy. After locating the position of the fetus in the uterus, the doctor inserts a sterile needle through the woman's abdomen into the amniotic sac that holds the fluid surrounding the fetus. Less than an ounce of this fluid is withdrawn and examined by a geneticist. By examining the chromosomal structures and enzyme components of the cells in the fluid, the geneticist is able to identify the presence of several kinds of inherited or metabolically derived defects that might be afflicting the fetus. He can also determine the sex of the developing infant, but this information alone is seldom of any real clinical significance.

Whether or not an amniocentesis should be done is based on a detailed family history and extensive blood studies. The risk of birth defects accelerates greatly with childbearing at an advanced reproductive age. For example, a twenty-year old woman has only a one in 2,500 chance of bearing a child suffering from Down's syndrome (an inheritable anomaly generally characterized by marked physical and/or mental retardation), at age forty that chance increases to one in 100, and at age 45 the risk further escalates to one in 40.

Sonography

With sonography the doctor is able safely to obtain an image of the fetus within the womb. Prior to the development of this technique, an image could be seen only by means of X-ray studies, which, because of their possible association with birth defects, now are utilized only in extreme situations.

Sonography employs sound waves to produce a two-dimensional image of the fetus. The image is projected on the screen of an oscilloscope and can be photographed for future reference. A long, thin transmitting apparatus called a transducer is placed in direct contact with the woman's abdomen; and electrical current is then introduced to cause the transducer to emit sound waves as it is moved back and forth across the abdomen. These waves traverse the abdomen and reverberate echoes from the various levels of the uterine cavity. The echoes pass back through the transducer and register as bright dots on the screen of the oscilloscope to form an image of the fetus.

Sonography can be used prior to performing an amniocentesis; it makes possible more accurate location of the position of the fetus and placenta in the uterus. It is also a valuable aid in differentiating between placenta previa and placental abruption. These conditions, which will be discussed later in this chapter, have similar symptoms but must be managed quite differently. Among its other uses, sonography may be used to pick up the outline of a pregnancy as early as 6 weeks, identify a multiple pregnancy within the same time span, and detect a stillbirth or deviation from normal fetal growth and development at any time after 12 weeks.

Abdominal Electrocardiography

In the past the only techniques available for monitoring the fetal heart rate were phonocardiography (an application of a microphone to the mother's abdomen to pick up the fetal heart sounds or Doppler ultrasound (the usage of a small transducer to direct a sound beam to the fetal heart so that its beat comes back in the form of an echo). The effectiveness of these methods is limited due to the presence of abdominal noise and to the characteristically excessive movement of the high-risk infant.

In abdominal electrocardiography two tiny electrodes are placed on the mother's abdomen, one to record her heart rate and the other to monitor the heart rate of the fetus. Outside interference can be filtered out and the greater strength of the mother's heart beat can be moderated so as not to obscure the weaker heart signal of the fetus.

Direct Monitoring of the Fetal Heart

In high-risk pregnancies, once labor has passed beyond the initial phase a catheter can be passed through the cervix and into the uterine cavity to record intrauterine pressure and a small electrode affixed to an accessible part of the fetus to monitor fetal heart rate. During labor the fetal heart rate normally fluctuates between 120 and 160 beats per minute. It is rapid with the onset of uterine contractions and decreases with the relaxation of the uterine muscles. If the heart rate ceases to fluctuate regularly, the fetus may no longer be able to compensate for the strains brought on by labor and immediate delivery may be necessary.

DISORDERS OF THE FEMALE REPRODUCTIVE SYSTEM

DISORDERS OF MENSTRUATION

Dysmenorrhea

Painful menstruation is called dysmenorrhea. Usually the discomfort is more severe if the woman is experiencing unusual tension, fatigue, or cold. Most cases of dysmenorrhea have no detectable organic cause; however, the patient should be examined by a gynecologist so that possible pathology can be discovered and treated. For example, malposition of the uterus, which can cause dysmenorrhea, can be surgically corrected.

If there is no organic cause for the dysmenorrhea, the application of heat to the lower abdomen, rest, and a mild analgesic such as aspirin will give symptomatic relief.

Amenorrhea

The absence of menstrual flow is called amenorrhea. This condition is normal after menopause, during pregnancy, and sometimes during lactation; however, it can also be caused by endocrine disturbances, chronic wasting disease (such as starvation and tuberculosis), and psychological factors. It is fairly common for young women who are in the process of making major changes in their way of life to miss periods.

Menorrhagia

Menorrhagia, or excessive bleeding at the time of normal menstruation, can be caused by endocrine imbalances and emotional upsets. It may also result from ovarian or uterine tumors and inflammatory diseases of the pelvic organs. It is difficult to describe how much bleeding is being experienced. A rough estimate can be made by the number of pads or tampons needed each day.

Metrorrhagia

Bleeding or spotting between periods is called metrorrhagia. The causes include the same conditions involved in menorrhagia; however, metrorrhagia can also be caused by cancer or by a threatened abortion. The amount of bleeding is not related to the seriousness of the underlying condition. Even when she experiences only slight spotting, a woman should have a gynecological examination.

ABORTIONS

Medically speaking, the term abortion is the termination of a pregnancy before the fetus is viable (about 28 weeks after conception). After this time and until the time of a full-term delivery, the expulsion of the fetus is called a premature birth.

Spontaneous Abortion

A miscarriage or spontaneous abortion usually occurs before the twelfth week of pregnancy. Maternal diseases may be the cause of the abortion; however, the most frequent cause of this type of an abortion is due to abnormal development of the fetus. Physical trauma or emotional shock do not usually cause an abortion.

Threatened Abortion

Bleeding or spotting during pregnancy may indicate that an abortion is likely to occur. About half of the women who have these symptoms lose their babies. Sometimes the doctor may prescribe bed rest in order to lessen the possibility of an abortion.

Incomplete Abortion

In this instance, some of the products of the pregnancy are expelled and some are retained. The retention of a portion of the placenta or other tissues can result in a serious infection. For this reason, all expelled tissues must be examined by a doctor, who can determine whether the abortion was complete. If it is an incomplete abortion, the woman may need a D and C to remove the retained parts.

Missed Abortion

A missed abortion is one in which the fetus has died but is not expelled. It may be retained for two months or longer. The fetus must be removed surgically or sometimes it is possible to induce uterine contractions with medications and cause expulsion of the fetus.

Therapeutic Abortions

Depending upon the religious beliefs of the physician and the patient, a physician may perform an abortion if the pregnancy might endanger the mother's mental or physical health. A therapeutic abortion might also be done if it seems certain that the baby will have serious defects. In many states, it is now legal to perform an abortion if the woman does not feel that she is physically, emotionally, or financially able to raise the child properly. Although in many instances abortion may be a desirable solution to a problem, it is certainly not a sensible substitute for instructions in how to prevent unwanted pregnancies.

ECTOPIC PREGNANCY

In an ectopic pregnancy, the fertilized ovum is implanted someplace outside of the uterus. The most common site is the fallopian tube. Because the tube has so little room for expansion, the growing fetus and placenta will rupture the tube. The woman will experience severe sharp pain and probably shock. Profuse bleeding occurs both vaginally and in the pelvic cavity.

As soon as possible the woman must have a salpingectomy, surgical removal of the tube. Because the rupture of the tube and the internal bleeding will very likely cause peritonitis, immediate and appropriate treatment of this serious complication is necessary.

INFECTIONS

Pelvic Inflammatory Disease

Inflammation of the organs of the pelvis may result when pathogens enter these structures through the vagina, the lymphatics, or the bloodstream. Symptoms include a foul vaginal discharge, backache, and abdominal and pelvic pain, in addition to fever, nausea and vomiting, dysmenorrhea, and menorrhagia.

Treatment includes bed rest, antibiotics, and warm sitz baths. Tampons should not be used as they may obstruct the flow of the discharge.

Because adhesions can block the fallopian tubes, pelvic inflammatory disease may result in sterility.

Vaginitis

The normal acidity of the vaginal secretions provides a natural defense against most organisms. However, a variety of pathogens can infect the vagina. The most common organisms that cause vaginal infections are a protozoa, Trichomonas vaginalis, and a fungus, Candida albicans.

Both of these pathogens cause a vaginal discharge called leukorrhea. The discharge is irritating and causes severe itching. Often the urinary meatus is affected and the woman will have frequency of urination and burning on urination.

Treatment involves the use of vaginal suppositories. Flagyl is an oral medication specific for Trichomonas. Usually, the patient should avoid douching because the unsterile equipment may cause her to reinfect herself.

In menopausal vaginitis, the atrophied vaginal membrane is easily traumatized and infected. Pruritis and vaginal discharge occur. Intercourse is painful. Estrogens—both oral and in the form of vaginal creams—are used in the treatment of this condition.

Puerperal Infection

Infections following childbirth are usually caused by streptococci or staphylococci. Although these infections are not common today, they can be very serious. They can cause generalized sepsis if the organisms enter the bloodstream. Possible causes of puerperal infections include rupture of the membranes several days before delivery, postpartum thrombophlebitis, and delivery of the baby under unsterile conditions.

TUMORS

Endometriosis

In this condition there is tissue that resembles that of the endometrium outside of the uterus. It may be found on the ovaries or elsewhere in the pelvic cavity. This tissue menstruates when the uterus does. Because there is no outlet for bleeding, adhesions may develop. Patients often have severe dysmenorrhea, menorrhagia, metrohagia, and pain on defecation. They are also often anemic.

The condition is relieved by menopause, either natural or surgical. Sometimes the condition can be treated successfully by hormone therapy. The objective of this medical treatment is to keep the woman in a nonbleeding phase of her menstrual cycle for a prolonged period of time.

Ovarian Tumors

Cysts are the most common tumors of the ovaries. Often women with ovarian cysts have no symptoms and no treatment is required. Occasionally, they must have the cysts removed surgically.

Some ovarian tumors may have masculinizing effects, including atrophy of the breasts, hirsutism (excess body and facial hair), and sterility. Usually, when the tumors are removed, a gradual reversal of these symptoms takes place.

Cancer of the ovaries is very dangerous because it is frequently not detected until the tumor has spread to other organs. Radiation or chemotherapy is used in conjunction with surgery.

Uterine Tumors

Fibroid tumors of the uterus are benign growths. These tumors vary greatly in size and may be single or multiple. Often women with fibroids are asymptomatic; however, if symptoms are present, menorrhagia is the most common. If the tumor is large, the woman may feel pressure in the pelvic region; if it causes pressure on the urethra, she may experience retention of urine. In such cases, the surgical removal of the tumor or perhaps the uterus may be necessary.

Cancer of the Uterus

The most common malignancy of the female reproductive tract is cancer of the cervix. Cure is possible only if the disease is discovered before it has spread. The importance of routine examinations and Papanicolaou tests cannot be over emphasized. Cancer of the cervix as well as cancer of other parts of the uterus is treated surgically, and with radiation and chemotherapy.

UTERINE DISPLACEMENTS

The uterus may be displaced backward (retroversion) or bent forward at an acute angle (anteflexion). Sometimes no symptoms occur; however, frequently the displacement causes dysmenorrhea. If the displacement causes severe discomfort, surgery may be necessary to move the uterus into a more natural position.

When the muscles of the pelvic floor are weakened, the uterus may herniate downward. This condition is called a prolapsed uterus. Symptoms include backache, fatigue, and pelvic pain. The woman may experience stress incontinence (a little urine seeps out when she coughs).

Often a rectocele or cystocele occurs when the uterus is prolapsed. A rectocele is a protrusion of a part of the rectum into the vagina, and a cystocele is a protrusion of the urinary bladder into the vagina.

Surgical repair of the prolapsed uterus and of a rectocele or cystocele is a relatively simple procedure. Generally, even women of advanced age tolerate the surgery quite well. If, however, the woman is a poor surgical risk, displacement can be reduced by inserting into the vagina a pessary, which will hold the uterus in a more normal position. If a pessary is used it should be removed, cleaned, and replaced about every six weeks.

VAGINAL FISTULAS

An opening between the urinary bladder and the vagina (vesicovaginal fistula) or between the rectum and the vagina (rectovaginal fistula) may develop. These fistulas may be congenital or develop as a result of an obstetric or surgical injury. The most common cause, however, is breakdown of tissue due to cancer or irradiation. Symptoms are most distressing, as the vagina is constantly being irritated by urine or feces.

Although surgery may help correct vaginal fistulas, it can only be done when inflammation and edema have been controlled. Unfortunately, surgery is not always successful.

DISEASES OF THE BREASTS

Cystic Mastitis

Cystic mastitis is characterized by lumps in the breasts. There may be tenderness of the breasts, especially a few days before the menstrual period. Often a well-fitted brassiere is all that is necessary to relieve the symptoms. In the case of multiple cystic disease, surgery may be necessary.

Benign Tumors

Fibroadenoma is a type of benign tumor of the breast. It is less common than are cystic diseases. The fibroadenoma is removed surgically.

Breast Malignancies

Cancer of the breast is the most common type of malignancy in women. Although the disease can occur at any age, it usually occurs after menopause. Successful treatment depends upon early detection. From the local American Cancer Society office or your doctor, you can obtain a pamphlet that gives directions for self-examination of the breasts. Medical assistants can also give a woman instructions on how to do this simple examination. Routine monthly breast examinations are the most valuable aid in detecting an abnormal lump in the breast.

As with most other types of cancer, treatment includes radical surgery, radiation, and chemotherapy. Following a radical mastectomy, the patient is often bothered by swelling of the arm on the affected side, which occurs because the axillary lymphatics have been removed. Exercises not only help in reducing the edema but are important in helping the woman regain full range of motion of the arm. "Help Yourself To Recovery," a pamphlet published by the American Cancer Society, is available at local Cancer Society offices. It describes exercises to be done following mastectomy.

Breast Abscess

Abscesses of the breasts occur most frequently following pregnancy. A localized abscess can be incised and drained. Antibiotics and warm compresses may also be ordered.

COMPLICATIONS OF PREGNANCY

Toxemia

Toxemia of pregnancy may occur in the last three months of pregnancy. It is characterized by an elevation of blood pressure, headache, edema, and proteinuria. In severe cases, it may lead to convulsions, coma, and death. If convulsions occur, the disease is called eclampsia.

Low sodium diets are prescribed. Diuretics may be used to relieve the edema. Sedatives and bed rest are helpful in lowering the blood pressure. Parenteral injections of magnesium sulfate lower the blood pressure and also act as a central nervous system depressant.

Infections

Urinary tract infections are common during pregnancy. Cystitis (infection of the urinary bladder) is characterized by urinary frequency, urgency, and burning. Acute urinary infections are treated by bed rest and specific antibacterial therapy.

With thrombophlebitis, an inflammation of a vein and clot formation within the vein, the patient will have a high fever and frequent chills. Small emboli of infected fragments of the thrombus may be discharged into the venous circulation and cause obstruction in the pulmonary circulation. Large pulmonary emboli may completely obstruct the pulmonary artery.

In femoral thrombophlebitis, the affected leg is warmer than normal and is swollen. The leg is tender, and dorsiflexion of the foot while the leg is extended causes pain.

Usually a patient with thrombophlebitis is put on bed rest with the legs elevated. Some doctors permit the patient to exercise the affected leg, in the belief that exercise facilitates circulation. The affected part should not be massaged as this might cause the clot to become dislodged. Anticoagulants prevent further clot formation.

PROBLEMS ASSOCIATED WITH LABOR AND DELIVERY

Prolonged Labor

Labor in excess of 24 hours is classified as prolonged labor. Prolonged labor increases the risk of postpartum shock, hemorrhage, and fetal mortality. The most common causes of prolonged labor are uterine inertia (weak, poorly coordinated uterine contractions), pelvic disproportions (a disparity between the size of the fetus and the space available for its emergence through the pelvis), and abnormal fetal position. It may be possible to rotate the fetus so that it can be delivered normally with the top of the head emerging first.

Frequently, with prolonged labor a Caesarean section is necessary. A Caesarean section, perhaps the most dramatic of surgical operations, is technically simple and has a low mortality rate. The baby is delivered through an incision in the mother's abdomen and uterus.

Placental Abruption

The placenta normally separates from the uterus only after the birth of the infant. When the placenta separates earlier, bleeding and severe abdominal pain occur. If placental abruption takes place during labor, the uterus scarcely relaxes between contractions. Fetal heart sounds may be absent or slow and irregular. Fetal and maternal welfare are both served best by prompt delivery. The membranes can be ruptured and labor can be induced by use of an intravenous infusion of Pitocin, a drug that causes uterine contractions.

Placenta Previa

The fertilized ovum normally implants high in the uterine wall; however, implantation low in the uterus or covering the cervix results in placenta previa. The main symptom of placenta previa is painless vaginal bleeding late in pregnancy. This condition is hazardous to the mother because of the hemorrhage it produces and to the fetus because of hypoxia resulting from decreased placental function. The condition is usually treated by performing a Caesarean section.

Postpartum Hemorrhage

The most common cause of hemorrhage following delivery is retained placental tissue. In most cases, the bleeding can be checked by giving ergotrate, which causes uterine contractions. If this procedure is ineffective, curettage must be performed.

DISEASES OF THE MALE REPRODUCTIVE SYSTEM

BENIGN PROSTATIC HYPERTROPHY (BPH)

Enlargement of the prostate is a common disease in men who are past middle age. The symptoms—frequency of urination, difficulty in voiding, retention of urine, and nocturia (frequent voiding during the night)—develop gradually and are frequently ignored until the enlargement is fairly advanced. Because of the retention of urine, the patient may also develop cystitis. Symptoms resemble those of cancer of the prostate. Clearly, it is important that a diagnosis be established even though some men with BPH do not require surgery.

The disease can be diagnosed by a rectal examination or by a cystoscopic examination. X-ray examination of the kidneys (pyelography) will give information about the possible damage to the upper urinary tract due to backup of urine.

Prostatectomy

BPH is treated by the surgical removal of part or all of the prostatic gland. The simplest procedure is a transurethral prostatectomy. A cystoscope is passed through the urethra and a cutting apparatus is used to slice away small pieces of the hypertrophied gland (see Figure 20–3). These pieces of tissue are washed away with periodic irrigations through the cystoscope. Bleeding is controlled by an electric cautery; however, for a few days following the surgery some hematuria is normal. The patient uses a Foley catheter until the urine is clear of blood. Fluids should be encouraged and constipation must be avoided as straining at stool may induce bleeding. Following discharge from the hospital, the patient should continue to force fluids for several weeks and should return to the doctor if hematuria reappears.

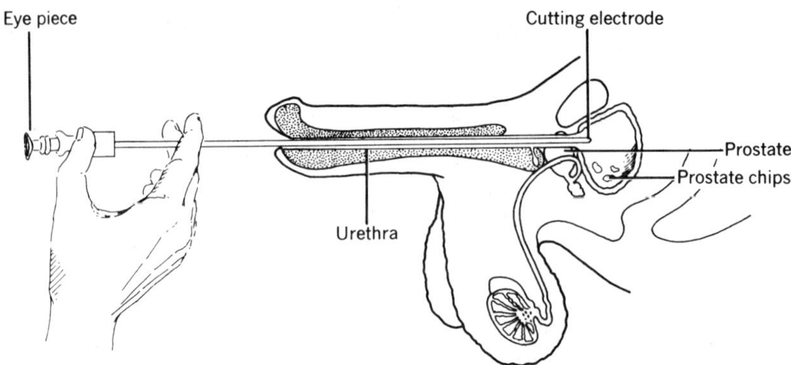

FIGURE 20-3
Transurethral resection of the prostatic gland being done through a cystoscope.

In some cases, more extensive surgery is necessary to treat benign prostatic hypertrophy. In a suprapubic prostatectomy, an incision is made into the bladder and the prostatic tissue is removed from above. The patient will have two catheters in his bladder, one leading from the suprapubic wound and one through the urethra. The suprapubic catheter is usually removed about 5 days following the surgery; however, the wound will continue to drain urine for a few days. These wounds heal slowly and unless great care is taken, they are likely to become infected.

In a retropubic prostatectomy the prostatic tissue is also removed through a low abdominal incision, but the bladder is not entered. This procedure can be used when the enlarged gland does not extend into the bladder.

The prostate can also be removed through a perineal incision (perineal prostatectomy). Recovery following a perineal prostatectomy is usually rapid. Obviously, the patient will experience some discomfort when he sits down. Warm sitz baths promote healing and help prevent infection.

CANCER OF THE PROSTATE

Prostatic cancer is fairly common in men over the age of 50. The symptoms are similar to those of BPH. Back pain may indicate that metastasis to the nerve sheaths has taken place.

If the tumor is still contained within the capsule of the prostate, the patient will probably be treated with radical surgery. When the tumor has metastasized, the treatment is palliative.

Female hormones, mainly estrogen, frequently relieve the symptoms temporarily. Radiation therapy may also give some relief from the painful metastases. If the tumor is obstructing the flow of urine, the surgeon may perform a transurethral resection.

INFECTIONS

Orchitis

Orchitis is an inflammation of the testes. It may be caused by a systemic infection such as tuberculosis or, more commonly, gonorrhea. Mumps occurring after puberty may cause orchitis. Orchitis can also result from trauma.

Symptoms include swelling, heat, and pain in the scrotum. The treatment is directed toward the underlying cause. In addition, the patient should be on strict bed rest with the scrotum elevated. An ice bag placed under the scrotum will help relieve the pain.

Epididymitis

Infections of the epididymis may be the result of the same conditions that cause orchitis. Symptoms are similar to those of orchitis; however, in addition there may be systemic manifestations such as chills, nausea, and vomiting. Antibiotics, bed rest with the scrotum elevated, and large fluid intake are usually ordered. If an abscess forms, it probably must be incised and drained.

HYDROCELE

A hydrocele is a large accumulation of fluid in the scrotum. It may result from infection or trauma, or it may occur without any known cause.

Although fluid may be removed by aspiration, recurrences are likely. A hydrocele may also be treated by open surgery.

VENEREAL DISEASES

Venereal diseases are infectious diseases that are transmitted by sexual intercourse with a venereally infected person. In some instances these diseases can be contracted in other ways, but such cases are so rare that they are of no major concern.

Around 1957 the reported incidence of new cases of venereal diseases was so low that many thought that these diseases were no longer a major public health problem. Effective antibiotics and intensive public health education programs were mainly responsible for the sharp decline. Unfortunately, shortly after 1957 a slight rise in the number of new cases was reported. This rise greatly increased in the 1960s and today venereal diseases are once again a major public health problem. Many strains of the organisms causing venereal diseases are resistant to the antibiotics that were once so effective.

Gonorrhea and syphilis are the two most common venereal diseases. Both are worldwide and are not limited to any socioeconomic group. Attitudes toward the diseases may vary according to socioeconomic level. Skilled venereal disease investigators concerned with case finding are often able to discover from even the most embarassed patient the names of persons with whom they have had sexual intercourse. Humiliating as this may seem, these people and those with whom they have had sexual intercourse must be examined and receive treatment. Unlike some other infectious diseases, natural immunity is not acquired after a person is infected with either gonorrhea or syphilis.

GONORRHEA

Gonorrhea is caused by the gonococcus, a gram negative diplococcus. The symptoms usually appear about 3 days to 2 weeks after intercourse with an infected person. The early symptoms are pain and burning on urination. Women may experience a yellowish vaginal discharge and abscesses of Bartholins glands. Even without treatment these early symptoms subside in a relatively short period of time. Unfortunately, the absence of symptoms does not mean that the disease has disappeared.

If untreated the infection moves up into the uterus and fallopian tubes in women, and into the prostate, epididymis, and seminal vessels in men. Infections and resulting adhesions in these structures can result in sterility.

More serious, the untreated infection may enter the bloodstream and cause septicemia. The patient will have fever, chills, and malaise. Once in the bloodstream the organisms have access to every part of the body. They can cause endocarditis (an inflammation of the lining of the heart) or meningitis. A late complication is arthritis caused by invasion of the joints by the gonococcus; this, however, is not a common complication today.

The mucous membranes of the eye are very susceptible to gonorrheal infections. If untreated, such infections can result in blindness.

If treatment is given in an early stage, cure is possible. Penicillin, if the organism is not resistant to the drug, is effective. If the organism is resistant to penicillin or the patient is allergic to it, other antibiotics such as tetracycline may be effective.

SYPHILIS

Syphilis is caused by a spirochete, the Treponema pallidum. This organism must stay wet to live. It is also sensitive to cold and is killed by soap. The transmission of syphilis must therefore involve bodily contact. Congenital syphilis is transmitted through the placenta of the infected mother to the fetus.

Persons with untreated syphilis are infectious for about 3 years. With early treatment, the patient is usually noninfectious within 24 hours after the start of therapy.

In the primary stage of syphilis, a painless lesion called a chancre develops on the mucous membrane where the spirochetes entered. In women this chancre may be inside the vagina and be completely unnoticed.

Secondary syphilis starts about 6 weeks after the initial infection. It is at this stage that serology tests become positive. Some patients, but not all, have a skin rash during this period. They may have some malaise or they may have no discomfort at all. During both the primary and the secondary stages the disease is curable. If untreated, the disease enters a latent period.

During the latent period the patient feels well and has no symptoms related to syphilis. The spirochetes are still present in the body and can be discovered by serology tests. The disease is usually not infectious during the latent stage.

The third and final stage of syphilis usually starts about 20 to 30 years after the initial infection. During this stage, evidences appear that the organisms have invaded the central nervous system or cardiovascular system.

In central nervous system syphilis, a deterioration of the dorsal columns of the spinal cord may take place. In this case, the patient is unaware of where his feet and legs are and must observe where he places his feet as he walks. This condition is called tabes dorsalis. Central nervous system syphilis can also cause insanity.

In the cardiovascular system, patches of necrosis may weaken walls of the aorta and cause aneurysms. As the aneurysm grows larger, the walls become thinner and thinner until it bursts and the patient bleeds to death. The organisms may also cause the heart valves to become incompetent and to allow regurgitation of blood. If the bundle of His is involved, it will cause heart block.

SUMMARY QUESTIONS

1. Discuss some of the tests that determine the cause of infertility.
2. How is a patient prepared for a gynecological examination?
3. For what purpose is a Papanicolaou Test done?

4. What symptoms necessitate a dilatation and curettage?
5. Differentiate between amenorrhea, menorrhagia, metrorrhagia, and dysmenorrhea.
6. During what time of pregnancy is a spontaneous abortion most likely to occur?
7. What is an incomplete abortion?
8. What is a placenta previa and what are its symptoms?
9. Discuss the events that normally occur during the three stages of labor.
10. What are the most common vaginal infections?
11. What are the symptoms of endometriosis and how is it treated?
12. What might cause a prolapsed uterus?
13. What is the treatment for breast malignancies?
14. What are the symptoms of eclampsia?
15. What is the most common cause of postpartum hemorrhage?
16. What are the symptoms of benign prostatic hypertrophy?
17. Discuss the surgical procedures used to treat BPH.
18. In the event of metastatic cancer of the prostate, what treatment might be given?
19. What are some of the causes of epididymitis?
20. How is epididymitis treated?
21. How is gonorrhea transmitted?
22. Discuss the symptoms that occur during the various stages of syphilis.
23. What is a hydrocele?

SUGGESTED READINGS

Gardner, A. F. *Paramedical Pathology.* Springfield, Ill.: Charles C Thomas, 1972.

Morton, R. S. *Venereal Disease.* Baltimore, Md.: Penguin Books.

Schifferes, L. J. *Healthier Living.* 3rd ed. New York, N.Y.: John Wiley & Sons, Inc., 1970.

Seward, E.M. "Preventing Post Partum Psychosis," *American Journal of Nursing,* March 1972.

21 Endocrine System

VOCABULARY

Amino acid
Corpus luteum
Gluconeogenesis
Hypothalamus
Involution
Ischemia
Lactation
Lymphocyte
Osteoblast
Pressor
Sella turcica
Steroid
Testes
Tropic
Uterus

OVERVIEW

I PITUITARY GLAND
- **A.** Adenohypophysis
  - **1.** Thyrotropic hormone
  - **2.** Adrenocorticotropic hormone
  - **3.** Gonadotropic hormones
  - **4.** Somatotropic hormone
  - **5.** Melanocyte-stimulating hormone
- **B.** Neurohypophysis
  - **1.** Vasopressin
  - **2.** Oxytoxin

II PINEAL GLAND
- **A.** Serotonin
- **B.** Melatonin
- **C.** Glomerulotropin

III THYROID
- **A.** Thyroxin
- **B.** Triiodothyroxin
- **C.** Calcitonin

IV PARATHYROID GLANDS

V THYMUS
- **A.** Lymphatic role
- **B.** Thymosin

VI ADRENAL GLANDS
- **A.** Adrenal Cortex
  - **1.** Glucocorticoids
  - **2.** Mineral corticoids
  - **3.** Sex hormones
- **B.** Adrenal medulla
  - **1.** Epinephrine
  - **2.** Norepinephrine

VII PANCREAS
- **A.** Insulin
- **B.** Glucagon

VIII KIDNEYS

IX GONADS
- **A.** Estrogen
- **B.** Progesterone
- **C.** Relaxin
- **D.** Testosterone

X LOCAL HORMONES
- **A.** Gastrin
- **B.** Secretin
- **C.** Pancreozyme
- **D.** Cholecystokinin
- **E.** Enterocrinin
- **F.** Entrogastrin
- **G.** Prostaglandins

Endocrine glands are ductless glands that pour their secretions directly into the bloodstream. These glands tend to be widely separated in the body. The actions of one gland often affects the actions of other endocrine glands. The endrocrine system works together with the nervous system in integrating most of the biochemical and physiological processes of the body.

PITUITARY GLAND

The pituitary gland is located in the sella turcica and is attached to the hypothalamus by a stalk (see Figure 21–1). The anterior lobe of the pituitary is called the adenohypophysis. Although it has no direct nerve connections with the hypothalamus, it is under its control. This control is maintained by means of blood neurohumors that circulate in the bloodstream between the hypothalamus and the adenohypophysis. The posterior pituitary is also under the control of the hypothalamus, however, by means of nerve connections.

ADENOHYPOPHYSIS

At the present time, at least seven hormones are known to be produced and secreted by the adenohypophysis. The function of most of these hormones is to cause some other endocrine gland to secrete hormones. The increased secretion of hormones from the target gland in most instances causes a decrease in the secretion of the tropic hormone from the adenohypophysis. This mechanism is called negative feedback. Essentially, it operates like a furnace thermostat. When the room temperature is lower than the thermostat setting, the furnace turns on; once the temperature is high enough, the furnace turns off.

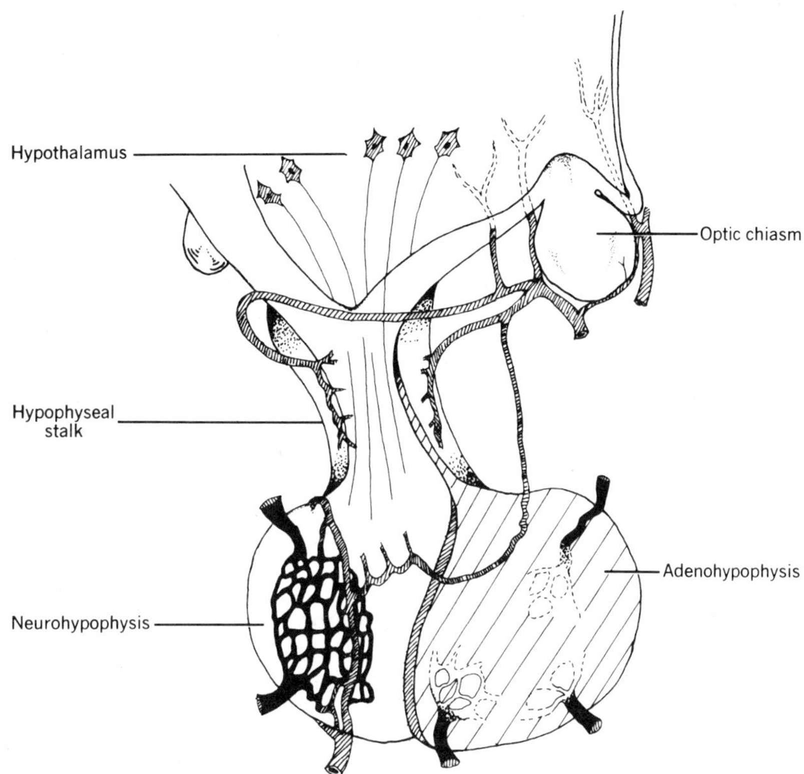

FIGURE 21-1
The pituitary gland attached to the hypothalamus by the hypophysial stalk.

Thyrotropic Hormone

The thyrotropic hormone (TSH, or thyroid-stimulating hormone) from the adenohypophysis stimulates the growth and the secretory activity of the thyroid gland. Increased secretions (chiefly thyroxin) from the thyroid will cause a decrease in the production of the thyrotropic hormone.

Adrenocorticotropic Hormone

The adrenocorticotropic hormone (ACTH) causes the adrenal cortex to secrete some of its hormones, chiefly the glucocorticoids (cortisol and hydrocortisone). Although many other hormones are produced by the adrenal cortex, it is mainly high levels of glucocorticoids that decrease the production of ACTH.

Gonadotropic Hormones

These hormones are concerned with the maturation and secretions of the gonads or sex glands. In the male the follicle-stimulating hormone (FSH) causes the maturation of sperm; in the female it stimulates the development of the ovarian follicles and causes the ovaries to produce estrogen.

The luteinizing hormone (LH) in the female causes ovulation and stimulates the formation of the corpus luteum and the secretion of progesterone. This hormone, called the interstitial cell-stimulating hormone (ICSH), in the male stimulates the secretion of testosterone.

The lactogenic hormone or luteotrophic hormone (LTH) stimulates the development of the breasts and following childbirth the production of milk. This hormone also causes the corpus luteum to secrete progesterone.

Estrogen, progesterone, and testosterone inhibit the production of the gonadotropic hormones.

Somatotropic Hormone

The somatotropic hormone has a growth factor that stimulates the growth of bone, muscle, and organs. Normally, its level is high in infants but by the age of 4 is at the adult level. The growth factor, first synthesized in 1971, is composed of 188 amino acids. Hopefully, this complex, synthetic growth factor will soon be in sufficient supply that it can be given to all children who have a deficiency of the factor and who without the synthetic hormone would become midgets.

The other part of the somatotropic hormone is the diabetogenic factor, which reduces peripheral glucose uptake by muscle and fatty tissue. It is an insulin antagonist; however, it stimulates the release of insulin from the pancreas.

Melanocyte-Stimulating Hormone

The melanocyte-stimulating hormone is responsible for normal pigmentation. The chemical structure of MSH is similar to ACTH. Both of these hormones, under certain circumstances, can cause increased pigmentation of the skin.

NEUROHYPOPHYSIS

The hormones released by the neurohypophysis are actually made by the hypothalamus and are merely stored in the neurohypophysis until the hypothalamus causes their release into the bloodstream.

Vasopressin

Vasopressin is also called the antidiuretic hormone (ADH). This hormone acts mainly on the collecting tubules of the kidney to increase their permeability and to cause reabsorption of water back into the bloodstream. The stimulus for the release of ADH is any condition that causes dehydration. Normally, a decrease in the production of ADH takes place after an increase in fluid intake.

Oxytoxin

Oxytoxin also has a mild antidiuretic effect; however, its main actions are the expulsion of milk from the lactating breasts and uterine contractions. Frequently, mothers who are breast-feeding their babies have fairly severe uterine contractions (afterpains) for the first couple of weeks following delivery when the infant is nursing.

PINEAL GLAND

The pineal gland, formerly called the pineal body, was thought to do nothing other than act as a radiological landmark because it calcifies soon after puberty. It is found just posterior to the third ventricle in the brain. Now there is considerable evidence that it produces several hormones and that these hormones do not decrease in amount or activity with the calcification. There is further evidence that the pineal gland inhibits the hypothalmic and pituitary production of all of their hormones except the somatotropic hormone.

SEROTONIN

Serotonin from the pineal opposes extremes in vascular diameter in the brain. For example, if there is too much vasocontriction in the cerebral vessels, serotonin causes these vessels to dilate. Serotonin levels in man are highest at noon and lowest at midnight. Serotonin, with the help of an enzyme, produces melatonin.

MELATONIN

Melatonin decreases ovarian activity. The action of this hormone is related to light and darkness; activity is greatest at night. Blind girls have their puberty somewhat earlier than the average, whereas albino girls usually have a delayed onset of puberty.

GLOMERULOTROPIN

The pineal produces glomerulotropin, which increases the adrenal cortical output of aldosterone for salt and water retention.

THYROID

The thyroid gland has two lobes that are lateral to the trachea and connected by an isthmus (see Figure 21–2). Hormones from this gland are influenced by TSH.

THYROXINE AND TRIIODOTHYROXINE

Thyroxine (T4) and triiodothyroxine (T-3) essentially perform the same function; however, the action of thyroxine is longer acting and less intense than that of T-3. Each increases metabolism and is essential for normal growth and development.

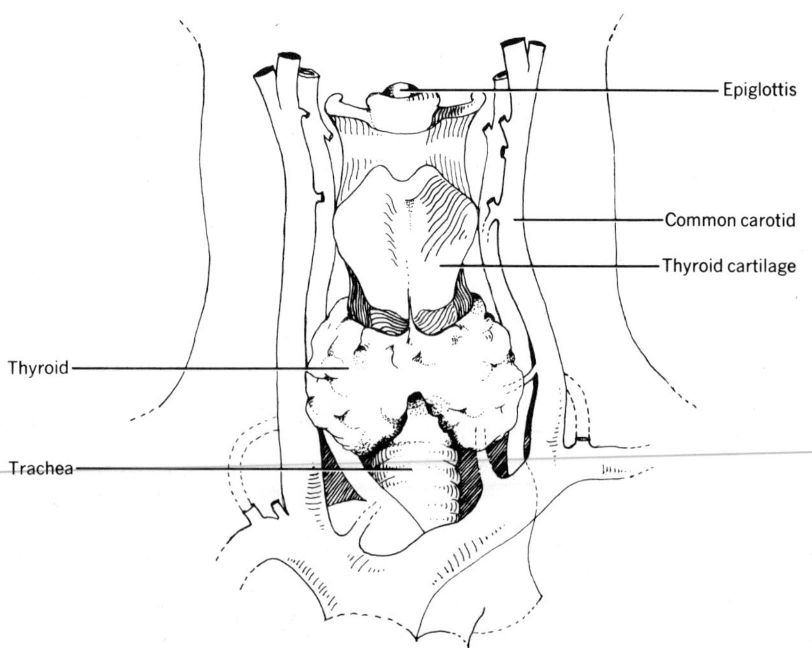

FIGURE 21-2
The thyroid gland and adjacent structures.

CALCITONIN

The thyroid also produces calcitonin. This hormone lowers the blood level of calcium by favoring the activity of the osteoblasts.

PARATHYROID GLANDS

The parathyroid glands are found behind the thyroid and in the same capsule with the thyroid. Usually four in number, the parathyroids secrete parathormone, which increases the blood level of calcium. The stimulus for the release of parathormone is a low blood calcium level.

THYMUS

The thymus gland has both endocrine and lymphatic functions. It is found inferior to the thyroid about the level of the second rib. It is conspicuously large in the infant and undergoes involution after puberty and under influence of stress. Involution, however, does not reduce its physiological importance.

In the child the thymus consists primarily of lymphocytes; in fact, in the fetus it is the only source of lymphocytes. It produces a mold for other organs (spleen and lymph nodes) to produce lymphocytes. The thymus produces thymosin, a hormone that enables lymphocytes to develop into plasma cells and to contribute to our immunological antibody-producing system.

ADRENAL GLANDS

The adrenal glands are small triangular yellow bodies located at the upper pole of each kidney. They have an abundant blood supply. The outer part of the adrenal gland is called the cortex, and the inner gray portion is called the medulla.

ADRENAL CORTEX

The adrenal cortex produces at least 30 different hormones that are essential to life. Most of these hormones are under the influence of ACTH. Circulating levels of these hormones vary over a 24-hour period. For a person on a normal daily schedule, maximum secretions occur between 2 AM and 8 AM; secretions are lowest in the late evening.

Glucocorticoids

There are several different glucocorticoids; however, their actions are similar. On protein they increase the amount of amino acids in the extracellular fluid and the rate of utilization of these proteins for tissue repair. On fat they enable muscle tissue to use fats for fuel. They increase the concentration of glucose in the blood as a result of gluconeogenesis. Although they increase the blood sugar level, they antagonize insulin by preventing muscle tissue from using the glucose. This action preserves the glucose for use by the brain. Glucose is the only form of fuel that can be used by the brain. The brain does not need insulin in order to utilize the glucose.

Glucocorticoids are anti-inflammatory in that they decrease vascular permeability so that fluid does not leak out and swelling is suppressed. They also decrease the activity of the fibroblasts that form scar tissue. For this reason, patients who have been on long-term steroid therapy may have poor wound healing.

The hormones also increase the gastric acidity. This fact, together with the fact that they are anti-inflammatory, explains why prolonged steroid therapy may cause gastric ulcers.

Glucocorticoids are anti-allergenic. Patients who suffer severe allergy may benefit from steroid preparations.

Glucocorticoids sensitize blood vessels to vasopressor substances. The release of these hormones at the time of serious injury helps to limit the degree of shock that usually occurs with severe trauma.

In general, glucocorticoids help you cope with stressful situations primarily by preserving your carbohydrate reserves. They are produced in response to pituitary ACTH. Under ordinary circumstances, high levels of the glucocorticoids will decrease the production of ACTH; under very stressful circumstances, however, this negative feedback mechanism is suppressed.

Mineral Corticoids

Although several mineral corticoids are produced by the adrenal cortex, the main one is probably aldosterone. Mineral corticoids act primarily on the distal kidney tubules but also to some extent on sweat and salivary glands. They cause the conservation of sodium and the elimination of potassium. Like the glucocorticoids, these hormones help us cope with stressful situations. Under stress, mineral corticoids help to preserve our fluid and electrolyte balance. These hormones are produced in response to glomerulotropin and to blood levels of sodium and potassium.

Sex Hormones

In both males and females, the adrenal cortex produces sex hormones: testosterone, estrogen, and progesterone. The male hormones from the adrenal cortex dominate in both men and women.

ADRENAL MEDULLA

The hormones from the adrenal medulla are not essential to life; however, they are important in our fight and flight responses. These hormones, called catecholamines, are norepinephrine and epinephrine. They do essentially the same thing as does the sympathetic nervous system (see page 182).

Norepinephrine dominates in anger and epinephrine in fear. Norepinephrine is a much more powerful vasoconstrictor than epinephrine and therefore elevates the blood pressure more. It does little if anything to increase the activity of the central nervous system. If you consider these two facts, the norepinephrine secreted in anger obviously does not help you think through the situation, and indeed it may raise the blood pressure to dangerously high levels. On the other hand, the actions of epinephrine secreted in fear are helpful in preparing you to cope with the danger.

PANCREAS

Because it has both an exocrine portion (see page 39) and an endocrine portion, the pancreas is actually a heterocrine gland. It is found in the abdominal cavity inferior to the stomach. The head of the pancreas is surrounded by the curve of the duodenum, and the tail extends over to the spleen.

The beta cells of the pancreas produce insulin to lower blood sugar. Factors that stimulate the release of insulin by the pancreas are high levels of blood sugar and the growth hormone. Glucose also stimulates insulin synthesis, but this process is relatively slow. Normally, you require about 50 units of insulin per day and you store a five-day supply. Insulin is degraded by the liver and to a lesser extent by the kidney. Half the circulating insulin is degraded in about 10 to 25 minutes. Insulin is antagonized by epinephrine, the glucocorticoids, the diabetogenic factor, and thyroxine. Thus, diabetes mellitus, which will be discussed in the following chapter, can obviously be a very complicated disease.

The alpha cells of the pancreas produce glucagon. Glucagon helps in the conversion of glycogen to glucose and raises the blood sugar. This hormone also stimulates the production of insulin by the pancreas.

KIDNEY

In addition to its excretory function, the kidney has a pressor system that resembles an endocrine function. The renal cortex, particularly if it is ischemic, produces an enzyme, renin. The substrate of renin is angiotensin, a hormone that has a pressor activity about 200 times that of norepinephrine. Angiotensin is also a tropic hormone for aldosterone.

GONADS

The gonads are the sex glands. These glands produce hormones that are essential to the normal functioning of the male and female reproductive systems. Chapter 19 discussed the sex hormones in greater detail. In the female the ovaries produce estrogen, progesterone, and relaxin. In the male the testes produce testosterone.

LOCAL HORMONES

Many hormones are produced by some of the structures of the gastrointestinal system. Because these hormones act in the organs of the gastrointestinal system they are called local hormones.

The stomach produces gastrin to cause the production of gastric juice. The duodenum produces secretin to cause the release of sodium bicarbonate by the pancreas. Pancreozymin is also produced by the duodenum to cause the release of digestive juices from the pancreas. Cholecystokinin is a hormone produced by the duodenum to cause bile expulsion from the gall bladder. The duodenum produces enterocrinin to cause the production of intestinal juices and to determine their makeup. Enterogastrone is a hormone produced by the small intestines to inhibit digestive secretions and the motility of the gastrointestinal tract.

Prostaglandins are hormones that act directly on the tissues that produce them. Sometimes the prostaglandins increase the activity of other hormones and sometimes they decrease their action. These hormones (there are at least 14) were originally thought to be produced by the prostate; however, it is now known that they are present in most, if not all, body tissues and fluids.

The prostaglandins specifically help to open closed airway passages in the respiratory tract and therefore lessen the severity of asthma. They function as nasal decongestants.

The prostaglandins also inhibit stomach secretions even in patients who are receiving large doses of cortisone. Increased intestinal contractions take place under the influence of these hormones. Excess amounts of prostaglandins prevent the normal breakdown of fat, a factor that may be of considerable importance in obesity.

In the reproductive system the prostaglandins play several roles. In females, they cause dilatation of the cervix and contractions of the uterus. These facts suggest that the hormones may be a factor in causing some miscarriages. They also cause the corpus luteum to regress. In males, the prostaglandins facilitate ejaculation. This action, together with increased uterine contractions, may help in sperm motility. In fact, it is known that infertile males have low prostaglandin content in their semen.

Prostaglandins regulate platelet aggregation and therefore prevent clot formation. Perhaps prostaglandins can even break up existing clots. In addition, they have a digitalis-like effect. The fact that prostaglandins lower blood pressure suggests that some types of hypertension may indicate a deficiency of the hormone.

SUMMARY QUESTIONS

1. Name the hormones produced by the neurohypophysis and discuss the actions of these hormones.
2. What gland produces calcitonin and what is the function of calcitonin?
3. What hormones are produced by the pineal gland and what are the functions of these hormones?
4. What hormones oppose the action of insulin?
5. List the hormones produced by the adenohypophysis and discuss the functions of each.
6. Name the three main classifications of hormones produced by the adrenal cortex.
7. What hormones are produced by the thyroid gland and what are the actions of these hormones?
8. What gland produces parathormone and what is the action of parathormone?
9. What hormones are produced by the gonads?
10. Compare the actions of epinephrine with those of norepinephrine.
11. Under stressful situations what hormones are primarily concerned with preserving our carbohydrate reserves?
12. Under stressful circumstances what hormones are primarily concerned with maintaining our normal fluid and electrolyte balance?
13. With respect to the endocrine glands, what is meant by the negative feedback mechanism?

14. Discuss the functions of the thymus.
15. What gland produces glucagon and what is the function of this hormone?
16. List the local hormones and give examples of their actions.

SUGGESTED READINGS

Berger, A. J. *Elementary Human Anatomy.* New York, N.Y.: John Wiley & Sons, 1964.

Best, C. H., and Taylor, N. B. *The Human Body: Its Anatomy and Physiology.* New York, N.Y.: Holt, Rinehart & Winston, 1963.

Laros, R. K., *et al.* "Prostaglandins," *American Journal of Nursing,* June 1973.

Lowenberg, Miriam E. *Food and Man,* New York, N.Y.: John Wiley & Sons, Inc., 1968.

22 Diseases of the Endocrine System

VOCABULARY

Anorexia
Benign
Hypertension
Hypoglycemia
Malignant
Polyuria
Trachea
Tracheostomy
Vertigo

OVERVIEW

I DIAGNOSTIC TESTS
- **A.** Tests for diseases of the thyroid
  - **1.** Basal metabolism
  - **2.** Protein-bound iodine
  - **3.** Radioactive iodine uptake
  - **4.** Thyroid scan
  - **5.** Blood cholesterol
- **B.** Tests for diseases of the adrenal glands
  - **1.** Twenty-four-hour urine specimen
  - **2.** Blood chemistry tests
  - **3.** Regitine test
- **C.** Tests for abnormalities of the pituitary
  - **1.** Twenty-four-hour urine specimen
  - **2.** Visual field examination
- **D.** Tests for diabetes mellitus
  - **1.** Urine examination
  - **2.** Blood sugar
  - **3.** Glucose tolerance test
  - **4.** Carbon dioxide combining power and CO_2 content

II DISEASES OF THE THYROID
- **A.** Hyperthyroidism
- **B.** Hypothyroidism

III DISEASES OF THE PARATHYROID GLANDS

IV DISEASES OF THE ADRENAL GLANDS
- **A.** Addison's disease
- **B.** Cushing's syndrome
- **C.** Pheochromocytoma

V DISEASES OF THE PITUITARY
- **A.** Acromegaly
- **B.** Diabetes insipidus

VI DIABETES MELLITUS

DIAGNOSTIC TESTS

TESTS FOR DISEASES OF THE THYROID GLAND

Basal Metabolism (BMR)

The basal metabolism test is performed to determine the rate at which the patient consumes oxygen under resting conditions. It is not as accurate as some of the other tests for thyroid function and is not done as frequently today as it was in the past. Because the test is relatively simple, it is sometimes used as an office procedure to help determine whether more elaborate tests for thyroid function are advisable.

Prior to the BMR, the patient should fast for about 10 hours and have rested as completely as possible. If it is done in the doctor's office, the procedure is usually scheduled for early morning. The patient's temperature is taken as soon as he arrives for the examination. His temperature must be normal because even a slight degree of fever will increase metabolism and the results of the test will be unreliable. The procedure should be explained to the patient and he should be assured that there is no discomfort involved in the test. He is then assigned to a quiet comfortable room and asked to rest and, if possible, to sleep for about 1 hour. Smoking is not permitted.

The procedure involves having the patient breath into a tube through his mouth. It may be necessary to put a soft sponge rubber clamp on the patient's nose in order to assure that breathing is done entirely through the tube. The machine attached to the tube measures the amount of oxygen consumed. Normal values are between minus 20 and plus 20. Values lower than this may indicate decreased thyroid function, and values above plus 20, an overactive thyroid.

Protein-Bound Iodine (PBI)

No special preparation for this test is necessary other than that the patient should not have ingested any unusual amounts of iodine for a few weeks before the test. Dyes used for X-ray studies of the kidneys, gall bladder, and bronchi contain iodine, as do some cough medicines. Thus, if the patient has had any of these examinations, or medications recently, the PBI should be deferred. Even antiseptic solutions of iodine on the skin should be avoided. PBI may also not be an accurate indication of thyroid function in women who are taking birth control pills.

The test involves the withdrawal of a sample of the patient's blood, which is analyzed for the amount of protein-bound iodine. Normal values for this test are 4 to 8 μg./100 ml. of plasma. As with the BMR, values below this level may indicate hypothyroidism and above, hyperthyroidism.

Radioactive Iodine Uptake Test

After fasting for 10 hours, the patient is given a capsule or a drink that contains sodium radioiodine^{131}. When the solution is used, it is tasteless and odorless. Twenty-four hours later a scintillator (an instrument that measures radioactivity) is held over the thyroid to measure the amount of radioactive iodine that has been taken up by the thyroid. The normal thyroid will have removed from 15 to 50% of the radioactive iodine during this period. An overactive thyroid will remove considerably more.

Thyroid Scan

The thryoid scan is similar to the radioactive iodine uptake test; however, the scintillator will record a graphic outline of the thyroid. This examination helps the physician to differentiate between benign and malignant growths of the thyroid.

Blood Cholesterol

A sample of the fasting patient's blood can be analyzed for cholesterol content. Normal values lie between 150 and 250 mg./100 ml. of blood. In hyperthyroid patients the values are often lower than normal. However, a lower level may also reflect the patient's dietary habits; for this reason it will be helpful to the physician if a simple dietary history is taken.

TESTS FOR DISEASES OF THE ADRENAL GLANDS

Twenty-Four-Hour Urine Specimens

The patient discards the first urine voided in the morning and records the time of the voiding. All urine voided from that time until the last voiding at the end of the 24-hour period is collected and refrigerated. Although the urine may be analyzed for many different constituents (see Table 18–1), the constituents of the 24-hour urine that are particularly important when diseases of the adrenal glands are suspected are chlorides, 17 ketosteroids, pressor amines, and aldosterone.

Blood Chemistry Tests

High blood potassium or low blood sodium and chloride may indicate Addison's disease. The finding of a blood sodium to blood potassium ratio of less than 30 is particularly significant.

Regitine Test

The Regitine test helps diagnose a tumor of the adrenal gland. The patient is given an intravenous infusion of 5 percent glucose in water. While the infusion is running, several blood pressure readings are taken and the blood pressure cuff is left in place between readings. After the patient is accustomed to having repeated blood pressure readings taken and to having the intravenous needle in his arm, a drug called Regitine is injected into the intravenous tubing. Immediately following this injection, blood pressure readings are taken as rapidly as possible for the next 5 to 10 minutes. A drop in the blood pressure probably indicates that the patient has a pheochromocytoma. This tumor secretes epinephrine and norepinephrine, which increases the blood pressure. Regitine neutralizes these hormones and the blood pressure falls.

TESTS FOR ABNORMALITIES OF THE PITUITARY GLAND

Twenty-Four-Hour Urine Volume and Specific Gravity

In diabetes insipidus, a disease of the posterior lobe of the pituitary, the 24-hour urine volume is much greater than normal and the specific gravity is low.

Visual Field Examination

A patient with a tumor of the pituitary gland will have reduced peripheral vision. The examiner covers his eye with one hand and sits facing the patient at a distance of approximately 2 feet. The patient, with the opposite eye covered, fixes his gaze on the uncovered eye of the examiner. The examiner then moves his finger inward from all sectors of the periphery. If the patient's visual field is normal, he will see the examiner's finger at approximately the same time as the examiner does. The test is done for each eye separately and then for the two together.

TESTS FOR DIABETES MELLITUS

Urine Examinations

A patient with untreated or poorly controlled diabetes mellitus has glucose and acetone in his urine. Although the urine appears dilute, the specific gravity is higher than normal. The total 24-hour volume of urine is also greater than normal.

Blood Sugar

A fasting patient with diabetes mellitus has a considerably higher level of glucose in his blood than the 80 to 120 mg./100 ml. of blood found in a normal patient. With a value above 130 mg./100 ml., under fasting conditions, diabetes mellitus may be present; however, the analysis should be repeated in order to confirm the diagnosis.

Glucose Tolerance Test

For a glucose tolerance test, blood sugar is measured in a fasting patient, after which he eats a normal breakfast. Two hours later another blood sugar examination is done. Normally, within this period of time, the blood sugar will have returned to the previous level. In the diabetic, the blood sugar level will remain elevated for longer than the 2-hour period.

Carbon Dioxide Combining Power and Carbon Dioxide Content

The carbon dioxide combining power and carbon dioxide content are blood chemistry examinations. Their results are usually lower than normal in a patient with diabetes. Normal values for the carbon dioxide combining power are between 21 and 28 mEq. per liter. Carbon dioxide content values are normally between 25 and 35 mEq. per liter.

DISEASES OF THE THYROID

HYPERTHYROIDISM (GRAVE'S DISEASE)

As the prefix "hyper" suggests, hyperthyroidism is characterized by increased thyroid activity. Although the cause of hyperthyroidism is unknown, doctors have found that conditions that increase the demand on the thyroid can precipitate hyperthyroidism. Physical or emotional stress, infections, and pregnancy all increase metabolism and therefore may predispose to hyperthyroidism.

The symptoms of hyperthyroidism include restlessness, tremors, emotional lability (laughing one minute and crying the next), increased pulse rate, increased systolic blood pressure, weight loss, intolerance of heat, and excessive sweating. The presence of exophthalmos, a bulging of the eyes, may also give the patient an appearance of being constantly startled. Sometimes a visible swelling of the neck due to the enlarged thyroid occurs. This swelling may be severe enough to cause hoarseness and difficulty in swallowing.

The medical treatment includes the administration of antithyroid drugs such as propythiouracil, methimazole, or Lugol's solution. Lugol's solution, which is potassium iodide, has an unpleasant taste and can stain the teeth. For this reason it is given in milk or fruit juice and is taken through a straw. Patients also need a very high caloric diet and frequent feedings.

If the condition does not respond satisfactorily to medical treatment, the patient may need surgery. Unless the hyperthyroidism is due to a malignancy of the thyroid, the surgeon will remove only a part of the gland. Following surgery, the patient must be observed closely for respiratory difficulties. Edema or bleeding can compress the trachea, an emergency that must be treated within minutes by the insertion of an endotracheal tube or by a tracheostomy. Obviously, the equipment to do this emergency treatment should be in the patient's room for the first few postoperative days.

An infrequent but life-threatening complication of thyroid surgery is damage to the parathyroid glands, which can cause tetany (muscular hyperactivity and spasm). The attendant must report immediately to the surgeon any complaint of muscle cramps or numbness and tingling in the extremities. The treatment of tetany is an intravenous dose of calcium.

HYPOTHYROIDISM

Hypothyroidism arising in an adult is called myxedema. A hypothyroid child has cretinism, a condition that will retard his growth, mentality, and sexual development.

A patient with myxedema is usually overweight, fatigues easily, and is sensitive to cold temperatures. His pulse and temperature have decreased. His tongue may enlarge and his lips swell. Nonpitting edema usually occurs about the eyes.

Patients are treated with thyroid extract, which they must take for the rest of their lives. Dramatic improvement takes place within a few weeks after treatment is begun. Because the patient feels so much better, he should be instructed that he must not discontinue his medication and that he must return to the doctor for periodic examinations.

DISEASES OF THE PARATHYROID GLANDS

Although disease of the parathyroid gland is not common, hyperparathyroidism results in an increased urinary excretion of phosphorus and calcium. Because these minerals are removed from the bones, patients may suffer from pathological fractures. They also frequently complain of fatigue and muscle weakness. Treatment is the surgical removal of the hypertrophied gland tissue.

Patients with hypoparathyroidism may develop tetany. If the parathyroid function is not markedly decreased, these patients may be helped with increased intake of vitamin D. If the condition is severe, they may need calcium salts and parathyroid extract.

DISEASES OF THE ADRENAL GLANDS

ADDISON'S DISEASE

Addison's disease results from hypofunction of the adrenal cortex. It may be caused by adrenal atrophy due to prolonged steroid therapy or destruction of the adrenal cortex by tuberculosis or cancer.

The symptoms of Addison's disease include decreased temperature, blood pressure, and metabolic rate. Patients have anorexia and weight loss, in addition to increased pigmentation of the skin and mucous membranes. Hypoglycemia is not uncommon, particularly before breakfast. The symptoms of hypoglycemia are hunger, weakness, sweating, trembling, and anxiety. If the hypoglycemia is severe enough, it may lead to confusion, coma, and convulsions. At the onset of these symptoms, the patient should eat candy or some other high carbohydrate food.

Patients with Addison's disease are treated by replacement of the deficient hormones. They may be given cortisone, hydrocortisone, or prednisone. An important aspect of the treatment is that the patient and his family understand the nature of the disease and the medications that are prescribed. Patients with Addison's disease may need a readjustment of the steroid dosage whenever they are threatened with stress of any type, infections, emotional strain, injury, or increased work load.

Ideally, a patient should have five or six small meals a day instead of three large meals. His diet should be high in proteins and low in fluids. He should carry candy with him to treat attacks of hypoglycemia.

The patient must also carry a card stating the fact that he has adrenal cortical insufficiency, the type and amount of medication he is taking, and his doctor's name and telephone number.

CUSHING'S SYNDROME (ADRENAL CORTICAL HYPERFUNCTION)

Cushing's syndrome may be due to a tumor of the adrenal cortex or of the pituitary gland. Progressive muscle wasting and weakness takes place. The bones lose calcium salts and are easily fractured; salts and water are retained. Patients will have peripheral edema and hypertension. In the event of infection, the symptoms are masked—perhaps to the extent that the infection may be very severe before it is noticed.

Depending upon its cause, adrenal cortical hyperfunction may be treated by X-ray therapy or by surgical removal of the adrenal gland. If surgery is performed, the patient should be treated as if he has Addison's disease.

PHEOCHROMOCYTOMA

Pheochromocytoma is a tumor of the adrenal medulla. The tumor secretes epinephrine and norepinephrine, which cause intermittent or persistent hypertension. Patients also have vertigo, nervousness, sweating, nausea, and vomiting. Treatment is the surgical removal of the tumor.

DISORDERS OF THE PITUITARY GLAND

ACROMEGALY

Acromegaly is caused by a tumor or hyperplasia of the anterior pituitary. It results in an overproduction of the growth hormone. In children, the condition produces giantism. In adults, the increased production of the growth hormone causes an increase in the size of the hands and feet. Patients have large, coarse features and a very large lower jaw.

The treatment for acromegaly is radiation of the pituitary gland. Even if the disease is successfully arrested, growth changes are irreversible.

DIABETES INSIPIDUS

Diabetes insipidus is a rare condition in which there is a deficiency of the antidiuretic hormone. Patients have excessive urinary outputs and may pass as much as 15 or 20 liters of urine in a 24-hour period. The specific gravity of the urine is very low. Patients are very thirsty and must consume large amounts of fluid to replace the body fluids being lost in the urine.

Treatment for diabetes insipidus is the administration of pituitrin. The drug can be given by injection or can be inhaled.

DIABETES MELLITUS

ETIOLOGY

Although diabetes mellitus is a very common disease, it is also an extremely complicated one. Indeed, it may not be due to a decreased production of insulin by the pancreas but to the many factors that oppose the action of insulin. Insulin is antagonized by epinephrine, glucocorticoids, thyroxin, and the diabetogenic factor from the pituitary. The disease may also be due to the overproduction of glucose by the liver or to poor storage of glucose by the liver.

Although diabetes sometimes occurs in individuals who have no family history of the disease, it is usually a hereditary disease. Its incidence varies among ethnic groups; for example, it occurs frequently in Jews. The Chinese and Japanese have a low incidence and usually have the disease in a relatively mild form if they have it at all.

SYMPTOMS

Some of the early symptoms of diabetes include thirst, polyuria, hunger, and loss of weight despite increased intake of food. Infections and wounds heal slowly. Patients are prone to infections such as boils and carbuncles. In severe cases of diabetes, they may experience impaired vision, neuritis, and peripheral vascular disease. Upon examination, patients will display increased blood sugar, and decreased carbon dioxide content and carbon dioxide–combining power. Glucose and acetone will be present in their urine.

TREATMENT

In mild forms of diabetes, dietary treatment and the practice of good health habits to prevent infections may be all that is necessary to control the disease process. In order to determine the appropriate diet for a particular patient, it is necessary to take into account the age, sex, activity, former dietary habits, and height and weight of the patient. If a diet is prescribed, the patient should thoroughly understand it. Some doctors allow their patients to eat whatever they like and to adjust the insulin dosage accordingly. This is frequently the practice for diabetic children because their activities tend to fluctuate from day to day much more than do those of adults.

The American Dietetic Association, the American Diabetes Association, and the Public Health Service have jointly prepared a valuable pamphlet for diabetics, *Meal Planning with Exchange Lists*. It contains six lists of foods. A food on one list can be exchanged for any other food in that list and will have approximately the same food value. The American Diabetes Association has also published a *Cookbook for Diabetics*. Many pharmaceutical companies distribute free pamphlets that are very helpful in teaching the patient about his diet as well as about other aspects of the disease.

In addition to diet instructions, the patient must know how to test his urine for the presence of glucose and acetone. Several different commercial preparations can be used to test urine. The patient can use clinitest tablets to determine the amount of glucose in the urine. He puts 10 drops of water and 5 drops of urine into a test tube. He adds one clinitest tablet and compares the color of the mixture with a color scale. A blue color indicates no glucose in the urine; green, a trace of glucose; yellow, 0.5 percent glucose; orange, up to 1.5 percent glucose; and red, more than 1.5 percent glucose. Acetest tablets are used to test the urine for acetone. The patient places drops of urine on an acetest tablet and compares the change in the color of the tablet with a color scale. If acetone is present in the urine, the tablets turn varying degrees of purple. The darker the purple, the more acetone is in the urine.

The patient also needs instructions concerning insulin. There are several types of insulin. Some are short acting (action starts within a half an hour after injection and lasts for 6 to 8 hours). More frequently, longer acting insulins are prescribed because these need only be taken once a day. As with diet prescription, the insulin dosage must be adjusted to the individual patient's condition and to his usual activity. The patient and some member of his family must be taught how to give the insulin. Instruction is usually given by the nurse in the hospital before the patient is discharged; however, after his discharge a Visiting Nurse should call on the patient in his home to ensure that he has really understood instructions.

If you consider that a diabetic patient has been confronted with the fact that he has a disease that will require treatment for the rest of his life and has had to learn about his diet, how to test his urine, how to administer insulin, and how to prevent complications of his disease, you will not find it surprising that many a patient is overwhelmed by his situation. An occasional visit by a Visiting Nurse not only is reassuring but also gives the patient an opportunity to ask questions that did not occur to him while he was hospitalized and under considerable stress.

In addition to insulin, several types of oral hypoglycemic agents exist. These drugs act either by stimulating the pancreas to secrete more insulin or by promoting the utilization of the glucose by the peripheral tissues. The oral hypoglycemic agents are used mainly for patients who have relatively mild diabetes.

COMPLICATIONS OF DIABETES MELLITUS

Diabetic coma and insulin reactions are both serious complications of diabetes. Table 22–1, outlines the causes, symptoms, and treatment of each of these complications. The patient and his family should be thoroughly familiar with each aspect of these complications. Of the two, insulin reaction progresses much more rapidly; therefore, prompt recognition and treatment are of utmost importance. Because a patient having an insulin reaction frequently appears to be intoxicated, the diabetic should always carry an identification card. The card, available from the American Diabetes Association, tells what should be done in the event of

TABLE 22-1

A Comparison of Diabetic Coma and Insulin Reaction

	Diabetic coma	Insulin reaction
Cause	Infection Insufficient insulin Dietary indiscretion Gastrointestinal upset	Excessive insulin Too little food Unusual exercise
Symptoms	Slow onset, hours to days Dry, hot, flushed skin Drowsy to comatose Hyperventilation, rapid shallow respirations Fruity odor to breath Thirsty May vomit Sugar and acetone in urine	Sudden onset, within minutes Pale, moist, cool skin Excited, possibly incoherent Possible convulsions and unconsciousness Hunger Second urine specimen free of sugar and acetone (first specimen unreliable)
Treatment	Give regular insulin Hydrate by giving fluids either by mouth or intravenously Make frequent blood and urine examinations Maintain patent airway Suction throat as needed Determine and treat the cause of the coma	Give high carbohydrate foods, either by mouth or intravenously Determine and treat the cause of the reaction, usually by means of reinforced health teaching

an insulin reaction and indicates the name, telephone, and address of the doctor as well as identification of the patient.

Vascular Disturbances and Infections

Diabetics are more prone to arteriosclerosis than are nondiabetics. The lower extremities are particularly vulnerable to a decreased blood supply. An ingrown toenail or any wound on the foot or leg is likely to become infected. Diabetes retards wound healing. Because a patient with an infection will experience increased metabolism, he will have to increase his insulin dosage. The diabetic should be instructed to wash his feet and legs daily in warm soapy water, rinse, and pat (not rub) dry. He should report to his doctor the slightest indication of infection any place in his body or the first sign of any illness.

The physician's assistant can be of great help to the diabetic patient and to the doctor. An assistant who can engage the patient in an unhurried conversation can often discover symptoms and complaints that he otherwise might not mention. Often a patient forgets an important symptom or feels that it is too insignificant to bother the doctor. Both the patient and the doctor will appreciate the assistant's concern and the extra time the assistant has taken to help the patient feel more like a person and less like a case of diabetes.

SUMMARY QUESTIONS

1. Describe several ways of testing thyroid function.
2. What instructions should be given to a patient who needs to collect a 24-hour urine specimen?
3. How is a glucose tolerance test done and what is the purpose of this test?
4. What are the symptoms of hyperthyroidism?
5. What measures might be used in the treatment of hyperthyroidism?
6. What is the difference between myxedema and cretinism?
7. What are the symptoms of Addison's disease and how is it treated?
8. How does Cushing's syndrome differ from Addison's disease?
9. What disorders might result from an overproduction of the growth hormone?
10. List the symptoms of diabetes mellitus.
11. Describe the type of health teaching needed by the diabetic patient and his family.
12. Differentiate between an insulin reaction and a diabetic coma.

SUGGESTED READINGS

McFarlane, L. "Children with Diabetes: Special Needs During Growth Years," *American Journal of Nursing,* August 1973.

Meloni, C. R. "Obesity or Cushing's Disease, "*American Family Physician,* Vol. 5, 1972.

Porter, B. M., Dentan, N. J., and Fisher, K. *Food and You.* Booklet for diabetics available from Communications Division, Intermountain Regional Medical Program, 50 North Medical Drive, Salt Lake City, Utah 84111.

Schultz, R. B., and Parra, A. "Relationship Between Body Composition and Insulin and Growth Hormone Responses in Obese Adolescents," *Diabetes,* July 1970.

23 Poisons

VOCABULARY

Analgesic
Anemia
Anoxia
Antagonist
Arrhythmia
Cyanosis
Diuretic
Emetic
Epiphysis
Erosion
Hematemesis
Hemodialysis
Hypnotic
Lavage
Nephrosis
Pylorus
Ulceration
Vertigo

OVERVIEW

The subject of poisons and the management of cases of poisoning are very complicated. One of the many reasons is the fact that two major unknowns often exist: the sensitivity of the patient to the poisonous substance and the nature of the poison. Of the thousands of Americans who die of accidental poisoning every year about one third are children under 5 years of age. Often it is difficult to determine what a child has ingested. Even when the nature of the poison is known, a child may be too ill or too frightened to cooperate with attempts to administer first aid. Even the most knowledgeable parents may be too anxious and distressed to be effective in their attempts to deal with the situation. Parents are very likely overwhelmed with guilt. In many situations, their guilt is not unrealistic because the tragedy might not have happened had they taken proper precautions in the storage of potentially hazardous substances. Clearly, a well-disciplined child is likely to be a safe child. A dramatic illustration is the case of a young mother who had only one child. She was concerned about raising an only child who was not obliged to share with siblings. For this reason she made a practice of requiring the child to share a part of a cookie or piece of candy with her. One day, when the child quickly put a handful of moth balls in his mouth, the mother (with remarkable calmness) requested that the child share them with her. Out of his mouth and into the mother's hand, came the moth balls.

Paramedical personnel who are knowledgeable about some of the general principles involved in the management of poison cases and who are able to act calmly and quickly in emergency situations are a valuable resource in a community. We shall first discuss the general principles for the treatment of poisoning and then consider in some detail the potential poisons that are likely to be found in any home as well as in some industrial setting. Particular attention will be given to the symptoms that are likely to give clues to the nature of the poison and to the appropriate first aid measures.

GENERAL PRINCIPLES FOR THE TREATMENT OF POISONING

INGESTED POISONS

In most instances, the unabsorbed poison should be removed as quickly as possible by inducing vomiting or by lavage. You perform lavage by inserting a stomach tube into the nose and on down to the stomach and irrigating the stomach with water or some solution that will inactivate the poison.

It is generally useless for you to induce vomiting or to lavage if the poison was ingested 2 hours before unless the victim is in shock, in which case at least some of the poison has probably not been absorbed. Warm mustard or salt water is often recommended to induce vomiting. Although such concoctions are nauseating and will indeed cause vomiting, most children will not swallow such a bad tasting fluid and you will lose valuable time. You should give the child a glass of milk or water. Most children will vomit if they quickly swallow large quantities of fluid. If not, you can gag the child with your finger or by stroking the posterior pharynx with a blunt instrument. To prevent aspiration of the emesis, you should invert the body and support the head. If available, syrup of ipecac in doses of 10 to 15 ml. repeated in 15 or 20 minutes will usually induce vomiting. If the ipecac is given and emesis does not occur, lavage is imperative because ipecac is irritating to the stomach and can cause serious problems if absorbed.

In certain situations you should not induce vomiting. If the poison is corrosive, it is an acid or alkali, it has done enough damage by burning the esophagus when it was swallowed. Clearly, you will compound the damage by causing vomiting. You should not induce vomiting if the victim is unconscious, because of the danger of his aspirating the emesis. If the toxic substance is oily, again the danger of aspiration contraindicates inducing vomiting. As we discussed earlier (Chapter 10), oil aspiration pneumonia can be very serious.

Neither should you stimulate vomiting, if the patient is convulsing, as it will increase the severity of the convulsions. If sedation and measures to control the convulsions are available, you can lavage the patient.

A second general principle to observe in dealing with a patient who has ingested poison, particularly if the unabsorbed poison cannot be removed, is to prevent the absorption of the poison. One or two ounces of olive oil or vegetable oil, by mouth but not forced, may help delay the absorption of poisons. Demulcents, such as a mixture of flour and water, beaten eggs, or mashed potatoes in water will not only delay the absorption of the poison but also soothe the irritated mucosa. However, you should administer only limited quantities of these solutions. One cup of the demulcent is sufficient. Excess amounts may open the pyloric valve and allow the poison to pass into the small intestine, where it is much more likely to be absorbed than if it remains in the stomach. You should recall that little absorption takes place in the stomach.

Activated charcoal, one or two tablespoons in eight ounces of water, is a potent absorbent and rapidly inactivates many poisons. You can substitute burned toast for activated charcoal, although it is not as effective.

You should identify the poison as soon as possible, so that

specific measures can be taken. Poison Control Centers, which exist in most major U.S. cities, can identify the poisonous ingredient in most household substances. They can also indicate over the telephone immediate first aid measures until the victim can be brought to a treatment center. You should keep the telephone number of the center in your locality in a prominent place near the telephone and you should know the location of the nearest Poison Control Center.

Once you have identified the poison, you should give the appropriate antidote or antagonist. Symptomatic treatment, such as measures to combat shock and to support respirations, should follow as needed.

When the poison is unknown, you should give the universal antidote. Available at most drug stores, it contains two parts activated charcoal, one part magnesium hydroxide, and one part tannic acid. Although the effectiveness of this antidote is debatable, it will do no harm and is certainly better than doing nothing until more specific treatment is available. The charcoal absorbs many poisons, the magnesium hydroxide neutralizes acids, and the tannic acid neutralizes bases as well as precipitates alkaloids and many metals and thereby prevents their absorption.

INHALED POISONS

In the case of poisoning from the inhalation of toxic gases or fumes, you should remove the patient from the source of the poison, remove any foreign objects from his mouth, and support respirations while transporting him to the hospital. The patient must receive oxygen as soon as possible.

INJECTED POISONS

To treat a patient suffering from an injected poison, you would apply a tourniquet proximal to the injection site. The pulse in arteries distal to the tourniquet should not disappear nor should you apply the tourniquet so tightly that it produces a throbbing sensation.

The local application of icc around the site of the injection will slow capillary blood flow in the area and thus limit the absorption of the poison to some extent.

SKIN CONTAMINATION

In cases of skin contact with poisonous substances, you should flood the contaminated area with water. You must remove the clothing while applying a continuous stream of water. The speed of washing is very important in reducing the extent of injury from corrosive or other agents that injure the skin. You must not use any chemical antidotes because the heat liberated by a chemical reaction may increase the extent of the injury.

EYE CONTAMINATION

You should immediately institute copious washing of the eye with water in cases of poisonous contamination to the eye. Delay of a few seconds will greatly increase the extent of the injury. You should continue washing for at least 15 minutes before you take the patient to a treatment center. As in the case of skin contamination, you must not use any chemical antidotes.

PREVENTION OF POISONING

HOUSEHOLD SOURCES OF POISONING AND PREVENTATIVE MEASURES

Proper storage of all potential poisons is of the utmost importance. All medicines should be properly labeled, dated, and stored in locked cabinets or at least well out of the reach of children. What is "out of reach" of the typical adventuresome 3 or 4 year old is not easy to determine. Children should never be told that any medicine is candy. They may believe you and act accordingly. All medicines including aspirin, iron tablets, and sedatives should have "Mr. Yuck" poison labels on them if a household contains young children.

Cleaning preparations, drain cleaners, polishes, and so forth should never be stored on a low shelf. Dangerous solutions should not be stored in pop bottles or in any container that resembles one ordinarily used for beverages or food.

Insecticides and rodenticides should be kept in a locked cabinet. Paint, paint removers, lacquers, and wood bleaches are all potential hazards. The storage of these substances should be given careful thought. As with other potential hazards children should be taught the possible dangers involved.

INDUSTRIAL SOURCES OF POISONING AND PREVENTATIVE MEASURES

Although the home is the most common site of accidental poisonings, we shall include a brief discussion of industrial sources of poisoning. Fortunately, most industries are well equipped to prevent or to deal with accidents that may occur in their day-to-day operations. You should know and strictly observe the safety rules of the institutions at which you practice.

Dust-forming operations should be conducted in closed systems with local exhaust ventilation. General room ventilation is not sufficient to control air contamination. Hoods for local exhaust ventilation should enclose the process as completely as possible.

Impervious floors and work tables should be used in areas where hazardous material are used. Such surfaces allow adequate cleaning and prevent accumulation of hazardous dusts and liquids.

Spilled dusts should be removed by vacuum cleaning. Sweeping should be done only with wet or oiled sweeping compounds. Spilled liquids should be removed by flushing.

Workers handling poisonous substances should be required to wash before eating or smoking. A change of clothing on leaving work should be required of these workers.

Protective clothing, gloves, goggles, and aprons should be used wherever necessary. If protective clothing is necessary, it should be laundered daily.

Eye fountains and showers must be provided for rapid removal of corrosive materials. Workers should be trained in the hazards involved in working with corrosive materials and understand the appropriate first aid measures to be taken in the event of an accident.

Facilities should be inspected frequently in order to detect failures or inadequacies in control methods. Adequate inspection in some instances may require continuous or intermittent sampling of air.

Workers in hazardous occupations should have preplacement physical examinations to detect chronic respiratory, kidney, or other systemic diseases. Workers with such conditions should not be exposed to any toxic fumes. Workers in hazardous occupations must be examined every 6 months to 1 year.

SPECIFIC TYPES OF POISONINGS AND SUGGESTED TREATMENTS

CLEANING AGENTS

Oxalic acid is a poisonous ingredient in many disinfectants and bleaching agents. The symptoms of this type of poisoning include gastrointestinal upset, erosions of the mouth, and hematemesis. You should give calcium in any form such as milk, chalk, lime water, or calcium gluconate. Calcium will cause the oxalate to precipitate and to prevent the absorption of the poison. You should then lavage the patient with a solution of calcium lactate.

Lye is a corrosive substance in many washing powders, drain cleaners, and paint removers. After ingestion the patient will suffer a burning sensation from the mouth to the stomach. The mucous membranes of the mouth first are soapy white and then become brown and edematous. Ulcerations and strictures of the esophagus are very likely to occur. You can give dilute vinegar or citrus juices, but you should not lavage or attempt to induce vomiting. Hospital treatment will include the passing of weighted esophageal catheters after the fourth day and the administration of cortisone to help prevent esophageal strictures.

When a patient ingests ammonia, he will experience a burning thirst, nausea, vomiting, diarrhea, and possibly respiratory failure. You should give citrus fruit juice or dilute vinegar. Olive oil helps to relieve the pain; however, you should not usually give analgesics, because most are respiratory depressants.

Hydrochloric acid is present in many metal cleaning preparations. This corrosive acid causes burning and erosions of the mouth and esophagus. Esophageal strictures will probably result from the erosions. The emesis has the appearance of coffee grounds. Kidney and liver damage are likely to occur. The patient should be observed for signs of respiratory failure and when necessary support respirations must be instituted. As immediate treatment to neutralize the acid and to prevent absorption, you can give milk of magnesia and egg whites. Later, measures to prevent or to lessen the degree of esophageal strictures are necessary.

Carbon tetrachloride, which is found in many cleaning fluids and paint solvents, is toxic when either inhaled or ingested. Patients will complain of headache, abdominal pain, and gastrointestinal disturbances. Patients often have hiccoughs and behave as though they are intoxicated. If the substance has been swallowed, you should give magnesium sulfate and lavage with large amounts of water. If it has been inhaled, you must remove the patient from the source of the fumes to fresh air and support respirations. You should limit fluids because renal failure may occur. Hospital treatment is primarily directed toward preventing or treating the renal damage. Late results of carbon tetrachloride poisoning are liver and kidney damage.

HYPNOTICS AND SEDATIVES

The initial symptoms of toxicity due to central nervous system depressants are excitement followed by progressive depression and possibly coma. The most serious problems are those of respiratory depression, shock, and a decreased urinary output. Emergency measures are directed toward the management of the shock and support of the respirations. Hemodialysis may be necessary depending upon the severity of the condition.

KEROSENE, GASOLINE, BENZENE, AND NAPHTHA

The symptoms of poisoning from kerosene, gasoline, benzene, and naphtha include central nervous system depression and pulmonary insufficiency. Death may result from pneumonia. You must not induce vomiting because there is likely to be aspiration of the emesis, which can result in oil aspiration pneumonia. A couple of ounces of mineral oil may help to prevent the absorption of the poison and to hasten the passage through the gastrointestinal tract.

CARBON MONOXIDE

With carbon monoxide poisoning the patient will have a bounding pulse and be flushed. He may experience vomiting, headache, and vertigo. If his pupils are dilated (increased in size), the prognosis is grave. You must resuscitate the patient and as soon as possible give a combination of 5 percent carbon dioxide (to stimulate respirations) and 95 percent oxygen.

ALCOHOL

The source of poisoning from alcohol may be that used for beverages, rubbing alcohol, medications that are in tincture form, or methyl alcohol used for fuel. The symptoms include a characteristic odor of the breath, coma, and respiratory depression. Poisoning from methyl alcohol may induce convulsions.

If the patient is not convulsing, you should induce vomiting or lavage to remove any unabsorbed alcohol. Because alcohol is one of the few substances absorbed by the mucosa of the stomach, the sooner emergency measures are taken the more likely they are to be effective. The patient should then be given symptomatic treatment.

STRYCHNINE

Some rodent baits contain strychnine. Unfortunately, strychnine is also used in tonics and some patent medicines used to improve the appetite. The symptoms of strychnine poisoning are those of extreme central nervous system irritability. Convulsions are very likely to occur. You must immediately give the patient activated charcoal by mouth. After you have controlled the convulsions, you should lavage the patient with a dilute solution of potassium permanganate or tannic acid.

IODINE

Unfortunately, iodine is still a fairly common antiseptic solution used in many homes. Soap and water are effective antiseptics and are hardly as toxic as iodine; thus, it is unnecessary and unwise for you to have iodine in your ordinary home medicine cupboard.

Following the ingestion of iodine the patient's lips and mouth are brown and he will complain of thirst. He will experience vomiting and possibly bloody stools. If the poisoning is severe, convulsions and circulatory collapse will probably occur.

If the patient is convulsing, you must immediately take measures to control the convulsions. Then you should lavage the patient with a soluble starch solution or egg whites. Measures to prevent shock or to lessen the degree of shock may be necessary. However, any stimulants should be given only with great caution because they may bring on convulsions.

IRON

Doctors often prescribe ferrous sulfate (iron tablets) for the treatment of certain types of anemia. Many tonics readily available at the drugstore contain iron. Although under some conditions a person may need additional iron in the form of medicine, dietary sources of iron and the body's natural conservation of iron are probably a good solution to simple iron deficiency problems. Adult daily doses of iron preparations such as ferrous sulfate are about 900 mg. As little as three grams (about a four-day supply for an adult) can be fatal to small children. Many iron tablets have a colorful coating and look very much like popular candies. The dangers involved are obvious.

Iron has a corrosive action on the mucosa of the gastrointestinal tract. Liver damage and circulatory collapse can result from iron toxicity. Circulatory collapse may develop very rapidly. Signs of poisoning include a weak rapid pulse, pallor, and cyanosis.

You should give demulcents and protective agents such as milk, magnesium oxide, or bismuth subcarbonate and should perform lavage as soon as possible. Plasma or plasma expanders given intravenously for shock and dehydration may be essential. Deferoxamine is a chelating drug that converts both free and tissue-bound iron to a harmless complex that can be eliminated by the kidneys.

HEADACHE REMEDIES

The most important symptom resulting from toxicity due to headache remedies is cyanosis. Vascular collapse may be caused by anoxia and direct central depression. These symptoms are due to the toxic properties of acetanilid or acetophenetidin, each of which is used in many headache remedies.

You should lavage the patient and give him an enema and a laxative to cleanse the intestinal tract. Oxygen and support of respirations may be necessary. Parenteral fluids and stimulants may be needed to combat shock.

Aspirin is responsible for over 20 percent of all poisonings reported in the United States. Sensitivity to the toxic effect of aspirin varies greatly. The early symptoms of aspirin poisoning are rapid, deep respirations, as well as vomiting, extreme thirst, fever, sweating, and confusion. If the intoxication is severe, circulatory collapse, coma, convulsions, and hemorrhaging will occur.

Parenteral fluids, oxygen, and measures to reduce the fever such as cold sponge baths will be necessary. Vitamin K will combat the bleeding tendency.

DIGITALIS

Digitalis is a drug commonly used in the treatment of heart disease. Patients with these conditions take the drug daily for long periods of time. These people are not usually the victims of digitalis poisoning. It is instead the curious youngster who finds a bottle of interesting tablets on the breakfast room table and consumes the entire bottle of tablets.

The symptoms of digitalis toxicity include gastrointestinal disturbances, visual disturbances, dizziness, and drowsiness. A wide variety of cardiac arrhythmias may develop.

You should lavage the patient immediately. Drugs used to treat the life-threatening arrhythmias include quinidine, procaine-amide, atropine, and potassium or magnesium salts. Each of these drugs has potentially dangerous side affects and should be administered with great caution and only under circumstances in which the patient can be kept under careful observation.

Many digitalis preparations are long acting and have an accumulative effect. The patient for whom the drug has been prescribed may easily forget whether he has taken a dose or not and may take additional pills. Clearly, you should instruct the patient to keep records of when he took his pill in order to prevent an inadvertant overdose.

HEAVY METALS

The most common source of poisoning by heavy metals is lead. Poisoning may be due to the inhalation of fumes caused by the combustion of storage battery casings or to the ingestion of lead. Children are the most likely victims of lead poisoning from ingestion. Although paint sold today does not contain lead, children not uncommonly have eaten flakes of old paint from the walls and woodwork of older homes. Their chewing on lead pipes and lead toys may also result in lead poisoning.

The symptoms of lead poisoning are gastrointestinal disturbances, anemia, a lead line on the gums, and increased bone density at the epiphysis. The latter is observable on X-ray.

The treatment for poisoning due to the ingestion of lead includes lavage with a dilute solution of sodium sulfate and the administration of demulcents or egg whites and saline cathartics. The child will have to be hospitalized for deleading of the bones. Vitamin D and C as well as increased intake of milk will be prescribed. Intravenous infusions of a dilute solution of calcium disodium versenate may be given daily or twice daily for a period of 5 days. Overdosage of this drug or prolonged treatment can cause nephrosis or potassium depletion. Although the child must be carefully observed for these untoward reactions to the treatment, they are not likely to cause any permanent damage.

Mercury poisoning is usually the result of ingestion of mercurial antiseptics or overdosage with mercurial diuretics. The symptoms are bloody diarrhea and kidney failure. Milk and eggs will inactivate the mercury. The patient should then be given copious lavage. Parenteral injections of BAL are effective when administered early and in adequate dosage.

ATROPINE AND BELLADONNA

Eye drops, nose drops, and the seeds of plants such as Jimsonweed occasionally cause poisoning in children. The symptoms include thirst and dryness and dilatation of the pupils. Sometimes the patient experiences confusion.

Withdrawal of the poison may be all that is necessary. If the toxicity is severe, you should lavage with dilute tannic acid, administer sedation, and keep the child in a quiet, dark room.

MOTH REPELLANTS

Toxic symptoms from the ingestion of moth repellants include gastrointestinal disturbances, motor instability, and urinary urgency. The urine appears brown or black. Convulsions and coma may occur.

If the patient is convulsing, you should induce vomiting or perform lavage after the convulsions are controlled. You should also give large quantities of fluid to decrease the concentration of the toxin in the urine.

INSECTICIDES AND RODENTICIDES

Most insecticides and rodenticides contain fluoride, phosphorus, arsenic, or cyanide. All are very toxic. The symptoms of these types of poisoning are gastrointestinal distress, bleeding, and circulatory and respiratory failure. Kidney and liver damage often result from poisoning with insecticides or rodenticides.

Many rodenticide manufacturers advertise that their products are safe around pets and children. They probably are reasonably safe if they contain an emetic that will cause the child or pet to vomit (rodents cannot vomit). Nonetheless, all should be considered potential hazards and therefore stored and used with caution.

When insecticides or rodenticides are ingested, immediate lavage or induced vomiting is indicated. Supportive treatment should be carried out as indicated. BAL is helpful in arsenic poisoning if it can be administered soon enough.

COMMON PLANT POISONS

The onset of symptoms of mushroom (toadstool) poisoning varies from a few minutes to several hours. If the symptoms do not appear for 6 or more hours, the condition is likely to be very serious. The patient's pupils are constricted and his eyes water. He will experience excessive salivation and gastrointestinal disturbances. Mental disturbances and circulatory failure occur with severe intoxication. You should lavage the patient with a tannic acid solution and give him saline purgatives. You must take measures to prevent or to treat the circulatory failure.

Ingestion of larkspur and March marigold will cause nausea, vomiting, diarrhea, weak pulse, burning and tingling sensations in the mouth, restlessness, and possibly convulsions. You should give the patient artificial respirations and injections of atropine. You must also take measures to control the convulsions.

Daffodil bulbs, iris roots, buckeye seeds, and the leaves and unripe berries of black nightshade will cause nausea, vomiting, and diarrhea when ingested. The treatment is removal of the poison by gastric lavage and administration of saline cathartics.

Several types of lilies are also poison. In addition to gastrointestinal disturbances, the symptoms of this type of poisoning include visual disturbances, a very slow pulse, and a low blood pressure (50 mm. Hg systolic or less). With very large doses the blood pressure may rise to 200 mm. Hg and there may be a rapid thready pulse. The treatment is removal of the unabsorbed poison and administration of injections of atropine.

Foxglove leaves contain digitalis. The symptoms of poisoning from the ingestion of foxglove leaves are the same as those due to intoxication from the medication. The patient requires similar treatment.

Mistletoe is also poisonous when ingested. Patients have diarrhea, vomiting, and a slow pulse. You should remove the unabsorbed poison by lavage or by inducing vomiting. You should also observe the patient for signs of cardiac arrhythmias; if they are present appropriate measures to correct the arrhythmia must be instituted.

All parts of black laurel are poisonous. Symptoms of poisoning from black laurel are salivation, tearing, nasal discharge, slow pulse, low blood pressure, vomiting and if severe paralysis. Treatment is administration of atropine by injection.

Injections of atropine also combat poisoning from betel nut seeds. The symptoms of this type of poisoning are gastrointestinal disturbances, dyspnea, impaired vision, and convulsions.

Rhubarb leaves contain oxalic acid. The symptoms and treatment of rhubarb poisoning are the same as those used for a person who has ingested cleaning agents that contain oxalic acid.

The principal manifestations of poisoning from contact with poison ivy or poison sumac are generalized edema and blisters. Acute poisoning can result not only from contact with these plants but also from ingestion or inhalation of smoke from the burning plants. In severe cases of poisoning of this type, laryngeal or pharyngeal edema, weakness, malaise, fever, and a decreased urinary output may occur. If the poisoning seems severe, the patient should have a tracheotomy and be observed carefully for signs of renal failure. (You should see Chapter 18 for a discussion of the signs and treatment of renal failure.)

Emergency measures to be taken when poisoning results from contact with poison ivy or poison sumac include thorough washing with strong laundry soap and water. You should launder the clothing and expose it to air and sunlight for 48 hours. Starch or oatmeal baths help to allay itching. The doctor may prescribe topical applications of cortisone cream or oral doses of 25 to 100 mg. of cortisone every 6 hours. He may also order antihistamines.

If the poison has been ingested, you should perform gastric lavage or induce vomiting as soon as possible. Saline cathartics also help to remove the unabsorbed poison.

SUMMARY QUESTIONS

1. Where is your nearest Poison Control Center?
2. What services are available by calling a Poison Control Center?
3. If a poison has been swallowed, what circumstances contraindicate inducing vomiting?
4. What are the symptoms of aspirin poisoning?
5. What should be the immediate treatment in cases of poisoning from aspirin?
6. What should be done in the event of eye contamination?
7. What should be done in the event of skin contamination?
8. What measures should be taken to help a person who has inhaled poisonous fumes?
9. Discuss measures that may delay the absorption of ingested poisons.
10. What is the most likely poisonous ingredient in metal polishes?
11. What are the symptoms of an overdose of sedatives?
12. What should be done for a person who has taken an overdose of sleeping pills?
13. What are the symptoms of carbon monoxide poisoning?
14. What are likely sources of strychnine poisoning?
15. What are the symptoms of strychnine poisoning and what measures should be undertaken for this patient?
16. How is a patient treated who has taken a toxic dose of an iron preparation?
17. What are the symptoms of digitalis poisoning?

SUGGESTED READINGS

Dreisbach, R. H. *Handbook of Poisons.* Los Altos, Calif.: Lange Medical Publications, 1955.

Liston, R. A. *What You Should Know About Pills.* New York, N.Y.: Pocket Books, 1970.

"Man—His Environment and Health," *American Journal of Public Health,* Part II, January, 1964.

Schifferes, J. J. *Healthier Living.* 3rd ed. New York, N.Y.: John Wiley & Sons, 1970.

Wilson, B. R. (ed.). *Environmental Problems: Pesticides, Thermal Pollution and Environmental Problems.* Philadelphia, Pa.: Lippincott, 1968.

24 Radiation

VOCABULARY

Anemia
Cardiac output
Chromosome
Erythema
Gene
Leukemia
Leukopenia
Malignant
Neoplasm

OVERVIEW

I HISTORICAL BACKGROUND
II ATOMIC ENERGY
III RADIATION SICKNESS
IV PROTECTION FROM RADIATION
V CLINICAL USES OF RADIATION
  A. Diagnostic procedures
  B. Therapeutic uses of isotopes
  C. Deep X-ray therapy

HISTORICAL BACKGROUND

What is matter? What forces bind matter together? Is there one kind of force or more than one? What are the properties of matter? For more than 2500 years man has been asking questions of this sort. We are still trying to answer them today.

The Greek philosopher Empedocles (500 B.C.) suggested that there are four basic types of matter: Earth, Air, Fire, and Water. These four materials combine to form all things by two universal powers, Love and Strife. Basically, this four element theory of matter survived for about 2000 years.

Although Empedocles's theory inspired the Medieval alchemists, they added mysticism to the explanation of the nature of matter. They believed that all kinds of matter have a common origin and possess a permanent "soul" housed in a temporary body. It was their belief that this body can be transformed from one form to another. This, in part at least, was the basis of Medieval belief that one could make gold from some other metal. All one needed do is to figure out how to effect this transformation. Although the solution to the problem is hardly practical, it has been found by twentieth-century scientists: Nuclear bombardment of platinum results in gold.

The first atomic theory of matter was developed by two Greek philosophers, Leucippus and Democritus, between 450 and 420 B.C. However, it was not until 1803 that John Dalton was able to show in a convincing way that atoms do exist and that they are a useful basis for understanding matter and chemical reactions. The ancient Greeks—and Dalton, too—believed that the atom was the smallest (and indivisible) particle of matter. We now know that atoms are divisible and that there are some 30 subatomic parts. In order to understand something of the nature of isotopes used in medical practice and the nature of radiation, we shall confine our discussion to only three of these subatomic parts: electrons, protons, and neutrons.

In 1913 Niels Bohr provided the theoretic and experimental justification of picturing the atom as a solar system. Electrons are small, negatively charged particles that revolve in shells around the nucleus of the atom. There are two electrons on the shell closest to the nucleus, 8 on the next shell, 18 on the next, and 32 on the next shell. The number of revolving, or planetary, electrons in an atom determines the atomic number of the atom. Ordinary chemical energy is possible because these electrons are not as stable as they might be. It is therefore the electrons that mainly determine the atomic character of the particular element.

Protons were discovered by Ernest Rutherford in 1911. These subatomic particles, which have a positive charge, are located in the nucleus of the atom. In any particular atom the number of protons is equal to the number of electrons revolving around the nucleus. Thus, ordinary matter is electrically neutral.

Neutrons were discovered by James Chadwick in 1932. These subatomic particles, which have no charge, reside in the nucleus. They provide the binding force that holds together all of the positively charged protons in the nucleus. The number of protons and neutrons in an atom determines the atomic weight of the atom.

Atoms of the same element (that is, those that have the same atomic number) can have different atomic weights because they have different numbers of neutrons. They are called isotopes. Of the well over 100 elements now known, some have as many as ten isotopes. Because the isotopes of a particular element have the same number of protons and electrons, their ordinary chemical behavior is the same. About 300 known isotopes occur naturally and are stable. However, by use of nuclear reactors, physicists and chemists have succeeded in making about 900 more, some of which are highly useful in medicine because of the energy they produce as they eject nuclear particles.

ATOMIC ENERGY

You should recall that ordinary chemical reactions occur because the electrons are not as stable as they might be. Atomic energy is possible because nuclear particles are not stable. Over 99 percent of the energy of an atom is found in the nucleus. If 4 grams of hydrogen are burned in the presence of oxygen, the result is the production of 120,000 calories. Four grams of hydrogen undergoing thermonuclear fusion produces 592 billion calories. This tremendous nuclear energy results when radioactive atoms attempt to achieve stability by throwing off nuclear particles. The change in the composition of the nucleus results in transmutation, the formation of an entirely different element.

Radioactive decay, arising from the instability of the nucleus of radioactive elements, is measured in terms of half-life. Half-life is the time it takes for an initial quantity of radioactive material to decay to one-half the original amount. The half-lives of different radioactive materials vary from billions of years (uranium U^{238} has a half-life of 5 billion years) to a matter of hours (bromine, Br^{82}, has a half-life of 36 hours). Some of the more recently discovered isotopes have half-lives of only a few seconds.

Rays are emitted in the process of decay. The naturally occurring rays are alpha rays, beta rays, and gamma rays. X-rays are manmade gamma rays. The penetrating ability of the alpha, beta, and gamma rays varies considerably. This fact is of importance not only in terms of the medical usefulness of the rays but also in terms of protecting vulnerable tissues from the harmful effects of the rays.

Alpha rays are two protons and two neutrons that move with a velocity equal to light. These rays are particles. Alpha rays cannot penetrate the epidermis (nonliving tissue) and can be stopped with cardboard. If alpha rays are ingested (swallowed or inhaled), they can penetrate up to a depth of five living cells. Uranium that is naturally radioactive is an alpha emitter. When the uranium decays it becomes thorium, which emits beta and gamma rays.

Beta rays are also particles. Electrons produced within the nucleus are thrown out and a neutron becomes a proton. Beta rays are much smaller than alpha rays and are more penetrating. They can "burn" the skin but cannot reach the internal organs from outside.

Gamma rays are not particles but are electromagnetic radiation. Gamma rays as well as the man-made X-rays are very penetrating. Because early workers with X-rays did not realize their danger, many physicists, doctors, and radiologists died over the years from excessive exposure to X-rays.

The three atomic radiations and X-rays are all ionizing radiations that convert atoms of matter through which they pass to ions. This procedure is not the orderly sort of ionizing that takes place when a sodium atom interacts with a chlorine atom; rather, it is a ruthless, indiscriminate knocking of electrons away from atoms and molecules. Although this type of behavior makes the radiations dangerous to human tissues, at the same time they can be very beneficial. First we shall consider the general symptoms of radiation sickness and then the valuable diagnostic and therapeutic uses of radiation.

RADIATION SICKNESS

The first symptom of radiation sickness is a decrease in the number of white blood cells. There may simply be a mild erythema of the skin; however, sometimes damage to the skin may be severe. There may be ulcerating wounds that do not heal. Loss of hair is fairly common. Gastrointestinal symptoms include nausea, vomiting, and diarrhea. Most patients who are given deep X-ray therapy or cobalt-ray treatments experience some gastrointestinal disturbances. Occasionally, internal bleeding occurs.

The cells that are most vulnerable to radiation damage are those that most frequently undergo cellular division. The primary site of radiation damage is the genetic material in the cell's nucleus. If the exposure to the radiation is very great, radiated cells are unable to divide when it is time to do so. If, as a result of deliberate exposure, the cells so damaged are cancer cells, the cancer is halted. If the exposure is relatively mild, the damage to the genes is less likely to be profound. Genes and chromosomes are capable of considerable self-repair. However, there are likely to be minor changes in the genetic material that are not serious enough to prevent cellular division but that can cause mutations. Future divisions of these cells may take the form of the erratic behavior characteristic of cancer. Thus, although radiation can possibly halt the growth of a cancer, it can also cause cancerous growths in otherwise healthy tissue.

PROTECTION FROM RADIATION

Measures for protection against radiation depend on the amount and kind of radioactivity present. They are based on the following principles: shielding around the source of radiation, distance from the source, and length of time of exposure.

Radioactive materials must be stored in lead-lined containers. The thickness of this lining is determined by the radiation physicist or safety officer. Depending upon the type of radioactive material being used, special techniques may be required to shield the body secretions of a patient.

Because it follows the inverse square law—that is, if the distance from the source of radiation is doubled, exposure decreases by a factor of four—distance provides great protection from the hazards of radiation. Generally, patients who are being treated with radioactive materials should be hospitalized in a private room and the bed should be as far from the door as possible. Many hospitals assign patients who are to receive this type of therapy to a corner room or to a room next to a linen or supply closet. Such practices provide protection for other patients by taking into account both distance and the shielding protection of the additional walls.

You should always minimize the time spent in the room of a patient who has been given radioactive substances. Those who attend the patient should be rotated so that no one attendant is exposed for unnecessarily long periods of time. The patient as well as his visitors should be advised of the nature and importance of these precautions. Their understanding of these precautions will not only help them protect themselves but will relieve them from unnecessary worry about exposure to radiation.

The length of time necessary to observe the precautions depends upon the type of radioactive material being used. In some instances no unusual precautions are necessary. For example, if the half-life of the radioactive material being used is only a matter of minutes or a few hours, there is little danger to those attending these patients.

CLINICAL USES OF RADIATION

The uses of radiation in medical practice are numerous and are constantly being revised and expanded. For this reason, we cannot attempt to make the following discussion comprehensive. If you are working with radioactive materials, you should constantly update your knowledge of the subject and be thoroughly familiar with the policies and procedures of the institution where you practice.

DIAGNOSTIC PROCEDURES

Radioactive isotopes are used to help in the diagnosis of many different conditions. In most of these conditions the isotope is given in what is referred to as a tracer dose. Few precautions or special handling procedures are necessary when diagnostic amounts of an isotope are administered. If the tracer dose is given orally, a nurse or technician usually administers the isotope. A physician usually performs intravenous administrations.

Long-handled forceps should be used to handle the capsule or vial containing the isotope. Although each dose gives off little radiation, the cumulative amount after several doses have been dispensed may result in considerable radiation exposure. When the dose is in liquid form, care should be taken to avoid spills. All spills should be immediately decontaminated. Paper cups or other dry wastes should be collected and disposed of according to the approved procedure in the institution.

Radioactive iodine (I^{131}) is used to diagnose diseases of the thyroid, kidney, and liver. It is also used to detect and to localize brain tumors. I^{131} is sometimes used to determine cardiac output.

An isotope of sodium is a beta and gamma emitter with a half-life of 14.8 hours. This isotope is given intravenously to help locate a blood clot. If circulation is poor (the probable location of the clot), there will be little radioactivity.

Radioactive phosphorus (P^{32}), a beta emitter, tends to concentrate in neoplasms. It is therefore helpful in detecting the location and determining the size of a neoplasm. Because this isotope inhibits the production of red blood cells, it may also be helpful in treating polycythemia vera, a disease characterized by an overproduction of red blood cells.

THERAPEUTIC USES OF ISOTOPES

Radiophosphorus is probably the least hazardous of the commonly used therapeutic radioisotopes. It may be given orally or intravenously for the treatment of certain types of leukemia and cancer of the prostate. Immediate vomiting by the patient after an oral dose and leakage from the injection site are the only potential hazards to hospital personnel when P^{32} is administered in therapeutic doses.

Radiogold (Au^{198}) is administered by injection into the pleural or peritoneal cavity for the treatment of metastatic cancer. This isotope, a gamma ray emitter, is given in fairly high dosages. These two facts necessitate cautious handling of the material and precautions to prevent excessive radiation to hospital personnel and visitors. For the first two days following the injection, strict precautions regarding the length of time spent by hospital personnel with the patient and the time allowed visitors should be observed. Because the isotope has a short half-life (2.7 days), the time spent with the patient after three days need not be restricted.

Therapeutic doses of radioiodine vary considerably depending upon the condition being treated. At low doses, the patient is not a radiation hazard and may be treated on an outpatient basis. At high doses, such as those given to treat thyroid cancer, the patient is a potential radiation hazard, particularly for the first 72 hours after administration of the isotope. Iodine is excreted mainly in the urine, although significant amounts can appear in saliva, sweat, and vomitus, and in the stools if the patient has diarrhea. Contamination of the patient or his surroundings from these sources is greatest during the first 24 hours after administration of the dose.

Implantation of radioisotopes is a means of treating malignant growths with high doses of radiation, while keeping the radiation dose to surrounding tissues relatively low. The implants may be temporary or permanent. Temporary implants are usually isotopes with a long half-life, and permanent implants have a relatively short half-life. The implants are performed surgically. All personnel involved in the care of the patient should be aware that he contains radioactive material.

DEEP X-RAY THERAPY

To prepare a patient for deep X-ray therapy, the radiation therapist examines the patient and takes measurements of the area to be treated. He marks the skin over the area with ink to make certain that the treatments will be delivered to an exact area during the entire course of therapy. These markings should not be washed off until the treatments are completed.

Before the treatment, the therapist positions the patient and aligns the X-ray beam to the area to be treated. The patient must remain very still, because any deviation from the prescribed position will alter the course of the rays. During the actual treatment, the radiation personnel remain outside of the treatment room and leave the patient alone for the prescribed period of time of the treatment. The treatment room is usually equipped with intercom systems and windows so that the patient can receive immediate assistance if needed. Once the X-ray machine is turned off, personnel can safely enter the treatment room, because there is no danger of radiation exposure.

Patients who receive deep X-ray therapy are frequently outpatients. They should be given instructions concerning possible radiation reactions, skin care, and diet.

It is not uncommon for the skin over the area being radiated to redden. You should not refer to this condition as a "burn" because the patient might misinterpret this as being the result of carelessness or mismanagement. You might use terms such as "radiation reaction" or "radiodermatitis". The patient should not apply ointments, powders, or cosmetics to the area being radiated, nor should he vigorously rub or wash the area with anything other than mild soap. The patient should not sunbathe or use any other form of rays such as ultraviolet or infrared lamps.

The patient's diet should be high in protein and carbohydrates and he should have additional vitamins. Increased fluid intake is usually recommended for patients receiving radiation therapy.

Some radiation reactions do not develop for weeks or months (or years, in some cases) after the treatments. The patient must be instructed to report any changes in his physical status to the doctor. He should have periodic blood counts because secondary anemia or leukopenia can develop.

SUMMARY QUESTIONS

1. What are isotopes?
2. List the symptoms of radiation sickness.

3. What are the chief measures to be used in protection against radiation damage?
4. How is iodine excreted from the body?
5. What precautions should you observe when you care for a patient who is receiving deep X-ray treatments?
6. What precautions should you observe when you attend a patient who has received a tracer dose of a radioactive isotope?
7. Explain how radiation can halt the growth of a cancer but can also cause cancer.
8. Discuss the recommended skin care for a patient who is being treated with X-rays.
9. Discuss the differences in the penetrating powers of the various rays.
10. What diet instructions would you give to a patient who is receiving deep X-ray therapy?
11. What is meant by half-life?

SUGGESTED READINGS

Arena, V. "Radiation Accidents: What You Need to Know About Them," *RN Magazine*, September 1973.

Chadwick, D. R. "Man—His Environment and Health," Part II, "Radiation and the Public Health," *American Journal of Public Health*, January 1964.

Gardner, A. F. *Paramedical Pathology*. Springfield, Ill.: Charles C Thomas, 1972.

Jordana, A. "Protection from X-ray Exposure," *The Professional Medical Assistant*, May/June 1973.

Simon, N. "Man—His Environment and Health," Part II, "Radiation in the Diagnosis and Treatment," *American Journal of Public Health*, January 1964.

25 Aging

VOCABULARY

Atrophy
Catabolism
Convolutions
Dysfunction
Endoplasmic reticulum
Hydrolysis
Lysosome
Mitochondria
Nephron
Proprioceptor
Sclerosis
Vital capacity

OVERVIEW

FREE RADICAL THEORY OF AGING

If Ponce de Leon were searching today for his magic spring, he would still be an explorer but the site of his expedition would be a biochemistry laboratory. He would probably search through the microscopic cells and the organelles within these cells.

The fact that in the United States the average life expectancy at the turn of the century was about 50 years and is now about 70 years probably obscures an important reality for most people who have reached adulthood. The longer life expectancy of today is largely due to decrease in infant mortality and better control of infectious diseases. In fact, the Biblical psalmist who advised that life span is three score and ten was apparently speaking not only for the population of his own day but also for ours.

It is a law of thermodynamics that the entropy (the relative amount of unavailable energy) of any physiochemical system tends to become more pronounced with the passage of time. This certainly is the case with the cellular systems of the human body. Consider the energy produced from a milkshake and a hamburger fed to a 4-year-old child and compare this with the available energy of a 60-year-old man resulting from the same caloric intake. A decrease in available energy with the resultant diminished capacity for spontaneous change would appear to be crucial to the physiological processes that are to be expected with the passage of time.

Because the relatively new science of radiation chemistry has had a profound impact on every other field in the natural sciences, those researchers and doctors pursuing the new and quite uncertain field of the biochemistry of aging are naturally looking in this direction to find answers to some of its questions. Many theories of aging exist but at the present time they all are oriented toward intracellular metabolism, which of course is dependent upon cellular nutrition and appropriate waste disposal.

The radiation theory of aging appears to concern itself with an intracellular tug of war that goes on between two factors and acts on a third—the intensity and duration of radiation-like effects, the polyunsaturated lipids upon which they act, and the antioxidants available to protect the lipids from excessive destruction. Longevity tables seem to substantiate that the balance is tilted in favor of radiation-like damage that finally succeeds after 70 or so years. Much evidence from what we now know is taking place within the cell makes this whole concept of longevity quite plausible.

Polyunsaturated fats may well be the primary source of free radicals within the cell that causes it to age. When radiant energy—which can penetrate throughout the body—strikes a polyunsaturated lipid that is present in the cell, one of two things may happen: If enough antioxidant is present the radiation has little effect or if there is an intracellular deficiency of antioxidant, the rays will strike the lipid and knock loose an H^+ atom. This procedure initiates the peroxidation of the polyunsaturated lipid which involves the direct reaction of oxygen and lipid to form free radical intermediates. The free radicals fly about in the cell until they strike other molecules and cause all sorts of damage.

We shall now consider how this free radical damage might bring about the decrease in available cellular energy, cause cellular catabolism, decrease cellular metabolism, and particularly decrease protein synthesis. All of these factors contribute to the decreased ability of the aged to adapt to change. The relative sensitivity to the free radical damage may explain why some people look older than their years. If enough cellular damage has taken place, the host looks old and cannot adapt to changes.

Should man make great efforts to slow the aging process? Apparently, this has been and is a universal desire, but perhaps this desire is of questionable wisdom. Laboratory evidence suggests ways of shielding against free radical damage by means of antioxidants. The antioxidants that have substantially increased the life span of mice and other laboratory animals are vitamin E and butylated hydroxytoluene.

Free radical damage to the polyunsaturated lipids is not an isolated incident but is occurring all of the time. Polyunsaturated lipids are necessary because they supply most of the materials from which membranous structures of the cell are built, particularly the membranes of the mitochondria, endoplasmic reticulum, and lysosomes.

When the mitochondrial powerhouses of a cell are destroyed by lipid peroxidation, cellular energy is diminished. The ensuing decreased activity of the cell, the tissues, the organ, and ultimately the organism results in the increasing entropy of the aged.

The endoplasmic reticulum of the cell is responsible for protein synthesis that is essential for tissue repair, enzyme production, and the other vital roles of body proteins. Cells with damaged endoplasmic reticula will be unable to carry out these essential functions. Obviously, cells of the aged individual do not have the capacity to heal wounds that once healed readily.

The lysosomes are also vulnerable to this destructive sequence of events. Lysosomes have the normal obligation of housing hydrolytic enzymes and dissolving cellular wastes in an orderly fashion. If the membrane of the cellular lysosome is punctured by free radicals, the hydrolytic enzymes are released and destroy the cell. The individual is one cell older. If the lysosome is not punctured by a free radical but destruction is taking place all around it, the lysosome goes about its job and attempts to clean up the cellular waste until the waste disposal job is simply too much and the cell drowns in its own wastes. This result is the tissue catabolism so evident in old age.

Different types of cells have different life spans. Those of the central nervous system do not reproduce and seem to drop out randomly, but, in general, functioning neurons last about the expected 70 years. White blood cells and cells of the lining of the gastrointestinal tract have a life span of only a few days and do reproduce. Clearly, lipid peroxidation does not operate at the same rate in all of the cells of the body. The amazing fact is that their life spans are somehow programmed in synchrony. The teeth of mammals are not reproduced; yet the mechanical wearing away of the teeth of herbivorous mammals, for example, horses and mice, proceeds at such a rate that the teeth last about as long as the rest of the animal. Much evidence substantiates that this synchrony is programmed in the genetic material of the particular organism. The synchrony of different aging changes may also be a function of the physiological interdependence of the various organ systems.

CONSEQUENCES OF AGING

When we consider the consequences of aging of the various body systems, we find an important common thread—aging systems have a decreased ability to adapt. The developmental—or maturing—process, on the other hand, is characterized by increasing effectiveness of the adaptive mechanisms. In addition, the higher the level of the organism and the more interdependent the systems, units, and subunits of which it is composed, the more striking will be the progressive interference with function. Small changes in structure of a unit can lead to big changes in performance. The more highly integrated the functions of the structures involved, the more vulnerable they are to the artillery of time. A striking example is the aging vision. Recall how complex and beautifully interdependent the processes are which produce perfect vision, refraction, accommodation, convergence, and pupillary action. Aging changes vision ability quite some time before the ravages of time may become evident in the other body-tissues.

We shall now take a system approach to the process of aging and explore the changes in each system. In doing this, however, we must not lose sight of the fact the the human body functions as a whole, not just as a group of isolated systems.

CENTRAL NERVOUS SYSTEM

In the central nervous system, total brain weight is decreased one sixth or more in the aged. The convolutional atrophy may be generalized but is most conspicuous in the frontal lobes, where the neurons are assigned the most complex thought processes. The fallout seems to occur in strips, suggesting some relationship to vascular dysfunction. Here, as elsewhere, vascular changes accelerate and accentuate the cellular loss. Recall that it is upon the vascular system that cells depend for their supplies and for the removal of the products of cellular metabolism.

The more intricate the task and the more judgment demanded for it, the greater the deficit in the aged. Generally, animals with organic brain disease perform simple tasks as well as do intact animals.

In the aged person, input and output are tightly coupled. Thus, an error made once or twice during the process of learning a new skill is difficult to eliminate. The old person needs continuous reassurance and information as he proceeds to do a task. This information may be hard to come by because vision and hearing may be defective. In addition, the information giver may unfortunately be impatient with the elderly learner.

Most of the delay in performing tasks is due to pauses between successive acts rather than to the slowing of the acts themselves. At any age it takes longer to aim an action than to perform it. This aiming takes progressively longer with age. A sequence of acts broken down into a series of individual ones with pauses between is also characteristic of interference with proprioceptive and cerebellar systems.

In general, what the novelist Henry James called "sagacity," or the ability to cope with life, deteriorates with age later than the mental abilities required to perform tasks. Apparently, the reason is that experience and wisdom continue to grow for a time after speed, memory, learning ability, and simple reasoning begin to fail.

Much has been written about the loss of recent memory in the elderly while their memory of past events seems to remain crystal clear. Whether this phenomenon is physiological or psychological (although it is difficult to separate the two) is not clear. Unfortunately, the events of yesteryear may well be worth remembering more than the day-to-day monotony and loneliness existent in the lives of so many aged people.

The deteriorating aspects of the central nervous system, and the consequences of this deterioration, obviously have something to do with the original equipment of the individual. If one has had an IQ of 130 or 140, a loss of 10 or 15 points is not going to make as inadequate an elder as a similar loss would in an individual whose original mentality was an IQ score of 90. Unattractive personality traits are likely to become more obvious. The youthful coping mechanisms to cover up negative features do not operate as effectively—in essence, we become who we really are.

Another aspect to the aging of the central nervous system that is interrelated with the other body systems is the marked decrease in the range of vital signs exhibited by the elderly. In disease, the elderly do not respond with great changes in body temperature, pulse rate, respirations, and blood pressure. Indeed, a fever of one or two degrees in an elderly patient may indicate a much more serious illness than does a fever of three or four degrees in a child.

CIRCULATORY SYSTEM

Some authorities attribute most, if not all, of the consequences of aging to circulatory changes. As has been mentioned, because human body cells depend upon the circulatory system for supplies, this hypothesis has some validity. However, were this dependence the entire story, other living organisms that do not have circulatory systems would not age, and they do.

Some of the main changes that take place in the human circulatory structures as they age include increased rigidity of the heart valves, thickening and roughening of the auricular endocardium, and atrophy of the apical portion of the left ventricle. The arteries harden, and the lining of these vessels becomes roughened. Destruction of the elastic tissue in the blood vessel walls takes place, and perfusion of blood through the circulatory system becomes more dependent upon the force of systole. Diastoltic pressure increases because peripheral resistance increases.

In the hemopoietic system, the total volume of red bone marrow decreases; however, no abnormality occurs in the blood cell counts of the aged. Anemias have the same etiology and consequences in the aged as they do in earlier years. Leukemias increase in frequency in advanced age but have no particularly distinctive features.

SKIN

Changes in the skin are most obvious and are directly the result of cell catabolism, decreased protein synthesis, and decreased available energy. The skin is dry and the hair loses its luster. Cells that once produced lubricants have been lost or have decreased their functional abilities. Wrinkled and characteristically thin skin is also a result of cell catabolism.

RESPIRATORY SYSTEM

Changes in the lungs that are strictly due to aging are less well defined than are those in most other tissues because these structures are also more exposed to environmental factors such as air pollution. The major changes that are related to aging are probably an increased susceptibility to respiratory infections and a decline in the efficiency of the bronchoeleminating system. These changes are due to the atrophy of columnar epithelium and the mucous glands of the lining of the bronchi.

Sclerosis of the bronchi and supportive tissues interferes with normal respiratory movements and a decrease in vital capacity. Certain deteriorations in the cardiovascular system also adversely affect the pulmonary system. Decreased perfusion and ventilation both contribute to an increase in the physiological dead space (the prealveolar respiratory tract structures) in the aged individual.

GASTROINTESTINAL SYSTEM

Some of the changes in the gastrointestinal system of the aged are similar to the changes that occur in the mucous membrane lining of the respiratory tract. These are particularly apparent in the mouth. The mouth is dry and parotitis (infection of parotid glands) may be associated with the decrease in saliva production. A dry mouth also leads to a diminished sense of taste, which will influence other digestive processes. It is common for the elderly to complain that food does not taste as good as it once did. Good mouth care can help reduce the problems related to this aspect of aging.

Atrophic changes in the glands of the stomach occur that may lead to achlorhydria and atrophic gastritis. The resulting indigestion will increase the nutritional problems of the elderly.

A decrease in protein synthesis causes a loss of muscle tone and predisposes the elderly to intestinal obstruction from hernia or from scarring adhesions or diverticula. These diseases are not much different in the elderly than they are in a young adult; however, fluid and electrolyte losses and toxic absorption associated with intestinal obstruction are more likely to be fatal in the elderly who are already debilitated. Pulmonary aspiration of vomitus in the elderly is more dangerous than it is in a youthful patient because of respiratory changes that we have already discussed.

Bowel atony (lack of muscle tone) may be responsible for chronic constipation in many elderly individuals. In addition to this, the elderly have a decreased sense of thirst and as a consequence are likely to be dehydrated. These factors will lead to hardening of the feces and constipation.

Although a decrease occurs in the size of the aged liver, no particularly unique features of the liver and biliary system predispose the elderly to diseases of these organs.

ENDOCRINE SYSTEM

Impairment of the adrenal cortex function has been suggested by doctors and researchers as a cause of senescence (aging). Reviewing the functions of two of the main groups of adrenal cortical hormones, you should recall that the glucocorticoids look out for the fuel and tissue repair supplies that the body needs to cope with emergency situations, and the mineral corticoids look out for the preservation of normal fluid and electrolyte balance. Considering this, the adrenal atrophy of aging certainly reduces the individual's ability to cope and to adapt.

Other endocrine changes also occur with advanced years. The ovaries atrophy and no longer produce hormones. The male probably experiences little, if any, decrease in the production of testosterone; however, spermatogenesis may decrease. Infants of elderly parents are more likely to have congenital defects than are the children of younger parents.

We have discussed menopausal changes and prostatic hypertrophy in Chapters 19 and 20. Both can cause aging individuals considerable difficulties.

Generally speaking, although diabetes mellitus that has its onset in middle age is not usually as serious or as difficult to control as is the diabetes of children and young adults, it can result in serious complications. The circulatory disturbances of the elderly, together with general tissue catabolism and the decrease in protein synthesis, contribute to the difficulty in management of complications of diabetes, such as leg ulcers.

MUSCULOSKELETAL SYSTEM

Osteoarthritis is so common in the elderly as to be regarded as physiological. Destruction of the articular cartilages and a decrease in the cartilages' ability to function as cushions takes place. Articular cartilage is at a great disadvantage, regarding healing, because cartilage normally has little blood supply.

Osteoporosis, which is an imbalance between bone formation and reabsorption, leads to diminution in bone density. Elderly individuals are more prone to fractures than are the youthful because of this change in the osseous tissue.

Skeletal muscles decrease in mass. Function of skeletal muscles is dependent upon accurate transmission of afferent and efferent nerve impulses; both are diminished in the aging individual. The cellular catabolism in the skeletal muscles and the increase in entropy with the passage of time help to explain the waning of strength of the aged person.

URINARY SYSTEM

The urinary system undergoes aging changes similar to those of the other body systems. Probably about 50 percent of the nephron units are lost in people over the age of 60.

Although kidney disease in the aged has no particularly unique features, clearly the aged individual is not as well equipped to cope with any additional kidney dysfunction. Loss of smooth muscle tone in the bladder may predispose to retention of urine. Loss of sphincter tone may cause incontinence (uncontrollable loss of urine).

Shakespeare admirably summarized all of the devastating effects of aging in *As You Like It:* "Sans teeth, sans eyes, sans taste, sans everything." Considering all the changes that the years bring about, perhaps an explanation is required for not what causes the body to age, but how it is able to last so long.

SUMMARY QUESTIONS

1. What is osteoporosis and what are the complications of this condition?
2. Describe the changes that aging brings about in the cardiovascular system.
3. What factors predispose the elderly to chronic constipation?
4. Why might complications of diabetes mellitus be more difficult for an elderly diabetic than for a younger diabetic?
5. Why are elderly people more prone to respiratory infections than young adults?
6. Describe the changes that aging brings about in the urinary system.
7. Discuss some of the consequences of adrenal atrophy.
8. What factors should you keep in mind when giving health instructions to an elderly patient?

SUGGESTED READINGS

Journal of the American Geriatrics Society, numerous articles.

Lowenberg, Miriam E. *Food and Man.* New York, N.Y.: John Wiley & Sons, 1968.

Macleod, C. D. "Dietary Intake of Older People," *Nutrition,* Spring 1970.

Schultz, T. "Who Needs The Aged Anyhow?" *Today's Health,* June 1973.

26 Stress

VOCABULARY

ACTH
ADH
Aldosterone
Atrophy
Epinephrine
Glucocorticoid
Glucagon
Homeostasis
Hypertrophy
Norepinephrine
Phagocytosis
Pineal
Serotonin
Syndrome

OVERVIEW

I THE NATURE OF STRESS
II FACTORS DETERMINING THE DEGREE OF RESPONSE TO STRESS
III "ADAPTIVE" PATTERNS
IV ALARM AND RESISTANCE

THE NATURE OF STRESS

We all have some preconceived ideas about what "stress" means. For this reason many an investigator in the field of stress has regretted that he ever started using the word. We emphasize that stress as being discussed in this chapter is not necessarily bad. Although the stressor can be a serious trauma or illness, it can also be a wonderful party, an interesting challenge, a wedding, or a licensing examination. Regardless of the nature of the stress agent—whether it acts on the personality of the individual involved, traumatizes body cells, or alters established living patterns of the individual—many of the body's responses to the stress situation are much the same.

Numerous investigators have contributed to the basic concept of the stress syndrome as we presently understand it. Doctor Hans Selye began serious investigation of the subject in 1936. The subject had occurred to him several years before when he was a medical student at the German University of Prague. As a novice, he attended lectures by a professor who presented many patients and enumerated signs characteristic of their various diseases; yet he could not see these subtle signs. What impressed him, but seemed to be completely ignored by the professor, was the fact that the patients had many symptoms in common. To him the patients all appeared to be "just plain sick," but such an appearance seemed to be too vague to interest the professor.

In Dr. Selye's early investigations he defined stress as the rate of wear and tear. Later, it became apparent that the body reactions that tend to repair the wear and tear—that is, the responses to the stress—are hardly distinguishable from the stress itself. In our discussion, we shall consider stress as a specific syndrome that has nonspecific causes. We shall divide the responses into two phases: the initial alarm reaction and the resistance phase. Classically, there is a third stage, that of exhaustion or death.

Early investigations resulted in specific, observable changes due to stress that was induced by a variety of means. These changes included adrenal hypertrophy, atrophy of the thymus, and changes in the lining of the gastrointestinal tract. That these changes occur is well established; however, our discussion of the topic will center around other alterations in the anatomy, physiology, chemistry, and psychosocial aspects of an individual who is attempting to adapt to a stressful situation.

In preceding chapters of this text we have been predominantly concerned with the uniqueness of particular disease processes. The chest X-ray of a patient with fractured ribs is different from that of a patient with pneumonia; however, both patients have much in common. They are experiencing stress.

It is therefore appropriate that we now consider what is common to any disease process or stress. Much of this is the essence of homeostasis—the adaptive mechanisms of the body functioning to maintain equilibrium. Ironically, the more the body is able to change, the better able it is to stay the same: The stability and equilibrium of an individual depend greatly upon the rapidity and efficiency with which change can be accomplished. The ability to adapt or to change determines the ease or disease of the individual. The skill and knowledge with which the stressed person's helpers support the appropriate body defense mechanisms may spell the difference between recovery or exhaustion.

FACTORS DETERMINING THE DEGREE OF RESPONSE TO STRESS

We shall now consider some of the factors involved in determining the degree of response to stress. Obviously, the intensity of the stress has something to do with the response. However, this is not the only factor, nor is it necessarily the major factor. How the individual perceives the situation is of great importance. Probably none of you consider your last flu shot a particularly stressful situation, yet your reactions to the injections you received when you were three or four years old evoked quite different reactions. At that time, you could find no reasonable explanation of why the usually kind, trustworthy adults suddenly held you down and stuck needles into you. You naturally responded to the situation in the way appropriate to how you perceived it. The reactions of an elderly patient with impaired sight and hearing who finds himself in strange surroundings and attended by strangers may be another example of how faulty perception of a situation may evoke responses that seem inappropriate. In *Midsummer Night's Dream,* Shakespeare explains this point beautifully:

> —or in the night with some imagined fear
> how easy a bush a bear doth appear.

Age is another factor that influences the degree of reaction to stress. Generally speaking, the younger the individual the greater the response and the more likely the individual will be able to adapt. You might observe the resiliency of the healthy youth. He undergoes much turmoil during the process of his adaptation to stress. You should remember that we are talking about nonspecific causes of stress but a specific syndrome. The delight of the 4-year-old on Christmas morning when he sees all of the the wonderful gifts is probably no greater than that of his grandmother, but the observable reactions are quite different. If the stress agent is a disease—for example, an infection—the vital signs of the 4-

year-old will change markedly, whereas the changes in the vital signs of an aged individual with a similar infection change very little. The aged are not able to change as much or as rapidly; as a result their stability—their life—may be sacrificed.

Previous exposure to the same or similar stress agents tends to decrease some of an individual's body responses but at the same time equips him to deal more effectively with the stress situation. During some of the early studies of stress, investigators exposed rats to thermal stress. They were placed in cold environments but those not cold enough to be a threat to their lives. On subsequent occasions, they were exposed to colder and colder temperatures. Finally, the rats who were so conditioned were dipped in cold water and confined to even lower temperatures. Simultaneously, rats that had not been conditioned to the stress of lower and lower temperatures were subjected to the same wet, cold environment. The rats who had the previous exposure to the thermal stress survived the experiment but those who had not had the previous exposures died. The investigators found on examination that there was considerable adrenal hypertrophy in the conditioned rats but none in the rats who had not been conditioned to the stress and were unable to adapt to it.

Perhaps this previous-exposure aspect of determining the degree of response to stress helps to explain some of the amazing accounts of escape by prisoners of war. When an opportunity to escape arose, prisoners who had been kept on near-starvation rations, tortured, poorly clothed, and caged in most unsanitary quarters were able to accomplish astonishing feats that you would assume could be done only by a person in the peak of physical condition.

Undoubtedly, other factors influence the degree of response to stress. What is important is your realization that response is a composite of many factors. Some factors you can do little about in your efforts to help the stressed individual, but the fact that you are giving them some thought may increase your sensitivity to the needs of people in stressful situations.

"ADAPTIVE" PATTERNS

Before going into the specifics of the physiological, anatomical, chemical, and psychosocial changes that take place in response to stress, we shall consider some general stress-adaptive patterns that all of you may have developed to one degree or another. Under some circumstances, these responses will help you cope with stressful situations. Depending upon the nature of the stressor, however, these patterns may not be helpful and indeed may compound your problems.

We shall consider the response of the mucous membranes of the respiratory tract when they are exposed to irritating gases. What happens when you inhale ammonia fumes? The mucous membranes increase their production of watery mucus, which dilutes the fumes and helps to reduce the injury to the more vulnerable tissues lower in the respiratory tract. Swelling of the mucous membranes probably will decrease to some extent the amount of irritating fumes that can pass to the lower respiratory tract. We shall assume that we have been unfortunate enough to develop a "nose pattern" of reacting to stress. Now the stress is not an irritating gas but perhaps an irritating supervisor who is constantly demanding of you more than you feel you are able to do. A watery nose will not increase your ability to meet the demands. Further, the increased mucus may provide a nutrient media for the growth of bacteria. You might very well develop a respiratory infection.

Do you remember the increased skeletal muscle tone that occurs in the fight or flight reactions? This reaction is clearly adaptive if the threat or stress is one that requires skeletal muscle activity. What happens, however, if you happen to be a "skeletal muscle reactor" and the stress is not one that can be solved by fighting or running away? You can feel the increased muscle tension in your back and neck. Indeed, your situation may be a real pain in the neck!

If you eat spoiled food, your gastrointestinal tract responds appropriately by eliminating it as promptly as possible. You vomit or have diarrhea. Suppose, however, that the stress agent is not spoiled food but rather the fact that your plans have been spoiled by some unfortunate turn of events. If you happen to be a "GI reactor" you are likely to respond to this stress by having a gastrointestinal upset. Such a response will hardly help you to straighten out the situation.

Many times in the preceding chapters you have seen how automatic vasomotor activity adjusts the blood flow to various parts of the body and enables the different organs to adapt to changing needs for action. If you are a "vasomotor reactor" and the stress happens to be an embarassing situation or a slip of your tongue that you hope will go unnoticed, your vasomotor adaptive pattern will probably cause you to blush and will compound your embarassment.

What can be done about the unfortunate patterns that you have developed over the years? No simple solution is readily available to you, but you may be helped just by being able to recognize these patterns. If you consider the extensive neural connections between the frontal lobe of the brain and the hypothalamus that is responsible for controlling many unconscious body functions, you should not be surprised that your thoughts and emotions do influence body functions.

ALARM AND RESISTANCE

Now we shall turn to some of the specific aspects of the alarm and resistance phases of stress. You should remember that the causes of the stress are nonspecific and that the degree of response depends upon many factors.

Initially, a decrease in the blood sugar level occurs because glucose is being used for the energy needed to meet the emergency. Energy needs of the body are paramount in the process of adapting to the stress situation. In response to the decreased blood sugar, glucagon is released to cause the liver to convert glycogen to glucose. Another aspect of the resistance phase is the release of glucocorticoids. These hormones from the adrenal gland cause gluconeogenesis (the manufacture of glucose by the body from noncarbohydrate sources), and the blood sugar level increases. The glucocorticoids antagonize insulin and force muscle and most body tissues to utilize fats for fuel. This procedure preferentially saves the glucose for use by the brain. Glucose is the only fuel that the brain can use, and brain cells do not need insulin in order to utilize glucose. Normally, increased production of glucocorticoids will function as a negative feedback mechanism and depress the pituitary production of ACTH, but stress depresses this negative feedback and the glucocorticoids are produced in quantities greatly above their ordinary levels.

Another characteristic of the alarm phase of stress is a loss of sodium and water from the vascular compartment, which lowers the blood pressure. In stress situations, not only those associated with hemorrhage, the patient suffers some degree of shock. Many compensatory mechanisms help the stressed individual to resist this shock.

A decrease in capillary filtration of fluid into the interstitial spaces of the peripheral body tissues will save the remaining vascular volume to supply vital organs, heart, and brain. The relatively greater osmotic pressure in the peripheral capillaries helps to pull interstitial fluid back into the capillaries, thereby increasing the circulatory volume at the expense of the peripheral tissues whose nutrition is less vital in this emergency than that of the heart and brain.

The resulting dehydration of the peripheral tissues stimulates the pituitary release of ADH. This hormone causes a decreased urinary output and the conservation of fluid. Mineral corticoids, chiefly aldosterone, are released to effect the kidney conservation of sodium and water. To compensate further for the lowered blood pressure, the kidney produces the enzyme renin. This enzyme enters the plasma and causes the production of the hormone angiotensin, a powerful vasoconstrictor, which helps to elevate the blood pressure.

Hormones from the adrenal medulla also help to elevate the blood pressure by causing vasoconstriction. The increased levels of glucocorticoids sensitize the blood vessels to all of the vasopressor substances. Serotonin from the pineal opposes the vasoconstriction in the vessels of the brain.

In the alarm phase, an increase in the blood potassium level takes place. High levels of blood potassium depress cardiac action, which will also contribute to the shock picture of stress. Under the influence of aldosterone the normal kidney will compensate by excreting the excess potassium.

During the alarm phase, the stressed individual has an increased white blood cell count. The increase may be due to an actual increase in the manufacture and release of new white blood cells. If this is the source of the increase, the cells being released are probably immature cells (band cells, or metamylocytes) that are not capable of phagocytizing bacteria if the nature of the stress is infection, nor are they capable of removing the tissue debris that results from trauma.

The compensatory mechanisms of the latter aspect of the alarm phase are the production of the anti-inflammatory glucocorticoids and the atrophy of the thymus and other lymphatic tissues. If the increase in the number of circulating white blood cells is due to the release of splenic stores of mature white cells that can be helpful, then the mineral corticoids that are proinflammatory will favor the actions of these white blood cells.

You are all familiar with the neural response to a threatening situation. This is the fight or flight response mediated by the sympathetic nervous system. If you review the widespread actions of the sympathetic nervous system (page 182) you will understand clearly that all of these are intended to help the body cope with stress.

Society, however, does not generally approve of your making a habit of striking just anyone who happens to anger you; in fact, an angry fighter is likely to be charged with disturbing the peace. Likewise, you can imagine what might happen if, while you are being reprimanded by someone in authority, you run and hide. In spite of the fact that your sympathetic nervous system equips you to deal with anger and fright in these ways, society does not ordinarily allow you to cope with feelings of fear and anger in such a way.

If you consider the above factors, you can see that many psychosomatic illnesses such as those evidenced by increased skeletal muscle tone (chronic, functional low back pain) or essential hypertension may have a fairly sound physiological basis. You might think about the implications of this on the health of an individual whose life situation is dominated by either fear or anger. In *Die Fledermaus,* Johann Strauss wrote, "Happy [and healthy] is he who forgets what cannot be altered."

You have all been angry and you have all experienced fear. You know perfectly well that these are not the same emotions even though they are both the result of stimulation of the sympathetic nervous system and the increased production of epinephrine and norepinephrine. Perhaps at least part of what makes these emotions different is the fact that in anger the dominant hormone is norepinephrine and in fear it is epinephrine. Norepinephrine is a much more powerful vasoconstrictor than is epinephrine. The common expression, "I was so mad I thought I would blow my stack", may unfortunately have some sound physiological basis. Because norepinephrine also does not stimulate the cerebral cortex, the angry person does not always exercise the best judgment.

The initial neural response to the stress situation is obviously energy consuming. The individual is anxious, excited, and depending upon the nature of the situation may also have expended a considerable amount of motor energy as well as emotional energy. Rest is clearly indicated and the body knows this. Unfortunately, the urgent need for rest is not always as obvious to well-meaning friends and relatives who view convalescence as an occasion for visiting. Naturally, most people who are ill appreciate the visits of friends and family, but few people are able to visit the sick without unintentionally making certain demands on the stressed individual. They naturally expect the patient to engage in conversation and to express enthusiasm about the activities reported to them. It would be much more supportive of the normal adaptive mechanisms if the visits were short and the accounts of, for example, an exciting ball game, were saved for another occasion.

Many ritualistic hospital practices also may run counter to the patient's urgent need for rest. You might consider how you in your own specialized area might effect at least a little change in the system.

In concluding the discussion of stress and the way it affects patients, we shall give some thought to the stresses in our own lives and the lives of our associates and coworkers. All of us experience some degree of stress when we enter a new situation or begin a new phase of our lives. Perhaps you experienced a bit of stress when you began this course. However, as the weeks passed and you became more certain of what was being expected of you, the stress level declined. You may be surprised and amused to learn that teachers experienced this stress too. Even the most experienced teachers feel the stress of the start of the school year. Provided that the stress level is not so high as to interfere with effective teaching and learning, such stress is good. The feeling of challenge is exciting and equips you to put your best efforts to the task ahead. You are rewarded for these efforts by the satisfaction of having learned and increased the scope of your perspectives.

However, for everyone there are times that stress levels can get so high that they are overwhelming. Such situations usually arise not from one difficult aspect of life but from the compounding of many difficulties. This naturally can make everyday tasks not only unpleasant but also so difficult as to be beyond capability.

No one plays a single role in life. Not only are you a student, you are someone's relative or friend, you may have a job, and you surely have many responsibilities other than your classroom ones. All are important. What do you do when things start going wrong in one sphere of your life after another? Although true mental illness, which must be dealt with only by qualified professionals, is often evidenced by such circumstances, most of you at some time or another may find your situation not to your liking. We shall consider some possible ways of dealing with such circumstances.

Dissect your situation. Attempt to clearly identify the stressor and all of its various parts. Concerning each of these parts, ask yourself why this must be done. If some parts of your burden are not really necessary, discard them. If all have a valid reason to exist, then ask yourself if each needs attention now. Some may just as well be taken care of at another time. (However, this suggestion does not apply to preparation for your final examinations!)

Having analyzed the various components of the stress, you should also look critically at your resources—that is, in the context of the stress syndrome, your resistance or defense mechanisms. List these resources and give careful thought to how effectively they are being utilized to help you resist the stress. In taking inventory of your resources, you may very well uncover some that have not been utilized at all.

Another major component of stress is worry. Textbooks of psychosomatic medicine are full of case reports describing many disease conditions that result from chronic worry. Even the worried person usually recognizes that worry alone will not solve his problems. Unfortunately, well-meaning friends who tell him not to worry often only sharpen his focus of attention onto his problems. One of the best remedies for worry is diversion. He should undertake a strenuous task that needs all his attention. He may not forget his worries, but they will fade.

The diversion required to dissipate the devastating fatigue of worry is not always as simple. Everyone knows that all work (or worry) and no play is harmful, but what is work and what is play? Obviously, the nature of the diversion must accommodate the particular individual. A brisk walk in the fresh air might be relaxing for a student, but the postman would probably not find it a satisfactory diversion. The key point is that the activity—be it athletic, artistic, or social—must consume considerable energy. The energy so consumed will not be used to worry.

SUMMARY QUESTIONS

1. Discuss the ways in which the kidney helps to compensate for the shock of stress.
2. What hormones are involved in the compensatory mechanisms of stress and what is the specific function of each of these hormones?
3. Discuss some of the factors that determine the degree of response to stress.
4. What is gluconeogenesis?
5. In the resistance phase of stress, by what means is the blood sugar level increased?
6. Why do the glucocorticoids antagonize insulin but also favor an increased blood sugar?

SUGGESTED READINGS

Cannon, W. B. "Organization for Physiological Homeostasis," *Physiological Review,* July 1929.

Cannon, W. B. "Stresses and Strains of Homeostasis," *Physiological Review,* January 1935.

Cannon, W. B. *The Wisdom of the Body.* New York, N.Y.: W. W. Norton, 1932.

Galdston, Iago (ed.). *Beyond the Germ Theory: The Roles of Deprivation and Stress in Health and Disease.* New York, N.Y.: Health Education Council, 1954.

Selye, Hans. *The Physiology and Pathology of Exposure to Stress.* Montreal, Ont.: Acta, Inc., 1950.

Selye, Hans. *The Stress of Life.* New York, N.Y.: McGraw-Hill, 1956.

Selye, Hans, *et al.* "Adaptation Reaction to Stress," *Psychosomatic Medicine,* Vol. 12, No. 3, 1950.

Appendix A

Glossary*

abduction. moving an extremity away from the midline

abortion. the expulsion or removal of the embryo or fetus from the uterus any time before the twenty-eighth week of pregnancy, either by natural or artificially induced means

abrasion. the scraping away of the superficial layers of skin, a brush burn

abscess. localized collection of pus

accommodation. power of the lens of the eye to focus

acetabulum. cup-shaped cavity on the lateral surface of the innominate bone

acetone. a colorless liquid (dimethyl ketone) having a characteristic odor. Acetone may be present in the urine of patients with diabetes mellitus.

achlorhydria. absence of hydrochloric acid in the stomach

acidosis. an abnormally high amount of acid in the bloodstream or a decrease in the amount of base

ACTH. (adrenocorticotropin) a pituitary hormone that stimulates the release of some of the hormones from the adrenal gland

acute. sudden, severe, or sharp

Addison's disease. a disease resulting from a decreased function of the adrenal glands

adduction. moving an extremity toward the midline

ADH. (antidiuretic hormone) a pituitary hormone that causes the conservation of fluid by the kidney

adhesions. scar tissue binding together tissues that are not normally joined

adipose. fat or fatty tissue

adrenal gland. an endocrine gland located superior to the kidney

adrenergic. pertaining to the sympathetic nervous system

aerobe. an organism that requires oxygen

aerosol. a fine mist, frequently a drug used for inhalation

afferent. a structure such as a nerve or blood vessel leading from the periphery to the center

agglutinate. a clumping together

albuminuria. the presence of protein or albumin in the urine

aldosterone. a hormone from the adrenal gland

allergen. an antigen that produces an allergy

*This glossary is not intended as a substitute for a dictionary. The author recommends that students have access to *Taber's Cyclopedic Medical Dictionary,* C. L. Thomas (ed.), 12th ed. (Philadelphia, Pa.: F. A. Davis Co., 1973).

allergy. a hypersensitivity to an allergen
alveolus. a small space or cavity such as the alveoli of the lungs
amino acid. the basic structure of proteins
amniocentesis. the removal of a sample of amnionic fluid from the amniotic sac
amnionic fluid. fluid within the amniotic sac surrounding the fetus
amniotic sac. a sac in which a growing fetus is contained within the uterus. Sometimes called the "bag of waters"
analgesic. a drug used to relieve pain
anaphylaxis. a severe allergic reaction
anastamosis. the joining together of structures
androgen. a male hormone
anemia. a condition in which the blood is deficient in quality or quantity
anesthetic. an agent used to produce insensibility to pain
aneurysm. a localized dilatation of the walls of a blood vessel
angina. severe pain, frequently in the chest
angiogram. an X-ray procedure used for visualization of blood vessels
angiotensin. a plasma hormone
anaerobe. an organism that does not require oxygen
anorexia. loss of appetite
anoxia. lack of oxygen
antagonist. one that acts in opposition to another, for example, a muscle or a drug
anthelmintic. a remedy for worms
antiarrhythmic. a drug used to treat cardiac arrhythmias or irregularities of the heart beat
antibiotic. a drug that stops the growth of microorganisms in the body
antibody. a substance produced in the body that protects against specific infectious diseases
anticoagulant. a drug that decreases blood clotting
anticholinergic. a drug that blocks the passage of impulses through the parasympathetic nerves
anticonvulsants. drugs used to treat convulsions.
antiemetic. drug used to relieve nausea and vomiting
antihistamine. a drug used to counteract the effect of histamine, commonly used in the treatment of allergy
antimetabolites. drugs used in the treatment of cancer to decrease cellular division
antipyretic. a drug used to reduce fever
antispasmodic. a drug used to decrease muscle tone and contractions
antitoxin. an antibody that neutralizes a toxin

anuria. lack of urine production
aphasia. inability to speak
appendectomy. the surgical removal of the appendix
aqueous humor. clear fluid found in the anterior and posterior chambers of the eye
arrhythmia. an irregularity of heart beat
arteriosclerosis. hardening of the arteries due to deposits of fatty plaques in the lining of the arteries
ascites. an abnormal collection of fluid in the peritoneal cavity
aseptic. sterile, free of microorganisms
asthma. paroxysmal (episodic) dyspnea caused by constriction of the bronchioles
astigmatism. a visual defect due to an imperfect curvature of the refractive surfaces of the eye
atherosclerosis. a form of arteriosclerosis or hardening of the arteries
bacillus. a rod-shaped microorganism
bacteremia. the presence of bacteria in the blood
bacteria. microbes, or germs
bacteriostatic. an agent that halts the growth of bacteria
barbiturate. a drug used as a sedative or a sleeping pill
basal ganglia. deep lying masses of gray matter in the brain
Bell's palsy. a condition resulting from injury to the facial nerve causing paralysis of the muscles of facial expression on the affected side
benign. harmless, not malignant
bifurcate. the division of a structure such as the branching of a large artery to two smaller arteries
bilateral. affecting both sides
biopsy. the removal of a specimen of tissue for examination
blepharitis. inflammation of the eyelid
brachial. the region between the elbow and the shoulder
bradycardia. an abnormally slow pulse
bronchi. major air passageways in the lungs
bronchiectasis. a chronic lung disease characterized by a widening of the air passageways
bronchogram. an X-ray examination of the lungs by means of which the bronchi and bronchioles can be visualized
bronchitis. inflammation of the bronchi
bronchoscopy. an examination used to visualize directly the bronchus
Caesarean section. delivery of an infant through a surgical incision in the mother's abdominal wall and uterus
calcification. a process by which tissue becomes hardened by a deposit of calcium salts within its substance
calcitonin. a hormone from the thyroid gland

calculus. a stonelike formation usually composed of mineral salts

calyx. a recess of the pelvis of the kidney

Cantor tube. a hollow tube used to remove fluids and gas from the small intestines

cannula. a small tube for insertion into a body cavity

carbaminohemoglobin. a combination of carbon dioxide and hemoglobin, one of the forms in which carbon dioxide exists in the blood

carbon dioxide. a gas formed in the body through oxidation of foodstuffs

carbon monoxide. a poisonous gas formed by incomplete burning of carbon

carbuncle. an inflammation of subcutaneous tissue, terminating in sloughing and suppuration

carcinoma. a malignant growth of epithelial tissue

cardiac catheterization. a procedure in which a tube is passed through a vein into the right side of the heart

cardiac output. the rate at which the heart pumps blood, normally about 5 liters per minute in a resting subject

catabolism. the phase of metabolism involving the breakdown of substances and the production of energy

catecholamines. hormones produced by the adrenal medulla

cataract. an opacity or clouding of the crystalline lens of the eye

catheterize. the introduction of a hollow tube into a body cavity, such as the urinary bladder, to draw off fluid

cation. a positive ion such as sodium or potassium

cellulitis. inflammation of cellular tissue, especially subcutaneous tissue

central venous pressure. the pressure of the blood as it enters the right atrium

centrosome. a cytoplasmic organelle that is active in cellular division

cerebral vascular accident (CVA). pathology of a blood vessel in the brain; either a rupture of the vessel or an occlusion of the vessel with a blood clot

cervix. neck

chalazion. an infection of a meibomian gland in the upper lid of the eye

chemotherapy. treatment of a disease by means of administering chemicals

chiasm. a crossing

cholecystectomy. the surgical removal of the gall bladder

cholecystitis. an infection of the gall bladder

cholecystogram. an X-ray procedure used to visualize the gall bladder

cholelithiasis. gall stones
cholinergic. pertaining to the parasympathetic nervous system
cholesterol. fatlike substances found in body tissues
chromosome. a rod-shaped body that appears in the nucleus of a cell at the time of cellular division
chronic. persistent or prolonged
cilia. microscopic hairlike projections on the free surface of some columnar cells
circumcision. the surgical removal of the foreskin or part of it
cirrhosis. hardening of an organ
coagulate. to congeal or to clot
cocci. spherical-shaped type of microorganisms
colitis. inflammation of the large bowel
colloid. a gelatinous substance; a particle that, instead of being dissolved, is held in suspension
collateral circulation. an alternative blood supply
colostomy. a surgical procedure to form an artificial opening into the large bowel
conception. fertilization of an ovum
concha. a structure resembling a shell in shape
condyle. a rounded articular surface at the end of a bone
congenital. existing at or before birth
conjunctiva. the tissue lining the inner surface of the eyelid
conjunctivitis. inflammation of the conjunctiva
contaminate. to soil or to make inferior by contact or mixture
contraception. the prevention of conception, birth control
contracture. a shortening or distortion such as that resulting from shrinkage of muscles or from scar formation
convolution. an elevated part of the surface of the brain
convulsion. violent, uncoordinated, involuntary contractions of muscles
cornea. transparent outer surface of the eyeball
coronary occlusion. the blockage of a blood vessel supplying the heart
coronary thrombosis. a clot in a blood vessel supplying the heart
corpus luteum. a yellow mass in the ovary formed after the rupture of a graafian follicle
cortisone. a hormone produced by the adrenal cortex. A drug used in the treatment of many diseases, particularly allergies and chronic inflammatory diseases
cranial. referring to the skull
craniotomy. an operation on the cranium
crepitus. a crackling sound made by the rubbing together of the ends of fractured bones, or the sensation produced by palpating tissues that contain air

cretinism. a chronic condition due to congenital lack of thyroid secretion
cryptorchidism. undescended testis
crystalloid. a substance that when in solution will pass readily through body membranes
culdoscopy. a visual examination of the organs of the pelvic cavity of the female through a small incision in the pouch of Douglas
cutaneous. referring to the skin
cyanosis. a blueness of the skin caused by insufficient oxygen in the blood
cyst. a sac, especially one that contains a liquid or semisolid
cystitis. inflammation of the urinary bladder
cystocele. a hernial protrusion of the urinary bladder
cystoscopy. a visual examination of the interior of the urinary bladder
cytology. the study of cells
cytoplasm. the protoplasm of a cell excluding that of the nucleus
deaminization. the removal of an amino group from an amino acid
decongestant. a type of medication, such as nose drops, used to relieve congestion of mucus
decubitus. a pressure sore or bed sore
dehydration. deficiency of body water
delusion. a false belief that is contrary to the evidence
demulcent. a soothing, bland substance used to treat inflamed or abraded surfaces
dendrite. a process of a neuron
dermis. the true skin
dermatitis. inflammation of the skin
desquamation. peeling of the skin
diabetes insipidus. a disease due to a deficiency of the antidiuretic hormone from the posterior pituitary
diabetes mellitus. a metabolic disease in which carbohydrates are poorly oxidized
dialysis. the separation of crystalloids and colloids by means of a semipermeable membrane
diaphoresis. excessive perspiration
diaphysis. the shaft of a long bone
diastole. the relaxation of the heart
digitalis. a drug used to increase the efficiency of the heart beat
diffusion. spreading out of particles
diplococci. spherical-shaped microorganisms that occur in pairs
disaccharide. a sugar with the formula $C_{12}H_{22}O_{11}$
distal. farther from the body or from the origin of a part

diuresis. increased urine production
diuretic. a drug used to cause diuresis
diverticulum. a pouch leading from a main cavity or tube
dorsal. the posterior aspect or back of the body or an organ
dorsiflexion. flexion or bending of the foot toward the leg
DNA (deoxyribonucleic acid). a spiral-shaped molecule that contains the hereditary material of the cell
dysfunction. disturbed or abnormal function of an organ
dysmenorrhea. painful menstruation
dysphagia. difficulty in swallowing
dyspnea. labored breathing
dysuria. difficult or painful urination
ecchymosis. a bruise, discoloration of the skin caused by the extravasation of blood
ectopic. out of place
eczema. a skin disease characterized by patches of vesicules and crusts
edema. an abnormal collection of fluid in the intercellular tissue spaces
efferent. leading from some central structure toward the periphery
effusion. the escape of fluid into a space such as the pleural cavity or tissues
ejaculation. the expulsion of semen
electrocardiogram. a graphic recording of the electric current produced by the contraction of the heart
electroencephalogram. a graphic recording of the electrical currents produced by brain action
electrolyte. a solution such as salt solution that can conduct electricity
electromyography. a graphic recording of the electrical currents produced by skeletal muscle
embolus. a blood clot moving in the bloodstream
embryo. the fetus before the end of the second month of intrauterine life
emesis. vomitus
emetic. a drug used to induce vomiting
emphysema. a chronic lung disease usually characterized by greatly distended alveoli
empyema. pus in the pleural cavity
encephalitis. inflammation of the brain
endocarditis. inflammation of the lining of the heart
endocardium. the lining of the heart
endocrine glands. ductless glands that produce hormones
endogenous. originating within the body

endometriosis. a disease caused by the presence of endometrial tissue outside of the uterus
endometrium. the lining of the uterus
endoplasmic reticulum. a cytoplasmic organelle concerned with protein synthesis
endoscope. an instrument used to inspect the interior of a body cavity such as the stomach
endothelium. a type of tissue such as that which lines blood vessels
enzyme. an organic catalyst
eosinophil. a type of white blood cell
epidermis. the outer layer of the skin
epididymitis. inflammation of the epididymis
epigastric. abdominal region located medial to the hypochondric regions
epiglottis. a leaf-shaped cartilage that covers the entrance to the larynx during the act of swallowing
epilepsy. a disease characterized by seizures or "fits"
epinephrine. a hormone produced by the adrenal medulla and sympathetic nerve tissue
epiphysis. the ends of long bones
epistaxis. nose bleed
epithelium. a type of connective tissue
erosion. wearing away of tissue
erythema. an unusual redness of the skin
erythrocyte. a red blood cell
eschar. a slough produced by burning; a crust
esophagoscopy. the visual examination of the esophagus
esophagus. the canal extending from the pharnyx to the stomach
estrogen. a female hormone
etiology. cause
euphoria. an exaggerated feeling of well-being
eustachian tube. the canal extending from the throat to the middle ear
evisceration. the protrusion of internal organs through a wound
exacerbation. an increase in the severity or intensity of the symptoms
excoriation. an area where the skin has been scraped away or chafed
exocrine glands. glands with ducts such as sweat glands
exogenous. originating outside of the body
exophthalmos. abnormal protrusion of the eyeball
extension. the increasing or straightening of the angle at a joint
extrasystole. a premature contraction of the heart; a type of cardiac arrhythmia
exudate. a substance thrown out, pus or serous fluid

fallopian tubes. oviducts, tubes leading from the ovaries to the lateral aspect of the uterus
fascia. a sheet of fibrous tissue that covers muscles and certain other organs
feces. excrement from the bowels
fetus. the unborn infant
fibrillation. a type of cardiac arrhythmia
fibroid. a benign type of tumor of the uterus
fibrosis. the formation of fibrous or scar tissue
filtration. a process by which water and dissolved substances are pushed through a permeable membrane from an area of higher pressure to lower pressure
fissure. a cleft or groove
fistula. a deep ulcer or abnormal passage often leading from a hollow organ to the body surface
flaccidity. poor muscle tone
flatus. gas or air in the stomach or intestines
flexion. the decreasing of the angle at a joint, for example, the bending of the elbow
flora. plant life; the bacterial content of the intestines
fluoroscope. a machine used to examine internal organs visually and to observe the movement and contour of the organs
Foley catheter. a tube used for the continuous drainage of urine
foramen. an opening
fossa. a shallow or hollow place in a bone
frontal. the region of the forehead or a plane that divides the body into front and back portions
fundus. a rounded base or part of a hollow organ most remote from its mouth
furuncle. a boil
gamma globulin. a blood fraction that carries antibodies
gangrene. death of tissue
gastrectomy. the surgical removal of a part or all of the stomach
gastritis. an inflammation of the lining of the stomach
gastroscopy. a visual inspection of the interior of the stomach
gavage. feeding by means of a stomach tube
gene. a hereditary unit of the chromosome
genetics. the study of heredity
genotype. the genetic makeup of an organism
gingivitis. inflammation of the gums
G.I. series. an X-ray examination of the gastrointestinal tract
glaucoma. an eye disease characterized by increased intraocular pressure and impaired vision or blindness
glomerulus. a coiled mass of blood capillaries within Bowman's capsule of the kidney
glucagon. a hormone produced by the pancreas

glucocorticoids. a group of hormones produced by the adrenal cortex
gluconeogensis. the manufacture of glucose by the body from noncarbohydrate materials
glucose. a simple sugar
glycogen. animal starch, the storage form of glucose
glycosuria. glucose in the urine
Golgi apparatus. a cytoplasmic organelle concerned with the export of substances manufactured by the cell
gonorrhea. a common venereal disease
gout. a metabolic disease characterized by excess amounts of uric acid in the blood and painful swollen joints
Graafian follicles. structures in the ovaries where the ova are formed
gynecologist. a specialist in diseases of women
hallucination. hearing, seeing, or feeling things that do not exist
hematemesis. bloody vomitus
hematinics. a type of drug used to treat diseases of the blood
hematology. the study of the blood
hematoma. a swelling that contains blood
hematuria. blood in the urine
hemiplegia. paralysis of one side of the body
hemodialysis. a procedure to remove wastes or other toxic substances from the blood which can not be eliminated by the kidney
hemoglobin. the oxygen-carrying substance of the red blood cells
hemolysis. destruction of red blood cells
hemophilia. a sex-linked, hereditary blood disease characterized by an increased bleeding tendency
hemoptysis. bloody sputum
hemorrhoids. varicose veins in the walls of the anus
hemostasis. stopping the flow of blood
hemostat. an instrument used to stop hemorrhage
heparin. an anticoagulant
hepatitis. an infectious disease of the liver
hernia. a protrusion of a loop or part of an organ through an abnormal opening
herpes. a virus infection characterized by blisters on the skin and mucous membranes
hirsutism. abnormal hairiness
histamine. a chemical agent thought to account for certain allergies. Also a drug that causes vasodilatation
histology. the study of tissues
HNP (herniated nucleus pulposus). a slipped disc, usually in the lumbar region of the spine

Hodgkin's disease. a malignant disease characterized by swelling of lymph glands
homeostasis. a mechanism by which the internal environment of the body tends to return to normal whenever it is disturbed
hordeolum. a stye
hormone. a chemical produced by an endocrine gland
hydrocele. an abnormal accumulation of fluid in the sac surrounding the testis
hydrocephalus. a condition characterized by an increased amount of cerebrospinal fluid and a dilatation of the cerebral ventricles
hydrocortisone. a hormone produced by the adrenal cortex; also a drug used to treat allergies and chronic inflammatory diseases such as arthritis
hydrolysis. a chemical reaction involving water
hydronephrosis. an abnormal collection of urine in the pelvis of the kidney
hydrotherapy. the use of water in treating disease
hyperopia. farsightedness
hypertension. high blood pressure
hyperthyroidism. overactivity of the thyroid gland
hypertrophy. overgrowth
hyperventilation. rapid breathing
hypnotic. a drug used to induce sleep
hypochondriac. concerning the regions of the abdomen lateral to the epigastric region; also a person with a morbid anxiety about health
hypoglycemia. an abnormally low blood sugar
hypotension. low blood pressure
hypothalamus. a part of the brain concerned with the regulation of many visceral functions
hypovolemia. a decreased amount of vascular volume
hysterectomy. the surgical removal of the uterus
iatrogenic. caused by the physician or his assistants
idiopathic. of unknown cause
idosyncrasy. a peculiar sensitivity to some drug or other agent
ileostomy. an artificial opening into the ileum
ileum. the distal portion of the small bowel
ileus. obstruction of the bowel usually as a result of the inhibition of nerve impulses necessary to the maintenance of normal peristalsis
illusion. a misinterpretation of a sensory impression
immunity. the ability of the body to resist an infection
impetigo. a contagious infection of the skin
incontinence. inability to hold urine or feces

induration. hardening
infarct. an area of necrosis due to lack of blood supply
inflammation. a response of the tissues to injury. The signs of inflammation are pain, heat, redness, and swelling
infrared. electromagnetic waves that provide intensive dry heat
inguinal. the region of the groin
inorganic. not organic
insemination. fertilization of an ovum by a sperm
insulin. a hormone produced by the pancreas, also a drug used to lower the blood sugar
integument. the skin
intercostal. between the ribs
internal capsule. white matter tracts lateral to the thalamus in the brain
interstitial. the spaces between the cells, intercellular
intervertebral disc. a cartilagenous structure between each pair of vertebrae
intracellular. within the cell
intracranial. located within the craniun
intraocular. located within the eye
intravenous. located within a vein
involution. the return of an enlarged organ to its normal size such as the uterus following the birth of the infant
irradiation. exposure to any form of radiant energy such as X-ray or radioisotopes
ischemia. decreased blood supply
isotonic. solutions that have the same concentration or number of particles
isotopes. elements that have the same atomic number but differ in their atomic weights
jaundice. a yellow discoloration of the skin
jejunum. the second portion of the small intestine
ketosis. a disturbance of the acid-base balance of the body
kyphosis. hunchback
laceration. a wound caused by tearing
lacrimal glands. the glands that produce tears
lactation. the act and time of the production of milk by the breasts
lactic acid. a waste product produced by active cells
lesion. any pathological or traumatic change in a tissue
leukemia. a disease of the blood-forming organs resulting in an overproduction of white blood cells
leukocyte. white blood cell
leukocytosis. an increased number of white blood cells

leukopenia. a reduction in the number of white blood cells
Levin tube. a type of stomach tube used to gavage (tube feed) or to remove by suction the contents of the stomach
lidocaine. a local anesthetic agent
ligament. a fibrous band of tissue that connects bones or supports viscera
ligation. the application of a tie around a vessel or hollow tube such as the uterine tubes
lipid. a fatty substance
lithiasis. stone formation
lithotomy. the surgical removal of a stone; also a common position for a gynecologic examination
lordosis. curvature of the spine with a forward convexity
lumbar puncture. a procedure used to withdraw cerebrospinal fluid
lymphocyte. a type of white blood cell
lysosome. a cytoplasmic organelle
maceration. softening of a solid by soaking
macule. a small discolored spot on the skin that is not elevated
malaise. a generalized discomfort or sick feeling
malignant. usually pertaining to cancer or other life-threatening conditions
MAO inhibitor. a drug such as an amphetamine which inhibits monamine oxidase
mastectomy. the surgical removal of a breast
mastitis. inflammation of the breast
mastoiditis. infection of the mastoid sinuses
meatus. passageway
mediastinum. the space in the middle of the thorax
melanin. a dark pigment in the skin and hair
melanoma. a malignant tumor of the skin
menarche. onset of menstruation
meninges. the coverings of the brain and spinal cord
meningitis. inflammation of the meninges
menopause. the cessation of menstruation at the end of the reproductive period of life
menorrhagia. profuse menstrual flow
menstruation. the periodic discharge of blood and endometrial tissue from the uterus
metabolism. the chemical processes of life
metastasis. the transfer of a disease from one part of the body to another
metastasize. to form by metastasis a new site of disease in a distant part of the body

micturition. the passing of urine
midsagittal. a plane dividing the body or an organ into right and left halves
mitosis. cellular division
mitochondria. a cytoplasmic organelle
mononucleosis. an infectious disease characterized by fever, malaise, and swelling of the lymph nodes
monosaccharide. a simple sugar such as glucose
mucous membrane. a type of membrane that lines body cavities that open to the outside of the body
mucus. a secretion produced by mucous membranes
multiple sclerosis. a progressive, crippling disease of the nervous system
muscular dystrophy. a progressive disease characterized by wasting of the voluntary muscles
myasthenia gravis. a disease characterized by progressive paralysis of muscles without any sensory disturbance
myelin. a fatty covering around certain nerve fibers
myocardium. the heart muscle
myositis ossificans. a disease in which calcium salts are deposited in muscle tissue
myoma. a muscle tumor
myopia. nearsightedness
myositis. inflammation of muscle tissue
myxedema. a disease caused by a deficiency of thyroid hormones
naline. a narcotic antagonist
narcotic. drugs that produce sleep and relieve pain
necrosis. death of tissue
neoplasm. a new growth such as a tumor
nephrectomy. the surgical removal of a kidney
nephritis. inflammation of the kidney
nephron. a microscopic functional unit in the cortex of the kidney
nephroptosis. a downward displacement of the kidney
nephrosis. a kidney disease
nephrostomy. a surgical procedure involving the placing of a tube into the pelvis of the kidney for drainage purposes
neuralgia. pain extending along the course of one or more nerves
neurilemma. a covering around certain nerve fibers
neuritis. inflammation of nerve tissue
neurodermatitis. a chronic skin disease due to a nervous disorder
neuroglia. supporting tissue around neurons
neuron. a nerve cell
neurosis. a psychic or mental illness usually characterized by

anxiety and difficulties in adjusting to new or stressful situations

nevus. a mole or birth mark

nocturia. excessive urination at night

norepinephrine. a hormone produced by the adrenal medulla and sympathetic nerve tissue

nucleus. a central part of a cell containing the hereditary material of the cell

obesity. the condition of being overweight

obstetrics. medical speciality that deals with pregnancy and childbirth

occipital. pertaining to the base of the skull

occlusion. a blockage

olfactory. pertaining to the sense of smell

oligemia. a deficiency in the volume of the blood

oocyte. an immature ovum

oophorectomy. the surgical removal of an ovary

ophthalmoscope. an instrument used to examine the interior of the eye

optic disc. the origin of the optic nerve, or the "blind spot"

orchitis. inflammation of the testes

organic. pertaining to living matter

orifice. the entrance or outlet of a body cavity

orthopedics. a medical speciality that deals with disorders of the skeletal system

osmosis. the passage of water through a semipermeable membrane from areas of lesser concentration to areas of higher concentration

osseous. pertaining to bony tissue

ossification. bone formation

osteoarthritis. a chronic degenerative disease of the joints

osteoblast. a bone-building cell

osteomyelitis. infection of bone

osteoporosis. a disease in which there is a decrease in bone density

otitis media. an infection of the middle ear

ovulation. the release of a mature ovum from the ovary

palliative. treatment that is directed toward the relief of symptoms but that has little or no effect on the cause of the symptoms

palpation. examination of the body by means of feeling with the hand

palpitation. rapid action of the heart felt by the patient

pancreatitis. inflammation of the pancreas

papillae. small nipple-shaped elevations
papilledema. edema around the optic disc
papule. a small, raised lesion
paracentesis. the removal of fluid from the peritoneal cavity
paraplegia. paralysis of the lower extremities
parasite. a plant or animal that feeds on another plant or animal
parasympathetic. pertaining to autonomic nerves that originate in the lower part of the brain and the sacral portion of the cord
parathyroid glands. four small endocrine glands located on the posterior surface of the thyroid
parietal. pertaining to the body wall; also the region of the head posterior to the frontal region and anterior to the occipital region
Parkinson's disease. a progressive disease of the central nervous system characterized by stiffness, slowed movements, and rhythmic, fine tremors of resting muscles
paronychia. an infected hangnail
parotid. saliva-producing glands in the back of the mouth
parturition. the birth of an infant
pathogenic. anything that causes disease
pathology. a branch of medicine concerned with the structural and functional changes caused by disease
pectoral. pertaining to the breast or chest
pediatrics. a medical speciality that deals with diseases of children
pediculosis. infested with lice
pellagra. a vitamin-deficiency disease
pelvimetry. the measurement of the dimensions and capacity of the pelvis
percussion. a diagnostic procedure in which a part is struck with short, sharp blows to aid in determining the condition of the parts beneath by the sound obtained
perfusion. a liquid pouring over or through something
pericarditis. inflammation of the sac that surrounds the heart
pericardium. the membranous sac containing the heart
perineum. the anatomical region at the lower end of the trunk between the thighs
periosteum. the outer covering of bone
peripheral. away from the center
peristalsis. contractions of smooth muscle causing a wavelike motion
peritoneum. a serous membrane that surrounds the abdominal organs and lines the abdominal cavity
peritonitis. inflammation of the peritoneum
petechiae. small hemorrhagic areas

phagocytosis. the engulfing and destruction of bacteria and other foreign particles by white blood cells and other cells of the reticuloendothelial system
pharyngitis. inflammation of the pharynx
pharynx. the part of the alimentary canal that connects the mouth and esophagus
phenotype. the organism's physical appearance
pheochromocytoma. an adrenal tumor
phlebitis. inflammation of veins
phlebotomy. inserting a needle into a vein
pigmentation. the coloration or discoloration of a part
pilonidal cyst. a sac containing hairs usually located at the base of the spine
pineal. an endocrine gland located posterior to the third ventricle in the brain
pitocin. a drug used to induce uterine contractions during and after childbirth
pituitary. an endocrine gland located at the base of the brain
placebo. a substance that has no pharmacological action which is given to satisfy a patient's desire for drug treatment. It is also used as a control in scientific medical experiments.
placenta. the nutrient and excretory organ of the fetus
plasma. the fluid portion of the circulating blood
platelet. one of the formed particles of the blood that is important in clot formation
pleura. the serous membrane that surrounds the lungs
pleurisy. inflammation of the pleura
plexus. a network of nerves or blood vessels
pneumococcus. a type of microorganism
pneumonectomy. the surgical removal of a lung
pneumonia. an infectious disease of the lungs
pneumothorax. air in the pleural cavity
polyarthritis. inflammation of several joints
polydipsia. excessive thirst
polyps. saclike growths of the mucous membranes
polysaccharide. a complex sugar such as starch
polyuria. a greatly increased urinary output
postpartum. the period following the birth of an infant
prednisone. a drug with actions similar to those of the glucocorticoids. It is used to treat allergies and chronic inflammatory diseases
prenatal. the period of life before birth
presbyopia. a type of farsightedness that comes with advancing years
pressor. a substance used to increase the blood pressure

procaine. a local anesthetic
proctoscopy. a visual inspection of the rectum
progesterone. a female hormone
prognosis. an opinion of the probable outcome of a disease or injury
prolapse. an abnormal downward displacement of an organ
pronation. turning the palm downward
prophylaxis. the prevention of disease
proprioceptor. end organ of a sensory nerve fiber located in muscles and joints
prostate. a gland in the male located below the bladder and encircling the urethra
prosthesis. an artificial replacement for a part of the body that has been lost
proteinuria. the presence of protein or albumin in the urine
prothrombin. a blood protein that is the precursor of thrombin necessary for clot formation
proximal. closest to the point of attachment
pruritis. itching
psoriasis. a chronic skin disorder
psychogenic. caused by emotional factors
psychoneurosis. a mental illness produced by conflicts in the unconscious mind
psychosis. a severe form of mental illness in which the patient may have hallucinations and personality changes
psychosomatic. concerning a type of illness in which thought processes may disturb organic functions
ptosis. a drooping or a prolapse of a part
pulmonary. pertaining to the lungs
purpura. a disease characterized by the formation of purple patches on the skin and mucous membranes due to subcutaneous bleeding
purulent. consisting of or containing pus.
pus. a product of inflammation consisting of fluid, white blood cells, and bacteria
pustule. a small elevation filled with pus
pyelitis. inflammation of the pelvis of the kidney
pyelogram. an X-ray of the ureter and kidney especially showing the pelvis of the kidney
pyelonephritis. an inflammation of the kidney and its pelvis
pyelonephrosis. any disease of the kidney and its pelvis
pylorus. the valve between the stomach and the duodenum
pyogenic. producing pus
pyorrhea. an infection of the gums
pyuria. pus in the urine
quadriplegia. paralysis of all four extremities

radiation. the emission of radiant energy
radioisotope. a radioactive isotope
radiopaque. not readily penetrated by X-rays
râles. an abnormal respiratory sound usually associated with fluid in the air passages
rectocele. a protrusion of part of the rectum
red blood cell. cells that contain hemoglobin for gas transport in the bloodstream
refraction. deviation of light rays as they pass from one transparent medium to another; also the testing of eyesight for abnormalities of the lens or cornea
remission. the disappearance of the symptoms of a disease
renal. pertaining to the kidney
renin. an enzyme produced by the kidney in response to ischemia
resuscitation. the restoration to life or consciousness
reticuloendothelial system. tissue of the spleen, lymph nodes, liver, and bone marrow engaged in phagocytosis
retina. the innermost coat of the eyeball which contains the end organs of vision
retroperitoneal. behind the peritoneum
rheumatic fever. an infectious disease that may damage the heart valves
rheumatoid arthritis. an inflammatory disease of connective tissue characterized by remissions and exacerbations of pain and stiffness of the joints
Rh factor. a substance in the red blood cells important in the typing of blood for transfusion and in obstetrics
rhinitis. swelling of the mucous membranes of the nose and an increased production of nasal mucus
ribosome. cytoplasmic granules composed of protein and ribonucleic acid
RNA (ribonucleic acid). molecules within the cells that carry the genetic pattern from DNA to the cells newly being formed
roentgenogram. X-ray
rugae. folds inside some of the hollow organs such as the stomach and urinary bladder
sacroiliac. the joint between the sacrum and ilium
salicylate. a class of drugs used to relieve pain and to reduce fever. Aspirin is a salicylate
salpingitis. inflammation of the ovarian tubes
sarcoma. cancer of connective tissue
scabies. a skin disease caused by a mite that bores beneath the surface of the skin
Schlemm, canal of. a canal for the drainage of aqueous humor from the chambers of the eye
sclera. the outer coat of the eyeball

scleroderma. a disease characterized by smooth, hard, and tight skin. Other organs such as the lungs, heart, and muscles may also become hardened.
sclerosis. hardening
scoliosis. a lateral curvature of the spine
scurvy. a disease caused by a deficiency of vitamin C and characterized by an increased bleeding tendency
sebaceous glands. glands that secrete sebum, an oily substance that helps to keep the skin soft
seborrhea. an excessive secretion of sebum from the sebaceous glands
sedative. a drug used to relieve anxiety and to promote relaxation
sella turcica. a part of the sphenoid bone that contains the pituitary gland
semen. a white fluid produced by the male sex organs which serves as a vehicle for sperm
sepsis. poisoning by bacteria
septicemia. blood poisoning
serology. the study of the blood serum
serosa. any serous membrane
serotonin. a hormone produced by the pineal gland and also found in high concentrations in the platelets
serosanguinous. pertaining to or containing both serum and blood
serous membrane. a type of membrane that lines the closed cavities of the body
sigmoidoscopy. a visual examination of the sigmoid
silicosis. a pulmonary disease caused by the inhalation of dust of stone, sand, or flint
sinus. a space such as the paranasal sinuses
sordes. dark-brown foul matter that collects on the lips and teeth in the absence of good oral hygiene
spastic. involuntary muscle spasms
sperm. the mature male sex cell formed in the testes
sphygmomanometer. an instrument used to measure blood pressure
spinous process. a more or less pointed projection of a bone
spirochete. a spiral-shaped microorganism
splenectomy. the surgical removal of the spleen
staphylococcus. a ball-shaped microorganism that grows in clusters
stasis. a stoppage of the flow of any body fluid
stenosis. a narrowing

sterilization. a process by which materials are made free of microorganisms; also a procedure that makes an individual incapable of reproduction.

steroid. a type of hormone such as those produced by the reproductive glands and some of the hormones produced by the adrenal glands

stethoscope. an instrument used to listen to sounds produced within the body such as the heart beat

stimulant. a type of drug that increases the activity of an organ or the body in general

strabismus. the condition of being cross-eyed

streptococcus. a ball-shaped microorganism that grows in chains or in pairs

stricture. the narrowing of a passageway

subarachnoid. beneath the arachnoid or middle covering of the brain and spinal cord

subcutaneous. beneath the skin

subluxation. an incomplete or partial dislocation of a joint

sudoriferous glands. sweat glands

supraorbital. above the eye

suprarenal. above the kidney; also another name for the adrenal glands

suture. material used for sewing up a wound; also the joints between the bones of the skull

symbiosis. the living together or close association of two dissimilar organisms

sympathetic nerves. the fibers of the autonomic nervous system which originate in the thoracic and lumbar portions of the spinal cord

symphysis pubis. the place where the pubic bones join together

symptom. subjective evidence of disease; something the patient feels but the doctor cannot see, hear, or feel

synapse. the microscopic gap between one neuron and the next

syndrome. a set of signs and symptoms

synovial membrane. a type of membrane found surrounding the freely movable joints such as the knee

syphilis. a contagious venereal disease

systole. the contraction of the heart

tachycardia. abnormally rapid heart beat

temporal. pertaining to the region of the body anterior to the ear; the temple

tendon. white, fibrous connective tissue that connects muscles to bones

testes. the male reproductive glands that produce sperm

thalamus. a part of the brain that serves as the relay station for sensory impulses to the higher centers of the cerebrum
therapy. treatment
thoracentesis. a procedure used to remove fluid from the pleural cavity
thoracotomy. an artificial opening into the chest usually for the purpose of drainage of fluid
thorax. pertaining to the chest
thrombocyte. one type of formed particles of the blood
thrombocytopenia. a blood disease in which there is a reduction in the number of platelets
thrombophlebitis. inflammation of veins and clot formation within the affected veins
thrombus. a stationary blood clot within a vessel
thymus. an organ within the thoracic cavity that has both endocrine and lymphatic functions
thyroid. an endocrine gland producing hormones that influence metabolism and blood levels of calcium
thyroxin. a hormone produced by the thyroid
tonometer. an instrument used to measure intraoccular pressure
tonsillectomy. the surgical removal of the tonsils
torticollis. wryneck; abnormal contraction of cervical muscles producing an unnatural position of the head
toxemia. a general intoxication due to the absorption of bacterial products. The condition may also occur in pregnancy for reasons not presently understood.
toxicity. the quality of being poisonous
toxin. a poison
trachea. the windpipe
tracheostomy. an artificial opening into the trachea
trachoma. a serious infectious disease of the conjunctiva and cornea
tranquilizer. a drug used to relieve anxiety and to promote relaxation
trauma. any type of injury
tremor. involuntary shaking or trembling
trichomonas vaginalis. a parasitic infection of the vagina
trochanter. large, bony processes on the upper end of the femur where certain muscles are attached
tubercle. a small, rounded projection on a bone where muscles are attached; also small, rounded nodules produced by the bacillus of tuberculosis
tuberosity. a large, roughened projection on a bone where muscles are attached

tuberculosis. an infectious disease produced by the tuberculosis bacillus, usually involving the lungs however, it may also affect other organs
tumor. any swelling or new growth
tympanic membrane. ear drum
ulcer. an open lesion
ulnar. pertaining to blood vessels, nerves, or one of the bones of the forearm
ultraviolet light. electromagnetic radiation
umbilicus. the navel
unilateral. occuring on only one side of the body
urea. a nitrogenous substance that is one of the end products of protein digestion
uremia. a condition in which there is an excess accumulation of waste products in the bloodstream that should normally be eliminated by the kidneys
ureter. the tube leading from the kidney to the urinary bladder
urethra. the tube leading from the urinary bladder to the outside of the body
urinalysis. an examination of the urine
urology. a medical specialty that deals with disorders of the urinary system
urticaria. hives
uterus. the womb
vagina. the birth canal and female organ of sexual intercourse
varices. enlarged and tortuous veins
varicosed veins. dilated veins usually in the legs
vasectomy. a surgical procedure in which the vas deferens is cut and ligated
vasodilatation. an increase in the diameter of blood vessels, particularly the peripheral arterioles
vasomotor. presiding over the expansion or contraction of blood vessels
vasopressor. any substance that will cause constriction of the blood vessels
vasospasm. spasm of the blood vessels decreasing the caliber of the vessels
venereal diseases. diseases spread by sexual intercourse
ventral. pertaining to the anterior surface of the body or of an organ
vernix caseosa. a white, cheesy substance that covers the skin of the fetus and the newborn infant
vertebral canal. the canal that contains the spinal cord
vertigo. dizziness

vesicle. a blister
viable. capable of living
virulence. the relative infectiousness of a microorganism
virus. an ultramicroscopic parasite
viscera. internal organs
visceral. pertaining to internal organs
vital capacity. the amount of air that can forcibly be expelled after the largest possible inhalation
vitreous humor. a clear jellylike substance that fills the eye posterior to the lens
xanthochromic. a yellow color
zygote. the fertilized ovum

Appendix B

Common Medical Prefixes and Suffixes

a absent or deficient
adeno glandular
algia pain
an absent or deficient
arthro pertaining to joints
brachio arm
cysto bladder
dynia pain
dys difficult or painful
ecto exterior
ectomy surgical removal
emia pertaining to a condition of the blood
endo interior
entero pertaining to the intestines
gastro pertaining to the stomach
hema pertaining to blood
hemo pertaining to blood
hydro pertaining to water
hyper above or having a greater concentration
hypo below or having a lesser concentration
inter between
intra within
itis inflammation
mal disorder
meno pertaining to menstruation
meta after or changing
myo muscle
nephro pertaining to the kidney
neuro pertaining to nerve
oma tumor
osis abnormal condition or process
osteo pertaining to bone
para beside or beyond
pathy abnormality
peri around
phlebo pertaining to vein or veins
phobia abnormal fear
pneumo pertaining to air, lung, or breathing
psycho mental
ptosis falling or drooping

testosterone. a male sex hormone
pyo pus
stomy surgical opening
sub under
super over
tomy surgical cutting
uria contained within the urine

Appendix C

Medical Abbreviations

aa of each
a.c. before meals
A/G albumin globulin ratio
b.i.d twice a day
BMR basal metabolic rate
BP blood pressure
BUN blood urea nitrogen
CBC complete blood count
cc. cubic centimeter(s)
cm. centimeter(s)
CNS central nervous system
CSF cerebrospinal fluid
CVP central venous pressure
D and C dilatation and curettage
dr. dram
ECG or EKG electrocardiogram
EEG electroencephalogram
G.I. gastrointestinal
gm. gram
gr. grain
gtt. drop(s)
Hg. mercury
Hct. hematocrit
Hgb. hemoglobin
h.s. at bedtime
I.M. intramuscular
I.P.P.B. intermittent positive pressure breathing
I.V. intravenous
kg. kilogram
mEq. milliequivalent
mg. milligram
ml. milliliter
oz. ounce
PBI protein-bound iodine
p.c. after meals
pCO_2 partial pressure of carbon dioxide
pO_2 partial pressure of oxygen
p.r.n. as needed
q.d. every day
q.h. every hour
q.i.d. four times a day

q.s. sufficient quantity
sp. gr. specific gravity
stat immediately
t.i.d. three times a day

APPENDIX D

WEIGHTS, MEASURES, AND EQUIVALENTS

Apothecaries' System

Weight			Volume		
1 dram (ʒ)	=	60 grains (gr.)	1 fluid dram	=	60 minims (m.)
1 ounce (℥)	=	480 grains	1 fluid ounce	=	8 drams
	=	8 drams	1 pint (pt.)	=	16 ounces
1 pound (lb.)	=	16 ounces	1 quart (qt.)	=	2 pints

Metric System

Weight			Volume		
1 gram (gm.)	=	1,000 milligrams (mg.)	1 liter	=	1,000 cubic centimeters (cc.)
1 kilogram (kg.)	=	1,000 grams			

Approximate Equivalents

1 teaspoon (tsp.)	=	4 cc.	=	1 fluid dram
1 tablespoon (tbsp.)	=	15 cc.	=	1/2 ounce
1 teacup	=	120 cc.	=	4 fluid ounces
1 tumbler	=	240 cc.	=	8 fluid ounces

Weights

Metric		Apothecaries'	Metric		Apothecaries'
0.4 mg.	=	1/150 grain	30 mg.	=	1/2 grain
0.6 mg.	=	1/100 grain	60 mg.	=	1 grain
1.0 mg.	=	1/60 grain	1 gm.	=	15 grains
10.0 mg.	=	1/6 grain	15 gm.	=	4 drams
15.0 mg.	=	1/4 grain	30 gm.	=	1 ounce

Linear Measures

1 millimeter (mm.)	=	0.04 inch (in.)
1 centimeter (cm.)	=	0.4 inch
1 decimeter (dm.)	=	4.0 inches
1 meter (M.)	=	39.37 inches

Celsius–or Centigrade–Fahrenheit Equivalents

Celsius	Fahrenheit	Celsius	Fahrenheit
36.0	96.8	39.0	102.2
36.5	97.7	39.5	103.1
37.0	98.6	40.0	104.0
37.5	99.5	40.5	104.9
38.0	100.4	41.0	105.8
38.5	101.3	41.5	106.7

Index

Index